Rock Damage and Fluid Transport, Part I

Edited by
Georg Dresen
Ove Stephansson
Arno Zang

2006

Birkhäuser Verlag
Basel · Boston · Berlin

Reprint from Pure and Applied Geophysics
(PAGEOPH), Volume 163 (2006) No. 5/6

Editors

Georg Dresen
GeoForschungsZentrum Potsdam
Telegrafenberg D425
D-14473 Potsdam
Germany

e-mail: dre@gfz-potsdam.de

Ove Stephansson
GeoForschungsZentrum Potsdam
Telegrafenberg D425
D-14473 Potsdam
Germany

e-mail: ove@gfz-potsdam.de

Arno Zang
GeoForschungsZentrum Potsdam
Telegrafenberg D425
D-14473 Potsdam
Germany

e-mail: zang@gfz-potsdam.de

A CIP catalogue record for this book is available from the Library of Congress,
Washington D.C., USA

Bibliographic information published by Die Deutsche Bibliothek:
Die Deutsche Bibliothek lists this publication in the Deutsche Nationalbibliographie; detailed
bibliographic data is available in the internet at <http://dnb.ddb.de>

ISBN 3-7643-7711-9 Birkhäuser Verlag, Basel – Boston – Berlin

© 2006 Birkhäuser Verlag, P.O.Box 133, CH-4010 Basel, Switzerland
Part of Springer Science+Business Media
Printed on acid-free paper produced from chlorine-free pulp
Printed in Germany

ISBN 10: 3-7643-7711-9
ISBN 13: 978-3-7643-7711-3

9 8 7 6 5 4 3 2 1

PURE AND APPLIED GEOPHYSICS
Vol. 163, No. 5–6, 2006

Contents

Pure appl. geophys. 163 (2006) 915–916
0033–4553/06/060915–2
DOI 10.1007/s00024-006-0062-x

❙ Pure and Applied Geophysics

Rock Damage and Fluid Transport, Part I

GEORG DRESEN,[1] OVE STEPHANSSON,[1] and ARNO ZANG[1]

Introduction

Mechanical properties and fluid transport in rocks are intimately linked as deformation of a solid rock matrix immediately affects the pore space and permeability. This may result in transient or permanent changes of pore pressure and effective pressure causing rock strength to vary in space and time. Fluid circulation and deformation processes in crustal rocks are coupled, producing significant complexity of mechanical and fluid transport behavior. In addition to the effect of transient changes in pore pressure, fluids also promote chemical reactions that in turn will affect the pore space, the permeability and the mechanical response of rocks when subjected to loading. For example, the depletion of hydrocarbon and water reservoirs leading to compaction may have adverse effects on well production. Solution/precipitation processes modify porosity and affect permeability of aquifers and reservoir rocks. Fracture damage from underground excavation will critically influence the long-term stability and performance of waste storages. Reservoir and geotechnical engineering projects involved in oil and gas production, CO_2 sequestration, mining and underground waste disposal constantly present a wealth of challenging and fundamental problems to research in geomechanics and rock physics.

Since 1998 a series of thematic Euroconferences exist with focus on field and laboratory research in rock physics and geomechanics from both industry and academia. This topical volume results from the 5th Euroconference on Rock Physics and Geomechanics, which was held in Potsdam, Germany from 19–23 September 2004. The conference addressed four main themes including rock fracture mechanics and creep, fluid transport in rocks and in particular, upscaling models and the integration of field and laboratory research. It included a workshop focussing on brittle failure processes and rock fracture toughness testing methods. About 90 oral

[1]Telegrafenberg, GeoForschungsZentrum Potsdam, 14473, Potsdam, Germany

contributions and posters were presented at the conference and workshop from which the articles arose that are being presented in the two parts of this volume.

Part I mainly contains contributions investigating the nucleation and evolution of crack damage in rocks. Several studies apply advanced acoustic emission techniques and monitoring of elastic wave velocities to characterize crack damage ensuing in laboratory rock specimens and in the field (*Nasseri et al., Schubnel et al., Stanchits et al., Thompson et al., Kaselow et al.*). *Turcotte and Sherbakov* study temporal variations in acoustic emission activity and earthquake seismicity using continuum damage mechanics. Recent advances in the development of new or modified techniques to measure rock fracture toughness are presented (*Fowell et al., Sato and Hashida, Wang et al.*), and the effect of the intermediate principal stress on rock fracture is reviewed (*Haimson*). Modeling studies of brittle deformation on the field and laboratory scale reveal recent advances and limitations of numerical techniques (*Alassi et al., Napier and Backers*). *Guéguen et al.* discuss upscaling techniques such as effective medium theory, numerical methods and fractal models that relate mechanical and fluid transport in rocks at different spatial scales. Part II of this volume will include studies investigating the coupling of rock deformation and fluid flow.

Finally, we would like to express our sincere thanks to the reviewers that contributed their time and efforts to the success of this volume: *M. Albrecht, T. Backers, J. Baumgärtner , P. Benson, Y. Bernabe, A. Bobet, M. Bohnhoff, E. Brueckl, S. Chanchole, M. Di Marzio, D. Elsworth, B. Evans, S. Fomin, J. Fortin, R. Fowell, J. Fredrich, P. Ganne, V. German, A. Gray, Y. Guéguen, B. Haimson, S. Hainzl, L. Jing, A. Kaselow, C. Kenter, P. Knoll, H. Konietsky, H.-J. Kümpel, C. Lee, B. Legarth, X. Lei, Y. Leroy, I. Main, A. Makurat, P. Meredith, H. Milsch, J. Napier, J. Renner, R. Roberts, F. Roth, F. Rummel, E. Rybacki, E. Saenger, K. Sato, A. Schubnel, J. Shao, S. Shapiro, S. Stanchits, B. Thompson, C.-F. Tsang, D. Turcotte, A. Vervoort, G. Viggiani, Q.-Z. Wang,* and *R. Zimmerman.*

Pure appl. geophys. 163 (2006) 917–945
0033–4553/06/060917–29
DOI 10.1007/s00024-006-0064-8

Pure and Applied Geophysics

Fracture Toughness Measurements and Acoustic Emission Activity in Brittle Rocks

M. H. B. Nasseri,[1] B. Mohanty,[1] and R. P. Young[1]

Abstract—Fracture toughness measurements under static loading conditions have been carried out in Barre and Lac du Bonnet granites. An advanced AE technique has been adopted to monitor real-time crack initiation and propagation around the principal crack in these tests to understand the processes of brittle failure under tension and related characteristics of the resulting fracture process zone. The anisotropy of Mode I fracture toughness has been investigated along specific directions. Microcrack density and orientation analysis from thin section studies have shown these characteristics to be the primary cause of the observed variation in fracture toughness, which is seen to vary between 1.14 MPa.(m)$^{1/2}$ and 1.89 MPa.(m)$^{1/2}$ in Barre granite. The latter value represents the case in which the crack is propagated at right angles to the main set of microcracks. The creation of a significant fracture process zone surrounding the propagating main crack has been confirmed. Real-time imaging of the fracture process and formation of fracture process zone by AE techniques yielded results in very good agreement with those obtained by direct optical analysis.

Key words: Mode I fracture toughness anisotropy, acoustic emission, fracture process zone, microcrack density, microcrack orientation.

1. Introduction

Fracture toughness (K_{IC}) of rocks, which is one of the basic material parameters in fracture mechanics, is defined as the resistance to crack propagation. It is considered an intrinsic material property that is finding increasing application in a number of fields such as stability analysis, hydraulic fracturing, rock fragmentation by blasting, and earthquake seismology. The uniqueness of the fracture toughness parameter applies to homogeneous and isotropic materials. For distinctly anisotropic or inhomogeneous conditions, that are characteristic of most common rocks, there may not be a unique value for this parameter. As a fracture property, it must depend on the nature of pre-existing microstructure in the rock that contributes to the latter's inhomogeneity. Among these are the extent and severity of microcracks and their

[1]Lassonde Institute, Department of Civil Engineering, University of Toronto, Toronto, ON, M5S 1A4, Canada. E-mail: bibhu.mohanty@utoronto.ca

alignment, and grain-size distribution; the boundaries among which can also be viewed as nascent microcracks.

To date, the role of microstructure on fracture toughness has been investigated strictly in terms of its global effect on the measured value of the latter. Scant attention has been paid to investigating the exact correlation between the two, especially where demonstrated fracture toughness anisotropy is concerned. The same applies to determining the interaction between microstructural properties and the propagating crack in the fracture toughness measurements. In the latter, the crack initiation point is predetermined as is the crack path. The interaction between these preexisting microcracks in the vicinity of the main propagating crack, and the possible formation of an extended fracture zone around the latter for example, are important considerations. The latter would essentially constitute the fracture process zone (FPZ) that may have significant bearing on foreshocks and aftershocks in earthquakes. Quantitative investigation of these phenomena and real-time monitoring of the fracture process by means of acoustic emission techniques constitute the essential goals of this study.

In a related investigation aimed at characterizing the microstructures and fracture toughness for a selection of granitic rocks, a very good correlation among microcrack density, microcrack length and fracture toughness has been demonstrated (NASSERI *et al.*, 2002, 2005a). This study further revealed that the combination of high microcrack density and microcrack length is responsible for lowering fracture toughness values. The average microcrack length shows a better correlation with fracture toughness than grain size in the rocks studied. It is evident from this study that fracture toughness is greatly influenced by microstructural properties of the subject rock. Mode I fracture toughness anisotropy as a function of layered microstructures in metals, alloys and composite materials has been investigated in the past (LI and XIAO, 1995; ENOKI and KISHI, 1995; KEVIN and BRIEN, 1998). LI and XIAO concluded that three main mechanisms, namely, weak interface cracking, fracture path deflection and delamination toughening, contribute to the strong anisotropy of Mode I fracture toughness for materials with layered microstructures. The effect of the anisotropic nature of coal on fracture toughness has also been studied both experimentally and analytically (KIRBY and MAZUR, 1985). The study concluded, as expected, that the fracture toughness was higher when measured orthogonal to bedding plane and was lowest for cracks propagating along the bedding plane.

The correlation between microcrack fabric in granitic rocks and anisotropy of physical properties such as seismic velocity, modulus, compressibility, uniaxial compressive and tensile strength, and fracture toughness has been studied by several authors (SANO *et al.*, 1992; TAKEMURA *et al.*, 2003). It has been reported that the orientation of physical anisotropy corresponded well with that of splitting planes in granites, (CHEN *et al.*, 1999). SCHEDL *et al.* (1986) upon studying Barre granite using optical SEM and TEM techniques, concluded that the splitting planes and anisotropy were mainly caused by microcracks.

In terms of real-time monitoring of the fracture process, the use of acoustic emission techniques (AE) has proven to be an excellent diagnostic tool. Extensive work has been done in laboratory and field locations using AE studies to understand fracture processes (LABUZ et al., 1987; LOCKNER et al., 1991; MOORE and LOCKNER, 1995; YOUNG and MARTIN, 1993; ZIETLOW and LABUZ, 1998; ZANG et al. 1998, 2000; LEI et al., 2000; YOUNG et al., 2004). The brittle fracture processes taking place at the grain size scale have a profound influence on the mechanical properties of rocks. The acoustic emission technique (AE) lends itself very well to studying this process in real time. The development of fast acquisition systems has enabled AE experiments to show fracture development under constant stress loading, thought to better approximate the low strain conditions in the Earth.

2. Fracture Toughness

The material property associated with the ability to carry loads or resist deformation in the presence of a crack is defined as the fracture toughness, (K_{IC}). This parameter can be used to predict the nature of fracture onset depending upon the crack size and its sharpness (leading to the stress concentration effects), the stress level applied and the material property. The International Society for Rock Mechanics (ISRM) has recommended three methods, namely, Chevron Bend (CB), Short Rod (SR) and Chevron Cracked Notch Brazilian Disc (CCNBD) methods, (ISRM, 1988, 1995) for determining the fracture toughness of rock using core-based specimens. However, comparison tests among the three methods have consistently yielded 30 to 50% lower values with the CCNBD method than the other two (DWIVEDI et al., 2000). This anomaly has been recently resolved, and suitable corrective procedures drawn up to yield fracture toughness values with CCNBD that are consistent with the other two methods (IQBAL and MOHANTY, 2005), provided special care is taken in the selection of specimen sizes and handling of anisotropy issues. The advantage of the CCNBD test method over the previous one is considered to be due to its easy adaptation to measuring fracture toughness anisotropy within one rock type. It requires much higher failure loads and exhibits reduced data scatter.

2.1 Chevron Cracked Notch Brazilian Disc (CCNBD) test

In this method (ISRM, 1995) the fracture toughness (K_{IC}) is calculated by the following formula:

$$K_{IC} = \frac{P_{max}}{B\sqrt{R}} Y^\star_{min},$$

(1)

where

$$Y^\star_{min} = \mu.e^{v.\alpha 1}$$

(2)

$Y^{\star}_{\min}$ = critical dimensionless intensity value

$P_{\max}$ = maximum load at failure

B = thickness of the disc

μ and v are constants determined by α_0, α_B.

The geometry of the CCNBD is illustrated in Figure 1. The experimental parameters for Lac du Bonnet and Barre granites are given in Tables 1 and 2, respectively. All the dimensions of the geometry should be converted into dimensionless parameters with respect to specimen radius R and diameter D.

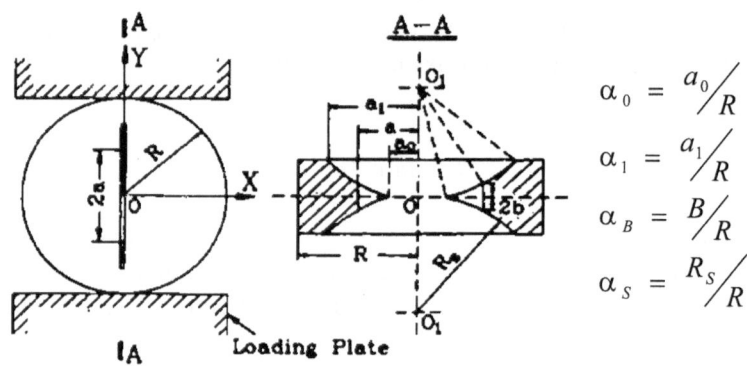

$$\alpha_0 = {a_0}/{R}$$

$$\alpha_1 = {a_1}/{R}$$

$$\alpha_B = {B}/{R}$$

$$\alpha_S = {R_s}/{R}$$

Figure 1
Geometry of the CCNBD and related parameters (after ISRM, 1995).

Table 1

CCNBD geometrical dimensions for Barre granite

Descriptions	Values mm	Dimensionless expression
Diameter D (mm)	75	
Thickness B (mm)	30	$\alpha_B = B/R = {\sim}0.79$
Initial crack length a_0 (mm)	8–9	$\alpha_0 = a_0/R = 0.21\text{–}0.25$
Final crack length a_1 (mm)	23.5	$\alpha_1 = a_1/R = {\sim}0.62$
Saw diameter D_s (mm),	50.0	$\alpha_S = D_s/D = 0.66$
Y*min (dimensionless)	${\sim}0.78$	

Table 2

CCNBD geometrical dimensions for Lac du Bonnet granite

Descriptions	Values mm	Dimensionless expression
Diameter D (mm)	194.8	
Thickness B (mm)	51.10	$\alpha_B = B/R = 0.52$
Initial crack length a_0 (mm)	16.67	$\alpha_0 = a_0/R = 0.17$
Final crack length a_1 (mm)	56.08	$\alpha_1 = a_1/R = 0.575$
Saw diameter D_s (mm),	147.4	$\alpha_S = D_s/D = 0.75$
Y*min (dimensionless)	0.76	

3. Selection of Samples

In this study Barre and Lac du Bonnet granites were selected for studying the effect of microcrack orientation on fracture toughness and fracture growth process under Mode I fracture condition. These rocks are characterized by their preferred microcrack orientations and widely different grain size distribution.

3.1 Microstructural Investigation

Microcrack observational techniques include 1) dye penetration prior to thin section preparation, 2) radiography and X-ray, 3) SEM and 4) TEM, (Kranz, 1983). In recent years the development of computer-aided image analysis programs has greatly facilitated microstructural characterization through analysis of digital images obtained from the thin sections (MOORE and LOCKNER, 1985; PRIKRYL, 2001; NASSERI et al., 2005a). These new techniques are based on direct measurements of crack length through line tracing, and grain boundary tracing from thin sections. These methods provide easier handling of larger amount of data collected and therefore provide a more representative assessment of microstructural properties. The process in this study consisted of the following stages: Image acquisition from thin sections, image preprocessing, microcracks and grain boundary tracing, measurements with automated image analysis programs (LAUNEAU and ROBIN, 1996) and data analysis. A number of scan lines at fixed angular intervals is drawn which intercepts microcracks. The latter are converted to bitmap lineaments with a minimum width of four pixels (Fig. 2a). The total number of intercepts is converted to microcrack density and expressed as microcrack length per unite area with an appropriate scale (i.e., cm/pixel). A rose diagram is used to represent the overall orientation of the lineaments with respect to particular axes on the plane of interest (Fig. 2b).

3.1.1 Microstructural characterization of Barre granite

Microstructural investigation in this study involved examination of thin sections along three orthogonal planes. Microstructural studies of this rock detailed the variation in terms of mineral size distribution, preferred microcrack orientation, microcrack density and microcrack length, measured along the three orthogonal planes. In our analysis one screen pixel represents 0.94 μm. Thus the shortest resolvable crack and grain size has a cut-off limit of approximately 1 μm. No attempt has been made in this study to measure the width of the microcracks, as some of them are finer than this cut-off limit. It has been shown that larger microcracks are the first to interact mechanically and thus dominate both the fracture process and the transport properties of rock (MADDEN, 1983; LOCKNER and MADDEN, 1991). Therefore limiting the observable crack and grain sizes to 1 μm or larger should not be considered a drawback in the present investigation. Correlation of the rock's microfabric elements with the mechanical properties such as that of fracture toughness requires knowledge of the original orientation of the rock or reference of orthogonal axis of the coordinate system (X, Y and Z axes) to the internal visible fabric of the rock. In the absence of any information regarding the natural *in situ*

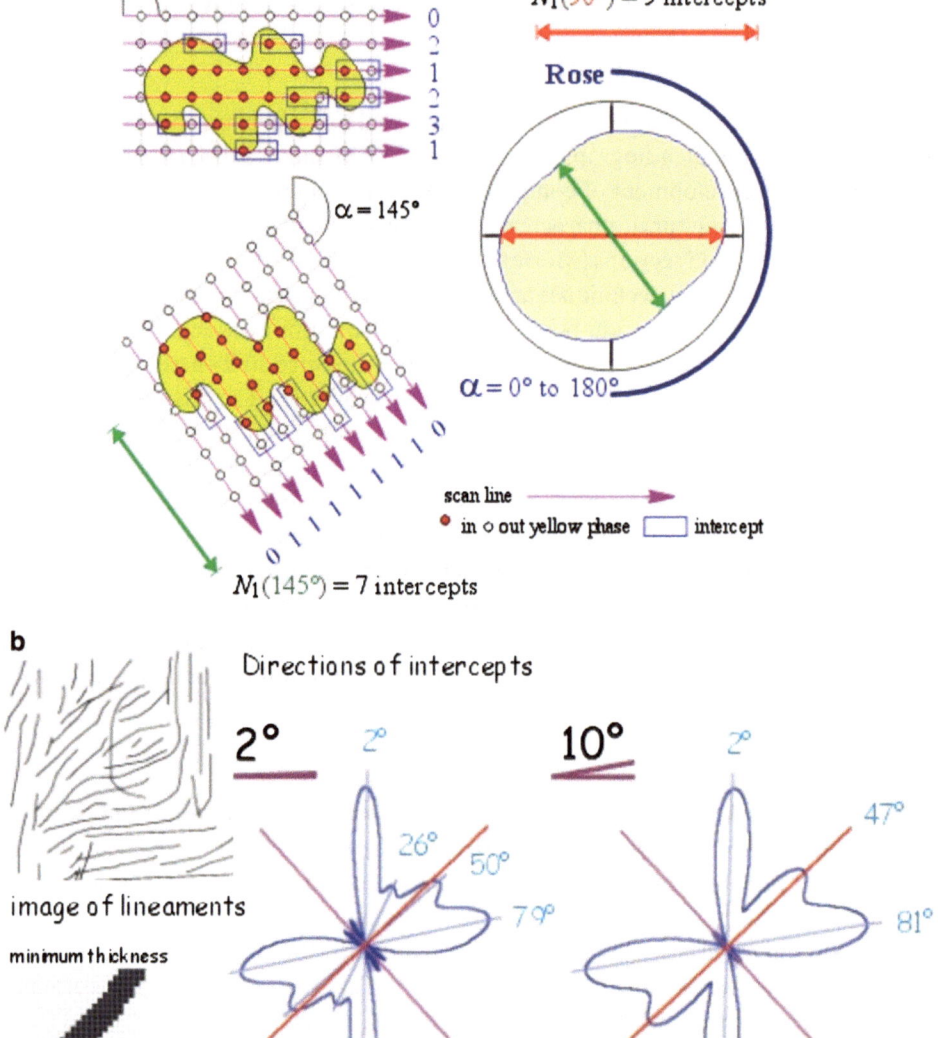

Figure 2
Illustration of scan lines intercepting a polygon or a lineament at different angular intervals (Fig. 2a), and rose diagram used to represent preferred orientation of lineaments with respect to particular axes on a plane of interest (Fig. 2b), (after LAUNEAU, P. and ROBIN, P-YF., 1996).

orientation of the granitic blocks and lack of visible internal rock fabric, thin sections were prepared normal to the three axes along which P-wave velocities were measured. Intermediate (3700 m/s) and slow (3300 m/s) and fast (4350 m/s) directions were assigned X, Y, and Z axes, respectively.

Barre granite is obtained from the southwest region of Burlington in the state of Vermont, USA. It is fine to medium grained rock with the mineral grain size ranging from 0.25 to 3 mm, with an average quartz grain size of 0.9 mm, which makes up 25% of this rock. The average feldspar grain size, the dominant mineral (70%), for this rock is 0.83 mm. The corresponding average biotite (6%) grain size in this rock is 0.43 mm. The microcracks are of intragranular type and are found in both the quartz, feldspar grains and along cleavage planes of biotite grains. The preferably oriented microcracks have an average length of 0.63 mm with maximum length of ~3.5 mm cutting through the larger quartz grains observed in the YZ plane. It is evident from the 3-D block diagram of microcrack orientation in Figure 3, that the larger microcracks (2–3 mm long) are seen to run parallel to the Y axis and the shorter ones (~1 mm) run at higher angles to the Y axis and are parallel to the Z axis. In the XY plane, intermediate size microcracks (~2 mm) are oriented parallel to the Y axis and shorter ones are again parallel to the X axis in this plane. The longer

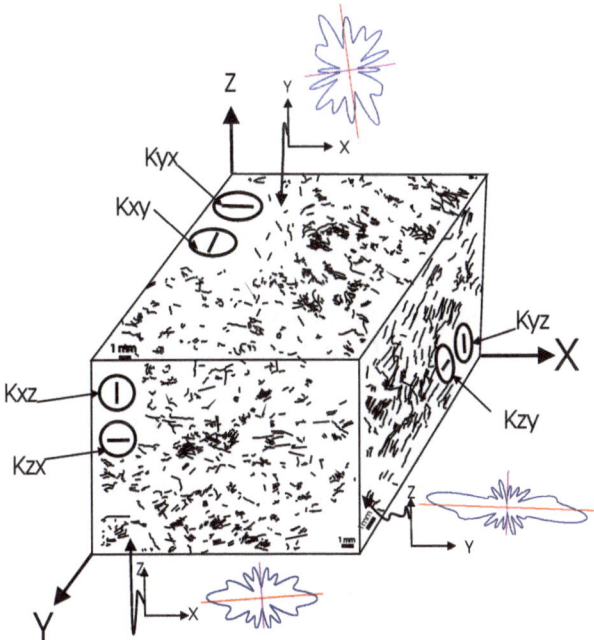

Figure 3
3-D block diagram showing location of CCNBD specimens prepared along each plane with respect to microcrack orientations in Barre granite; rose diagrams show the alignment of microcracks for each plane.

microcracks in the XZ plane are aligned parallel or at small angles to the X axis, whereas smaller microcracks are found to be parallel to the Z axis. The rose diagram representing the microcrack orientation and length along a certain direction for each plane is shown as well. Microcrack orientation analysis in Barre granite (Fig. 3) suggests a dominant alignment of pre-existing microcrack parallel to the XY plane. DOUGLASS and VOIGHT (1969) studied the microcracks orientation in Barre granite and demonstrated that a strong concentration of microcracks lies within the rift plane and the secondary concentration was found within the grain plane. XY plane is considered to be parallel to the rift plane and XZ parallel to the secondary concentration for Barre granite in this study.

3.2 Microstructural Characterization of Lac du Bonnet Granite

This rock is medium to coarse grained with the mineral grain size ranging from 0.5 to 8 mm. The average grain size for this rock is about 3.5 mm and is composed of 60% feldspar, 30% quartz and 8% biotite. The microcracks are of an intragranular type and found in both quartz and feldspar grains. The average length of the microcracks is 4 mm with a maximum length of $\sim$7 mm. The rock sample is from the Underground Research Laboratory, Atomic Energy of Canada, where the concept of deep radioactive waste disposal is being studied.

3.3 Fracture Toughness Anisotropy and Microstructure

The effect of microcrack orientation on fracture toughness of materials with layered microstructures and composites has been increasingly recognized as an important property. Similar investigations in rocks such as granite in which oriented microcracks have been found to be responsible for the splitting planes and anisotropy, help quantify the degree of anisotropy in otherwise isotropically and homogeneously treated rock. In the present study we use the CCNBD test method to evaluate this anisotropy in Barre granite. The Brazilian disk samples were prepared from three orthogonal planes with the notch orientation parallel to the main axis in each plane. The first index in CCNBD tests shown in Figure 3 represents the direction normal to the fracture plane, and the second index indicates the direction of crack propagation. The cracks were chosen to be parallel to one of the coordinate axes in this investigation and therefore CCNBD samples were prepared along six different directions with respect to microcrack orientation planes as shown in Figure 3.

Figure 4 illustrates the variation of fracture toughness measured along six different directions with the number of tests performed along each direction for Barre granite. Table 3 shows load at failure and calculated fracture toughness along six different directions including respective standard deviations, parallel to the axis of interest in three orthogonal planes. The average fracture toughness for the three fracture planes (i.e., YX, XY and YZ) reveal K_{IC} value of 1.9, 1.73 and 1.7

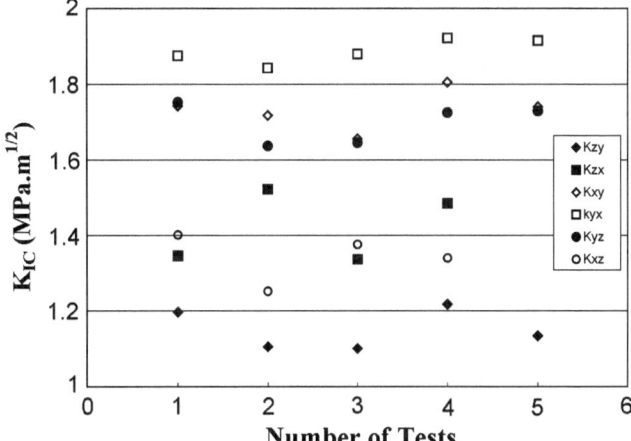

Figure 4
Variation of fracture toughness measured along six directions with the number of tests along each direction in Barre granite.

Table 3

Showing major mineralogical composition, measured load at failure and calculated fracture toughness along six different directions for Barre granite

Rock type	Quartz		Feldspar		Biotite		Load at failure	K_{IC}
	Av. grain % size (mm)		Av. grain % size (mm)		Av. grain % size (mm)		kN Std. Dev.	$(MPa.m^{0.5})$ Std. Dev.
XY Plane	0.93	25%	1.10	70%	0.46	4%	13 ± 0.46 14.3 ± 0.16	$\{$ Kxy = 1.73 ± 0.06 $\{$ Kyx = 1.89 ± 0.03
XZ Plane	0.95	23%	0.81	70%	0.43	6%	10.1 ± 0.54 10.3 ± 0.7	$\{$ Kxz = 1.34 ± 0.07 $\{$ Kzx = 1.42 ± 0.1
YZ Plane	0.94	26%	0.96	68%	0.40	4%	12.6 ± 0.42 8.5 ± 0.4	$\{$ Kyz = 1.7 ± 0.05 $\{$ Kzy = 1.14 ± 0.05

$(MPa.m^{0.5})$, respectively, whereas the other three fracture planes XZ, ZX and ZY exhibit average K_{IC} of 1.34, 1.42 and 1.14 $(MPa.m^{0.5})$, respectively. It is noted that the first index in describing the fracture planes refers to the direction normal to the fracture plane and the second index refers to the direction of crack propagation or notch's tip direction.

It is ascertained that K_{IC} measured along a direction normal to a pre-existing preferred microcrack plane (i.e., YX direction) is almost twice that measured in a parallel plane (i.e., ZY direction). Thin section studies and microcrack orientation analysis have been used to evaluate and explain these variations. Further analysis of morphology of the fracture forced to propagate along the direction that is perpendicular to the microcrack plane shows increased segmentation, roughness and creation of wing cracks (Fig. 5). This is in sharp contrast to the fracture morphology of a crack propagating parallel to microcracks in Barre granite (Fig. 6), where it is seen to be relatively straight and free of any irregularities on the crack surfaces.

4. Crack Propagation and Microstructure

4.1 Fracture Process Zone

The fracture process zone (FPZ) in rocks is defined as the region affected by microcracking and frictional slip surrounding the visible crack tip propagating under stress, (LABUZ *et al.*, 1987; VERMILYE and SCHOLZ, 1999). The width of the FPZ is

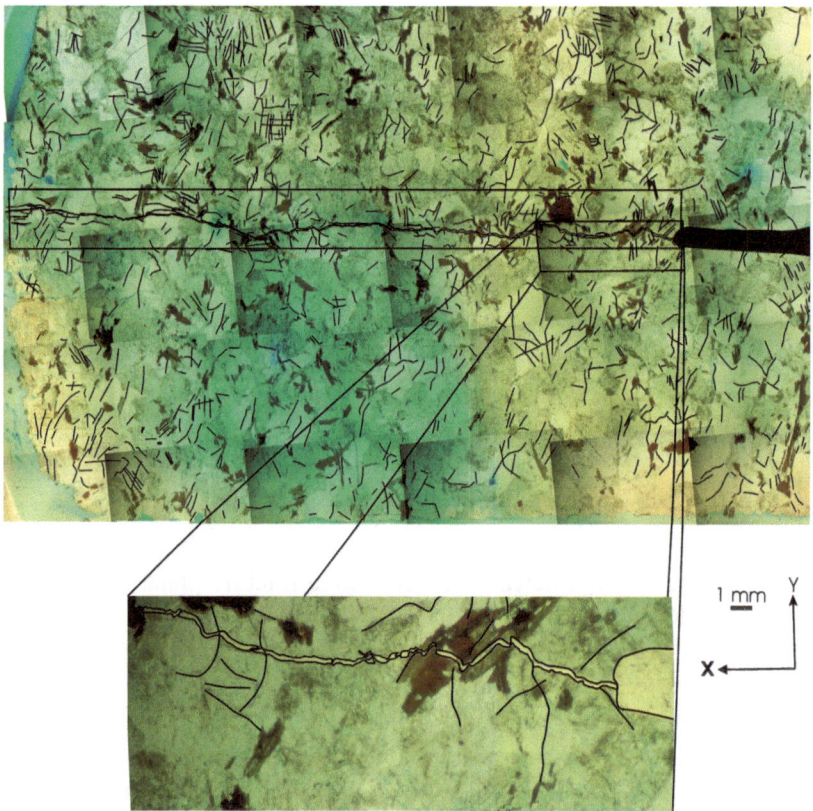

Figure 5
Fracture propagation (right side of image) normal to pre-existing preferably oriented microcracks in Barre granite. The close-up image shows the rough and segmented morphology of the fracture plane.

defined as the longest distance between visible cracks on either side of the main fracture and/or fault and its length is defined as the length between the fault tip and the crack with the greatest distance in front of the fault tip.

Dimensions of the fracture process zone (FPZ) based on the distribution of microcracks and acoustic emissions were investigated in various experimentally induced shear (LOCKNER, 1996; ZANG et al., 2000) and tensile fractures (ZIETLOW and LABUZ, 1998). These studies suggest that in a FPZ microcrack, density and number of AE show an exponential decrease as a function of distance from the main failure plane. Similar observations are reported in the field based on meso and microstructural examinations across the faults plane (SCHOLZ et al., 1993; WILSON et al., 2003). It is observed that microcrack density increases by an order of magnitude in the FPZ surrounding the fracture tip in comparison with the

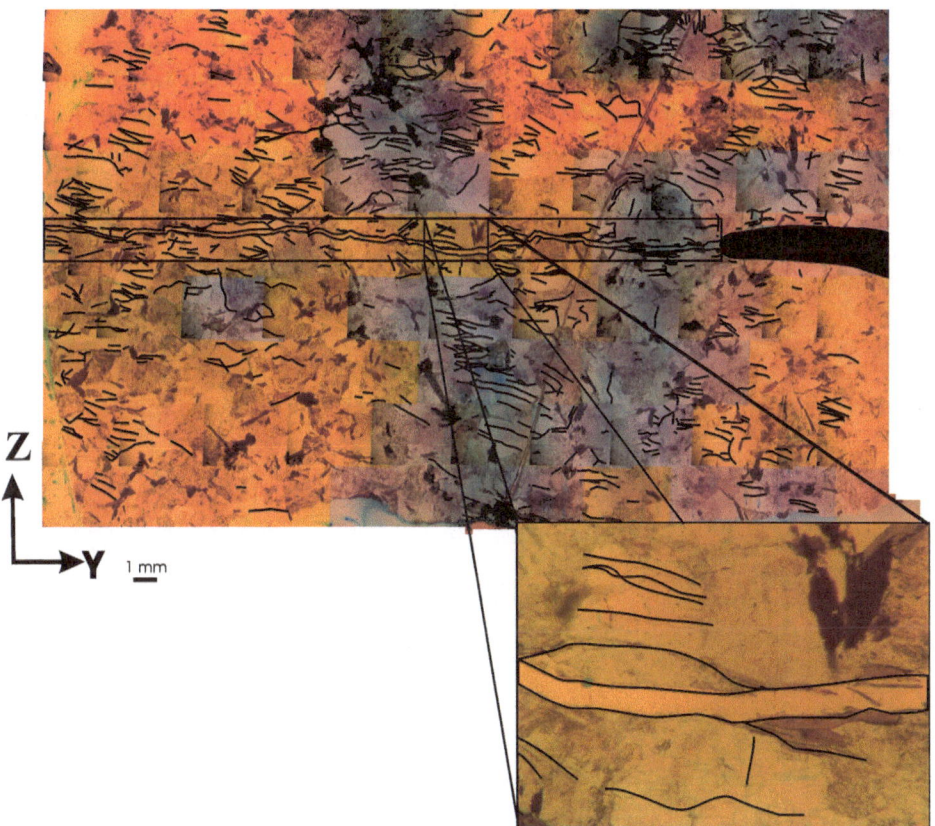

Figure 6
Fracture propagation (right side of the image) parallel to pre-existing preferably oriented microcracks in Barre granite. The close-up image shows the smooth morphology of the fracture plane. (The respective crack openings shown in close-up images in Figures 5 and 6 are not scaled and therefore, not comparable).

undamaged rock (NASSERI *et al.*, 2005b). In the present study the microcrack density profiles, as a function of distance on either side of the main fracture, have been taken as the basis for investigating FPZ. These are then compared with background or far-field microcrack density to delineate the width of the FPZ. Measurements of the longer transgranular microcracks on either side of the main fracture were used as additional guides. In this study an attempt is made to investigate the effect of oriented microcrack on FPZ width of Barre granite showing extreme values of fracture toughness. FPZ width analysis using a thin section technique for Lac du Bonnet granite for one plane is shown as well.

4.1.1 Barre granite

Figure 7 shows the microcrack density (Φ) and microcrack orientation comparison between the far-field and FPZ in Barre granite for a situation in which the fracture is propagated from the notch's tip along a direction normal to microcracks. Figure 7 is the same as Figure 5 but with the grains removed from view. Microcrack orientation in the process zone is found to be parallel to the fracture propagation plane, and its density was observed to be almost twice that of the far-field microcrack density. FPZ width varies between 2 to 3 mm and maximum deviation of the fracture plane from the main propagation path is 1.26 mm, and this deviation increases towards the end of the 3.12-cm long fracture. Fracture deviation is measured with respect to the hypothetical fracture plane originating at the notch tip, had the fracture remained perfectly planar.

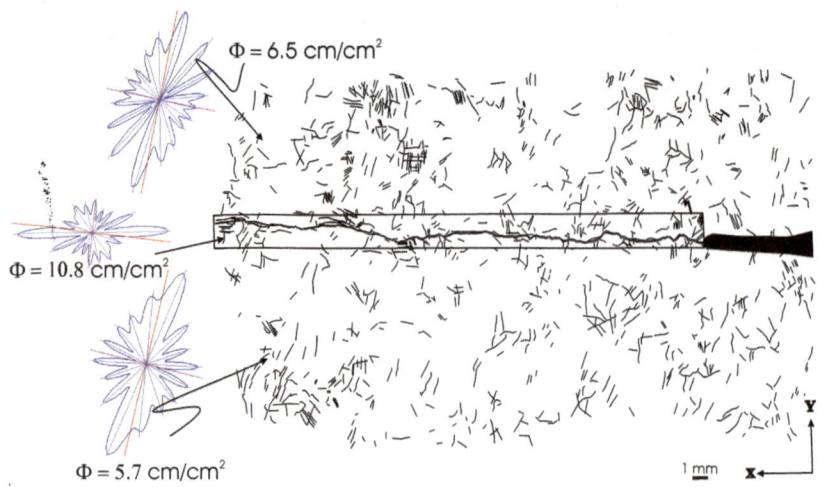

Figure 7

Microcrack density (Φ) and orientation (rose diagram) in the fracture process zone and far-field area in Barre granite (XY plane). Extended black box shows the width of FPZ.

Figure 8 shows the microcrack density and microcrack orientation comparison between the far-field and FPZ in Barre granite for a condition in which the fracture is directed parallel to microcracks. Figure 8 is the same as Figure 6, but with the grains removed from view. Microcrack density in the process zone is 12.6 cm/cm^2 where far-field microcrack densities on either side of FPZ are 6.7 and 4.5 cm/cm^2. The rose diagram showing the microcrack orientations in FPZ, unlike that of the far-field ones, does not show slight maxima perpendicular to the main flat maxima running parallel to the main fracture. FPZ width in this direction is 1.55 to 2 mm and the maximum deviation of the fracture plane from the main propagation path is 0.7 mm. This deviation is observed mainly at the mid section of the 2.85-cm long fracture.

4.1.2 Lac du Bonnet granite

In this experiment, only one side of the directed crack issuing from the notch broke through the edge of the disc sample; the other end of the crack stopped just short of the edge of the sample. These are designated as notch A and notch B, respectively. Comparison of microcrack density and orientation between far-field and FPZ along YX plane and from the notch A side is shown in Figure 9. FPZ width for this rock is ~10 mm. Microcrack density for far-field to FPZ varies from 4.4 cm/cm^2 to 10.7 cm/cm^2 and the rose diagram for the far-field area reveals several maxima parallel and normal to the fracture propagation direction. Whereas, the rose diagram for FPZ shows a single maxima parallel to the fracture direction. Figure 10 shows the microcrack density and the respective orientations measured in front of the arrested fracture (notch B) in Lac du Bonnet granite along the same plane. Microcrack density in front of the arrested tip reveals an order of magnitude increase in density (44 cm/cm^2) compared to far-field microcrack density. Microcrack orientations also reveal a sharp preferred direction in both FPZ and in front of the crack tip. These microfractures are also seen to be nearly parallel to the direction of crack propagation.

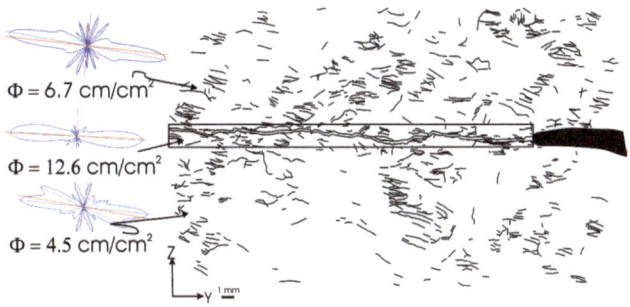

Figure 8

Microcrack density (Φ) and orientation (rose diagram) in the process zone and far-field area in Barre granite (YZ plane). Extended black box shows the width of FPZ.

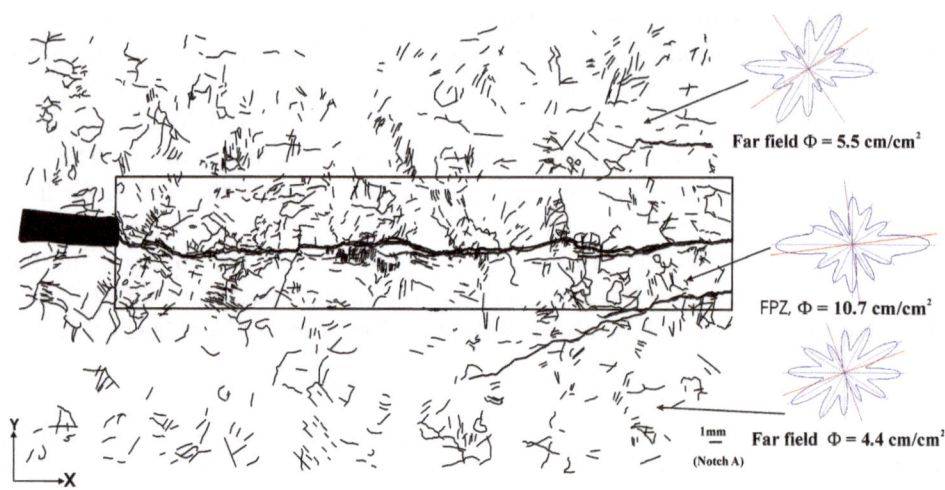

Figure 9

Microcrack density (Φ) and orientation (rose diagram) in the fracture process zone and far-field area in Lac du Bonnet granite (YX plane). Extended black box shows the width of the FPZ.

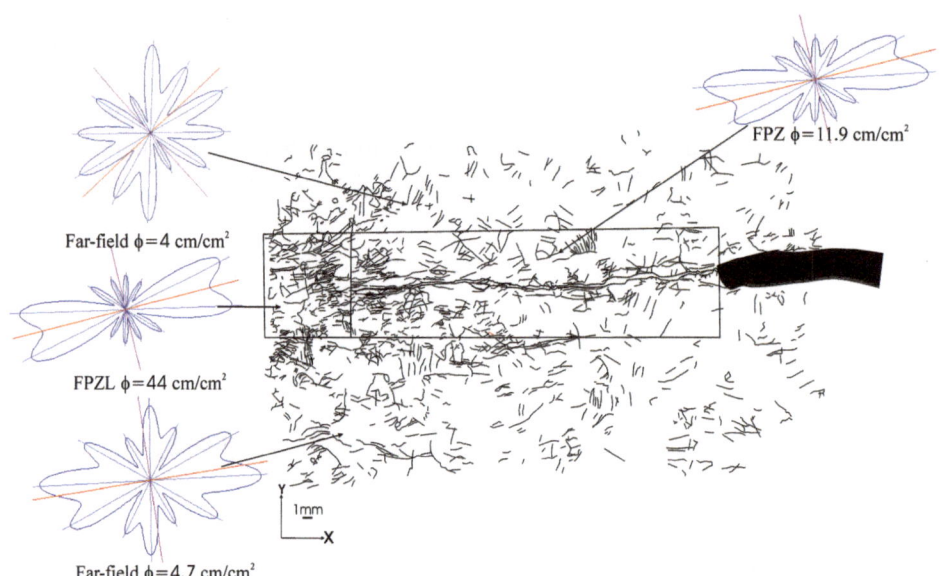

Figure 10

Microcrack density (Φ) and orientation (rose diagrams) in FPZ and far-field area in a seized fracture in notch B of Lac du Bonnet granite (YX plane). Extended black box shows the width of FPZ with the inset showing microcrack density ahead of arrested fracture.

5. Acoustic Emission and Crack Propagation

Both analytical and experimental AE studies have been carried out to demonstrate the usefulness of the former in core-based fracture toughness testing of rocks (HASHIDA, 1993). It has also been shown in plots of AE activity against time, that the steep rise of the former is representative of higher amplitude events than those generated at lower load levels (HASHIDA, 1993). Examination in thin sections of the test samples after the fracture toughness tests showed that the first region of the lower slope is associated with distributed microcracking around the notch tip, and that the steep slope corresponds to the localized macroscopic crack extension accompanied by the linking of microcracks.

The fracture toughness tests in Barre granite were carried out in 75-mm diameter samples. For reasons of assuring adequate precision in the location of acoustic emission events, detailed in the next section, considerably larger diameter samples are required. The AE portion of this study was therefore carried out in the Lac du Bonnet samples. The analogous nature of the fracture process in the two rock types, as demonstrated earlier, justified this selection. Except for scale, any conclusion based on the larger diameter samples should hold equally well in the smaller diameter sample with no loss of generality, especially for these two granitic rock types.

5.1 Experimental Procedure

5.1.1 Experimental setup

In this study, the sample was line-loaded in a MTS loading machine, at a displacement rate of 0.0005 mm/s for testing of Lac du Bonnet granite. A considerably slower rate than suggested in ISRM (1995) was selected to facilitate microcrack development imaging using AE techniques. The displacement, force and time base of the experiment were controlled and recorded with TestStar software. The diagram of the CCNDD specimen and sensor locations are shown along two different planes in Figure 11a. Figure 11b shows the conceptual diagram of experimental setup used to record ultrasonic data.

AE events were captured using the Hyperion Giga RAM Recorder, which stores continuous ultrasonic waveform data onto a 40 GB circular Random Access Memory (RAM) buffer at 14-bit resolution. In this experiment, the sampling frequency was 10 MHz, providing a 134 second segment of continuous waveform data on 16 channels. The system functions as a continuous circular buffer that can be transferred to disk following the recording of a significant event. Advances in technology have lead to rapid AE waveform acquisition systems, allowing new observations of rupture. However, we believe this to be the first system to record continuous waveforms for significant time periods, thus removing the effect of 'mask' time, during which no AE are recorded while events are written to permanent storage and removing the sampling bias imposed by trigger levels. In addition to continuous

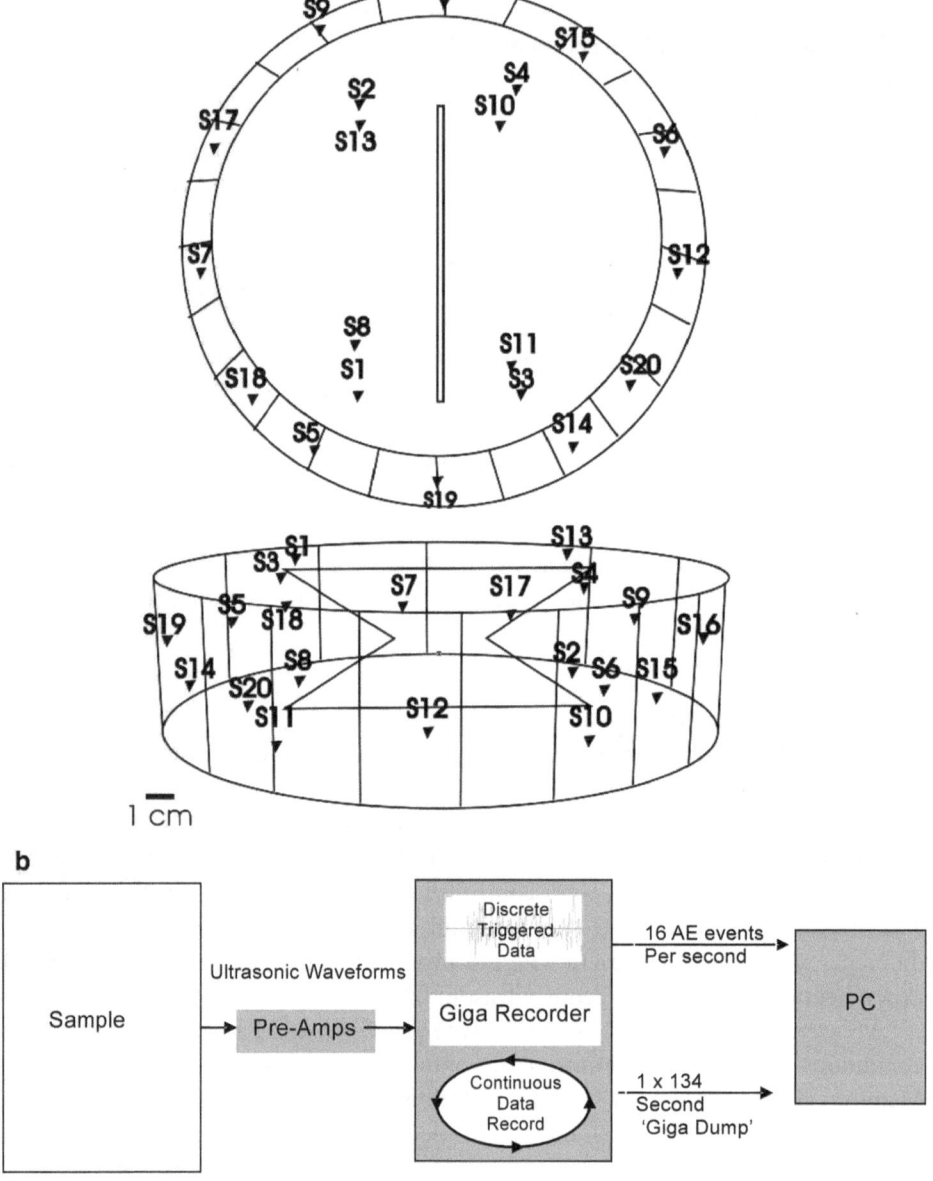

Figure 11

Schematic sketch of CCNBD specimen (Lac du Bonnet granite) and sensor locations on two different planes, (Fig. 11a) and experimental setup used for recording ultrasonic data (Fig. 11b).

data, the Giga RAM Recorder concurrently collects 'triggered' AE data during the entire experiment. Triggered data have identical recording parameters to continuous data. Each event has duration of 409.6 μs, and a maximum of 16 events per second are captured. The number of all AE counts per second is also recorded. The recorded ultrasonic signals passed through external 40 dB pre-amplifiers (PAC model 1220A). Four transmitters, dedicated as active sources for velocity surveys, were driven by a Panametrics 5072PR pulse generator module. The ultrasonic seismic velocity tests in the rock sample prior to notching exhibited anisotropic wave velocity behavior (Fig. 12), with the fast axis being 5290 m/s (AA' on Fig. 12) and the slow axis (EE' on Fig. 12) being 4550 m/s. The velocity anisotropy measurements were confined to prestressed conditions only, as such measurements during the actual loading process would have interfered with the accompanying acoustic emission events. The AE first arrivals were manually picked and source locations obtained using a Downhill Simplex algorithm (NELDER and MEAD, 1965), assuming a transversely anisotropy structure to an estimated accuracy of 3 mm. This necessitated the use of large diameter (195 mm) samples as mentioned earlier.

5.1.2 AE monitoring during crack initiation and propagation

Failure occurred 568 seconds after loading commenced, with the applied load reaching the peak value of 32.72 kN; the stress intensity factor K_I attains the highest value of 1.55 (MPa.m$^{0.5}$) representing the fracture toughness, K_{IC} of Lac du Bonnet granite. A total displacement of 0.28 mm was recorded in this experiment. Approximately 1000 AE events were captured, for which the first arrivals were picked automatically. It was found that the presence of the notch caused complications, introducing location error. For the purpose of this paper, we present 355 manually processed AE events using an optimized array of 8 receivers for one of

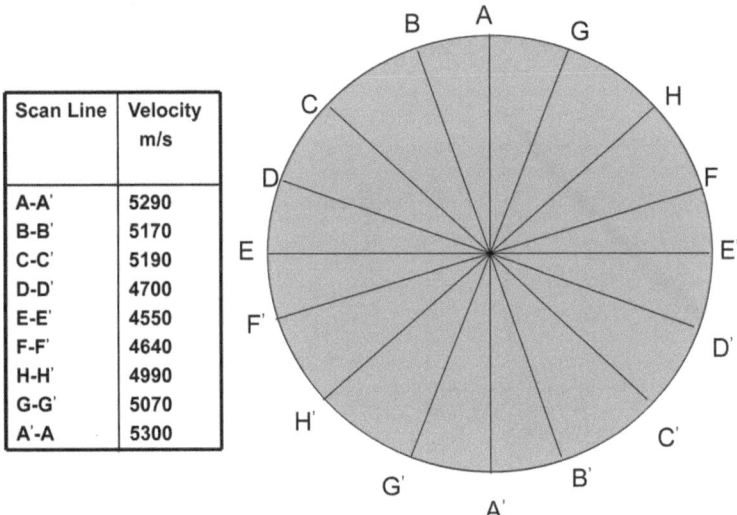

Scan Line	Velocity m/s
A-A'	5290
B-B'	5170
C-C'	5190
D-D'	4700
E-E'	4550
F-F'	4640
H-H'	4990
G-G'	5070
A'-A	5300

Figure 12
Sensor locations and measured *P*-wave anisotropy in Lac du Bonnet granite.

the notches of the sample (notch A). Such a technique improves location accuracy greatly by removing the effects of the notch and increasing the accuracy of the first arrival picks. This improved location accuracy is important in comparison with microstructural observations.

The results from this test are shown in Figures 13a–d. Figure 13(a) shows AE hits per second, along with force versus time of loading. A plot of the cumulative number of AE events and stress intensity factor (K_I) versus time is shown in Figure 13(b). The load versus displacement graph is shown in Figure 13(c). AE source locations are displayed from the final 17 seconds of the test, including post failure activity in Figure 14. These locations, divided into Stages A–E, are shown in Figure 13(d), which is an expansion of 13(a). AE locations are shown for each Stage as 1) discrete events, and 2) as a contour density plot. Views are shown along and perpendicular to the strike of the notch in Figure 14.

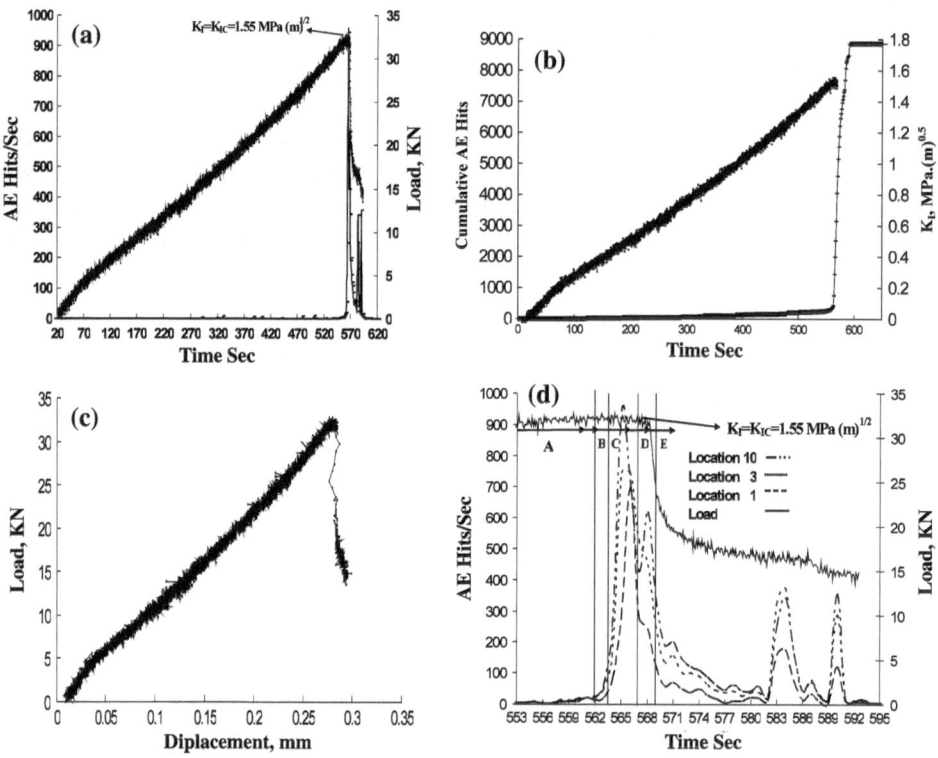

Figure 13
(a) Load and AE hits/second versus time (fracture toughness value for critical stress intensity factor is shown at top);. (b) SIF (K_I) and cumulative AE versus time; (c) Variation of load with displacement for CCNBD at failure; and (d) close-up window of load and AE hits (sensed at various locations) versus time around fracture initiation and failure.

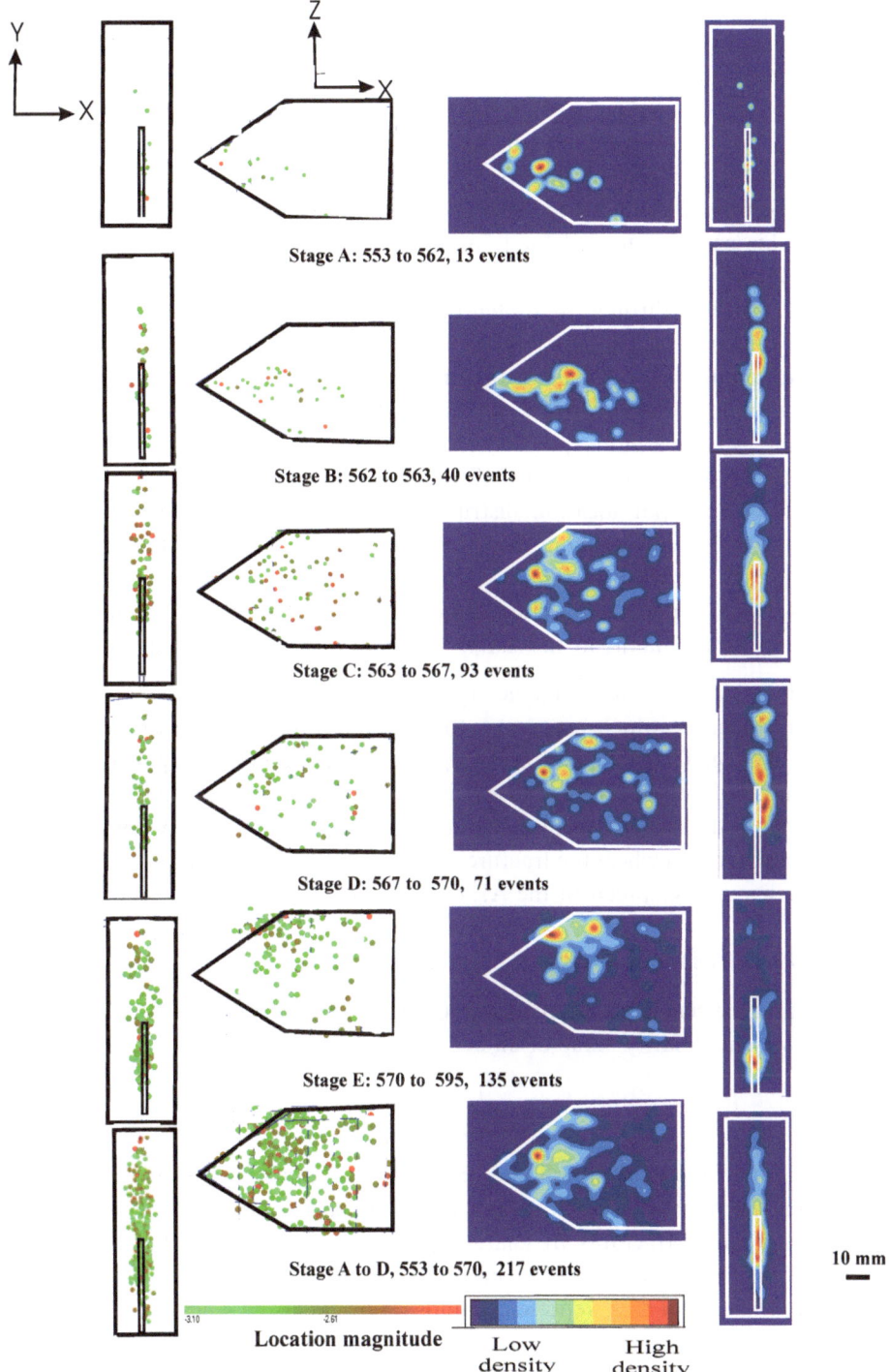

Figure 14
AE Source locations of 350 events (dots) and their equivalent density maps showing the final 17 seconds
(Stages: A-D) in the experiment. Stage E shows the AE source location for post-failure region.

5.1.3 AE hypocenter distribution and fracture growth

Stage A: This stage corresponds to the recorded activity during nine seconds between 553 to 562 seconds, during which 13 events locate in the triangle defined by the notch region (this region is subsequently referred to as the notch triangle).

Stage B: In this two-second period, from 562 seconds to 563 seconds, there is a marked increase in AE event rate. Forty events are located in a narrow zone extending from the notch tip; the events mainly locating in the notch triangle.

Stage C: This period extended for four seconds, from 563 to 567 seconds, during which 93 AE events were located. However, the distribution becomes more planar, spreading through the notch-tip triangle to the edge of the sample. This could be interpreted as the fracture front extending, which would suggest a fracture propagation velocity of the order of cm/s. This applies to stable crack propagation in the region of the notch tip. There is insufficient event coverage to allow one to have a more specific figure on this velocity.

Stage D: This three-second period (567 to 570 seconds) contains 71 located events which have a similar location distribution to the previous stage. During this period, failure occurs starting between 568 and 569 seconds. The time interval from the rapid increase in AE rate (Stage B) to failure is 6 seconds.

Stage E: A twenty-one second period of post failure activity is shown during which 135 events locate mainly at the sample boundary within the notch triangle area.

The AE locations exhibit a change in distribution, whereby a linear feature is observed in Stage B, which progresses to a planar feature intersecting the sample boundary in Stage C. This could be interpreted as the fracture propagating two or three seconds before the sample fails. It is interesting to observe that locations are concentrated in the triangle defined by the notch, and comparatively few events are observed in the region where the fracture must propagate to cause failure. The reason for this change in the spread of the AE events is unclear at the moment.

5.2 AE Focal Mechanism Analysis

AE source mechanisms are calculated using a first motion polarity method (ZANG *et al.*, 1998). The polarity value of an event is calculated by:

$$Pol = \frac{1}{K} \sum_{I=1}^{K} \text{sign}(A_1^I), \tag{3}$$

where A is the first pulse amplitude and K is the number of channels used for hypocenter determination. For the transducers used in this test, compressional pulses arising out of tensile fractures are characterized with positive polarity. Events in the range $0.25 < Pol \leq 1$ are defined as being tensile, $-0.25 \leq Pol \leq 0.25$ are defined as shear and $-0.25 \leq Pol \leq -1$ represent implosion or collapsed sources.

Analysis of 220 prefailure events revealed that 75% of the total events were of tensile type, 15% were of shear type and 10% had an implosion or collapsed source type. For 134 post-failure events, 40% were shear type, 39% were tensile and 21% were implosion type.

Figure 15 demonstrates AE hits/sec and the load curve during the evolution of failure in Mode I. The percentage of AE events type versus time, between 562 to 569 seconds, i.e., six seconds before failure and last second (568th to 569th seconds) during failure in notch A (Fig. 15a) is shown. AE activity during 562 to 568 seconds increased from 3 to 45 events. These were predominantly tensile in nature. At the 567th second, of the 52 events recorded, 60% were tensile, 25% shear and 15% implosion type. A total of 42 events were recorded one second (i.e., 568th second) prior to failure or stress drop. During the failure between the times 568th to 569th

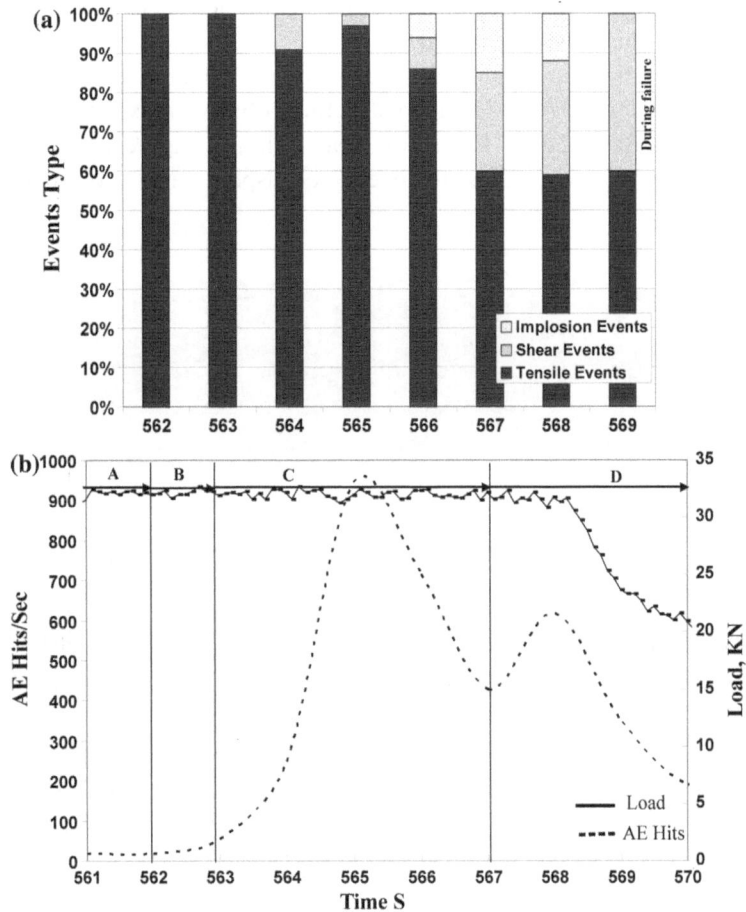

Figure 15

AE hits and the load curve during evolution of failure in Mode I. (a) Variation of event types (%) versus time at different stages of loading (A-D); (b) Variation of AE hits/sec versus time matched with the time in figure (a) to exhibit the relation between the event types (i.e., tensile, shear and implosion).

seconds, only ten events were recorded, in which 60% of the events were found to be of tensile and 40% proved to be of shear types. Variation of AE hits/sec versus time matched with time in Figure 15(a) is illustrated in Figure 15(b) to demonstrate the evolution of failure with time at various stages of loading and event types (i.e., tensile, shear and implosion).

6. Correlation of AE Technique with Optical Method

6.1 Microstructural Data from Notch A

A photo-mosaic showing the mineral fabric of notch A is presented in Figure 16; the notch being the black feature on the left side. The fracture plane resulting from this test is traced, and the microcracks are also marked. Superimposing AE source locations and optical images provides a unique opportunity to compare AE with microstructural observations. The lightly shaded region highlights the width of FPZ. For greater clarity, Figure 17 shows the same region with the mineral grains removed, to display only the fracture plane,

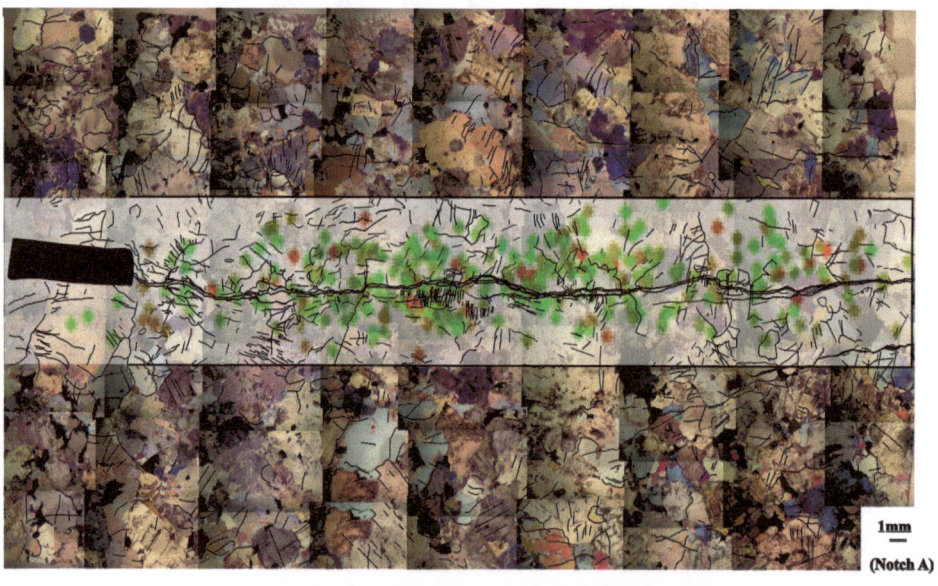

325 Events Scaled to Location Magnitude

-3.14   -2.35   -1.57

Figure 16
Tiled photo image of microstructure from thin section, around notch A (left side of image) with fracture trace and microcracks shown. 355 AE locations (dots) are superimposed.

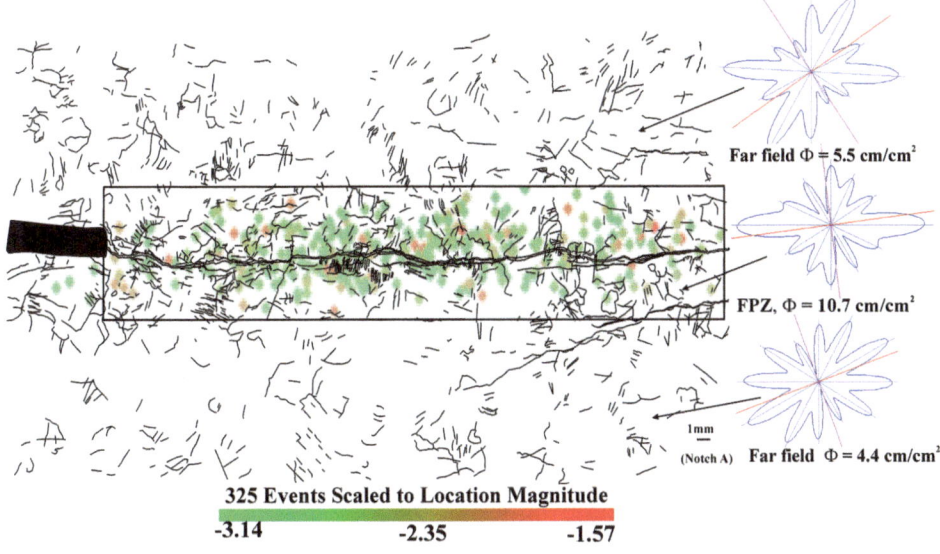

Figure 17

Microcrack density and orientation (rose diagram) in fracture process zone and far-field area in notch A.
Inset delineates the width of FPZ as derived from the superimposed AE events.

microcracks and AE events. In addition, Figure 17 shows the microcrack density and orientation (see the rose diagrams). The density and orientation of the microcracks were systematically calculated for consecutive 2.5 mm wide zones, orientated parallel to the fracture trace. The length of the fracture is measured to be 85 mm; the FPZ width is optically estimated to be 10–15 mm, which matches well with the width determined using AE techniques.

6.2 Comparison of AE and Optical Observations of the FPZ

Figure 18 is a comparison of microcrack density obtained from microstructural analysis, and AE source location density plotted against distance from the macro-fracture trace for notch A. The AE density was calculated equivalent to the microcrack density. For both the AE and microstructural densities, there is an exponential increase in density towards the fracture. The FPZ widths inferred from both AE and optical measurements are closely correlated. This is at variance with earlier work (JANSSEN et al., 2001) in which the FPZ resolution defined by AE was 2–4 times greater than the optical measurements. This discrepancy between these two studies could possibly be due to the fact that in the latter the width of FPZ would statistically represent the entire population of cracks and therefore be less dependent on some particular crack propagating under Mode I, as in the present study.

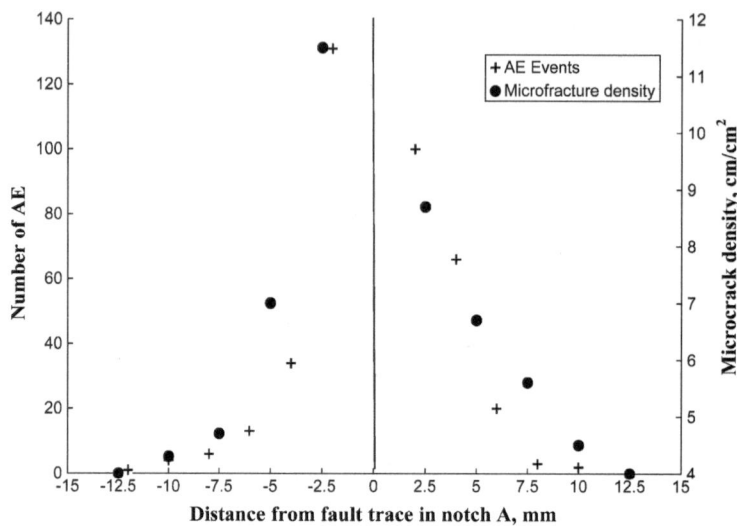

Figure 18
Correlation of AE events distribution with that of microcrack density as a function of distance from the fracture plane.

7. Discussions

The highest measured fracture toughness value i.e., $K_{IC} = 1.92$ (MPa.m$^{0.5}$) in Barre granite was observed to be along the X direction in ZX plane (Kyx, fracture plane), which is normal to the direction of the main set of oriented microcracks (i.e., XY plane). The lowest value of 1.1 (MPa.m$^{0.5}$) is found to be along the Y direction in XY plane (Kzy, fracture plane) where the fracture is propagated parallel to the Y direction in that plane along which delamination is easier. Two other planes with relatively higher average values i.e., 1.7 (MPa.m$^{0.5}$) were observed to be along directions that are normal to two orthogonal sets of pre-existing microcrack planes. The other remaining two directions (i.e., X and Z in YX and YZ planes) with average K_{IC} of 1.42 and 1.34 (MPa.m$^{0.5}$), respectively indicate the intermediate fracture toughness values for Barre granite. The crack propagation in the former direction is normal to YZ plane but parallel to YX plane. In the latter direction (in Z) the fracture is forced to propagate parallel to YZ and normal to XY planes. It is noted that the first index in describing the fracture planes refers to the direction normal to the fracture plane and the second index refers to the direction of crack propagation.

Further analysis of morphology of the fracture forced to propagate perpendicular to the microcrack plane also shows increased segmentation, roughness and creation of wing cracks (Fig. 5). This is in sharp contrast to the fracture morphology of a crack propagating parallel to microcracks in Barre granite (Fig. 6). This probably resulted in the different features associated with fracture process zone (FPZ), i.e., width and length of FPZ even within the same rock type. The width of the FPZ measured on ZY (lowest K_{IC}) plane (Kzy, fracture plane) is $\sim$ 1.55 to 2 mm and the maximum deviation of the fracture from the main path is 0.7 mm, whereas the width of FPZ measured on XY (highest K_{IC}) plane (Kyx, fracture plane) is 2.2 to 3 mm and the maximum deviation of the fracture from the main path is 1.26 mm. Microcrack orientation analysis of the far-field and FPZ in both the planes characterized with extreme K_{IC} values shows that within the width of the FPZ microcracks are parallel to the main fracture irrespective of the orientation of the microcracks with respect to propagation direction. However, it should be pointed out here that all reference axes in the experiments with test samples were assigned on the basis of measured P-wave velocities along three arbitrarily fixed orthogonal axes. Therefore, these directions may not necessarily represent the maxima or minima of the characteristics studied in the test rocks as whole.

Stress interaction between a fracture and a field of microcracks (or process zone) generate irregular fracture propagation front, which is generally referred to as rugosity front (the fracture does not propagate like a straight razor blade for instance, but more or less like a rugous wavefront). Considering the stress intensity factors in Modes II and III, the asymmetrical microcrack field generates stress perturbations in modes II and III, although the main fracture is loaded in Mode I. Consequently, stress perturbations in Modes II and III, when large enough (e.g., in XY plane, Fig. 5) can promote out of plane fracture propagation (KACHANOV, 1994).

In terms of AE activity, it is seen that the AE rate rapidly accelerates a few seconds before failure. Source locations suggest a migration of a fracture plane could be occurring, mainly in the triangular area defined by the notch. The velocity of this fracture growth would be of order cm/s. The lack of AE locations between this area and the loading platen, through which region the fracture must propagate, could be due to a bias in the sensor array geometry, preventing accurate locations in this area. It is also observed that the initial linear alignment of AE events near the notch tip spreads out to a more planar configuration, with major events aligned perpendicular to the propagation direction of the notch tip. The reasons for these observations are the subject of a continuing investigation. The majority of prefailure AE events are interpreted as tensile, whereas postfailure, an equal number (40%) are interpreted as due to shear and tensile fractures. This change in mechanism could be synonymous with movement of a fault plane accompanied by shearing of asperities. Focal mechanism analysis based on first motion results of shear failure experiments reveals the preponderance of quadrant type (double couple) and complex type events over tensile type events (reviewed by LOCKNER, 1993). Analysis of the first motion results of the events in the present study demonstrate the early predominance of tensile type events (85% to 97%) due to the nature of the test itself, changes to less dominantly tensile events (60%) compared to shear-type events (40%), 2~3 seconds prior to failure (Fig. 15). Generation of rugosity front can be equated with the creation of stress intensity factor in Modes II and III, as a result of stress interaction between the main fracture and the pre-existing field of microcracks.

A delay on the order of three seconds between failure of the sample (pick load drop) and peak AE activity has been noted. Similar observations have been made by previous researchers (LEI *et al.*, 2000). This could possibly be due to the decrease in the number of AE events that are recognized as separate events; for example, because of overlapping of AE events before failure. Analysis of cumulative AE energy during that instant could be useful for such clarification, which is a part of continuing research.

Microstructures analysis demonstrates that the width of the FPZ is between 10 and 15 mm while the fracture length is only 85 mm in Lac du Bonnet granite. In the FPZ, microcrack density is around 11 cm/cm^2, while the far-field background value is approximately 4 cm/cm^2. However, in front of an arrested fracture, the microcrack density reaches 44 cm/cm^2 which is consistent with theoretical investigations on the fractal nature of newly created fracture surfaces in brittle materials (CHELIDZE *et al.*, 1994). Similarly, the microcrack orientation is found to have a dominant plane in the FPZ, whereas in the far-field no dominant orientation is observed. Unlike previous studies, we find that both AE activities and microcrack densities increase exponentially towards the fracture plane. The two approaches also yield the same FPZ width, which suggests that the AE approach can be used as a diagnostic tool to investigate the formation of FPZ related to Mode 1 tests.

8. Conclusions

A systematic laboratory study involving tensile loading of homogeneous granitic rocks (Mode I fracture) under standard test procedures (i.e., Cracked-Chevron-Notch-Brazilian-Disc) and an advanced acoustic emission system (AE) has been carried out with Barre and Lac du Bonnet granitic samples. It was shown that the measured value of fracture toughness can vary by as much as 60% or more in a relatively homogeneous rock like Barre granite. This variation has been conclusively shown to be linked directly to the microstructural anisotropy present in the rock, as represented by its microcrack density distribution and its orientation with respect to fracture propagation direction in the fracture toughness test. A fracture propagating at right angles to the major set of microcracks yielded the highest fracture toughness value. Such fractures, in contrast to the fracture propagating parallel to the dominant microcracks, were characterized by increasing roughness, segmentation, and creation of wing cracks. All fractures issuing from the notch tip in the fracture toughness tests gave rise to significant fracture process zones (FPZ) surrounding the major crack. The microcrack density in FPZ was found to increase by a factor of two or more, compared to that of the prefractured state in both Barre and Lac du Bonnet granites. In both cases, the resulting microcracks in this zone were closely aligned with the major fracture plane. Real-time monitoring by AE evolution of the crack at the notch tip and that of FPZ with time showed that AE rate increased exponentially until a few seconds before actual failure of the sample. Analysis showed that the dominant fracture mechanism during the prefailure state was due to tensile mode, whereas, in the post-failure state it was divided equally between tensile and shear modes. Out of plane deviation of the main fracture may explain increased shear-type AE events just prior to final rupture loaded under Mode I. The formation of FPZ and the major fracture plane could be accurately tracked through the detected AE events, and the results on FPZ width agreed very well with that of the direct microscopic analysis of the failed samples.

The study reported here is part of a continuing investigation on fracture behavior of rock under varying loads for a range of parameters. This includes understanding the relationship between P-wave velocity anisotropy and crack density, AE focal mechanism analysis using the moment tensor method, cumulative AE energy, amplitude and duration of load as well as temperature. Establishment of a relationship between fracture toughness and ultrasonic velocities, which could allow the estimation of material toughness by performing P- and S-velocity measurements, would be of very significant interest in future research. Measurement of fracture toughness under varying strain rates under dynamic loading conditions, and the effect of the latter on formation of the fracture process zone would further our understanding of the fracture process in rock.

REFERENCES

CHELIDZE, T., REUSCHLE, T. and GUEGUEN, Y. (1994), *A theoretical investigation of the fracture energy of heterogeneous brittle materials*, J. Phys. Condens. Matter 6, 1857–1868.

CHEN, Y., NISHIYAMA, T., KUSUDA, H., KITA, H. and SATO, T. (1999), *Correlation between microcrack distribution patterns and granite rock splitting planes*, Int J Rock Mech. Min Sci. 36, 535–541.

DOUGLASS, P.M. and VOIGHT, B. (1969), *Anisotropy of granite: A reflection of microscopic fabric*, Geotechnique 19, 376–398.

DWIVEDI, R.D., SONI, A.K., GOELL, R.K. and DUBE, A.K. (2000), Fracture toughness of rock under subzero tempreture conditions, Int J Rock Mech. Min Sci. Geomech. Abstr. 30, 821–824.

ENOKI, M. and KISHI, T. (1995), *Effect of rate on the fracture mechanism of TiAl*, Mat. Sci. Engin. *A192/ 193*, 420–426

HASHIDA, T. (1993), Fracture toughness testing of core-based specimens by acoustic emission, Int J Rock Mech. Min Sci. Geomech. Abstr. 30, 1, 61–69.

IQBAL, M.J. and MOHANTY, B. (2005), *Fracture toughness measurement in brittle rocks*, Proc. 33rd Ann. Conf. Canadian Soc. Civil Engg., GC-220-1-7.

ISRM (1988), *Suggested method for determining the fracture toughness of rocks*, Int J Rock Mech Min Sci Geomech Abstr. 25, 73–96.

ISRM (1995), *Suggested method for determining Mode I fracture toughness using cracked Chevron notched Brazilian disc (CCNBD) specimens*, Int J Rock Mech. Min Sci. Geomech. Abstr. 32, 57–64.

JANSSEN, C., WAGNER, F.C., ZANG, A. and DRESEN, G. (2001), *Fracture process zone in granite: A microstructural analysis*, Int. J. Earth Sci. 90, 46–59.

KACHANOV, M. (1994), *Elastic solid with many cracks and related problems*, Adv. Appl. Mech. 30, 259–445.

KEVIN, T. and O'BRIEN, (1998), *Interlamiar fracture toughness: the long and winding road to standardization*, composites, Part B *29B*, 57—62.

KIRBY, G.C. and MAZUR, C.J. (1985), *Fracture toughness testing of coal,* 26 US Symp. on rock echancs, Rapic City, SD.

KRANZ, R.L. (1983), *Microcrack in* rocks: A review, Tectonophysics *100,* 449–480.

LABUZ, J.F., SHAW, S.P. and DOWDING, C.H. (1987), *The fracture process zone in granite: Evidence and effect,* Int. J. R. Mech. Min. Sci. Geomech. Abstr. *24,* 235–246.

LAUNEAU, P. and ROBIN, P-YF. (1996), *Fabric analysis using the intercept method,* Tectonophysics *267,* 91–119.

LEI, X., KUSUNOSE, K., RAO, M.V.M.S., NISHIZAWA, O. and SATOH, T. (2000), *Quasi-static fault growth and cracking in homogeneous brittle rock under triaxial compressive using acoustic emission monitoring,* J. Geophys. Res. *105,* 6127–6139.

LI, H.X. and XIAO, X.R. (1995), An approach on the Mode-I fracture toughness anisotropy for materials with layered microstructures, Engin. Fract. Mech. *52,* 4, 671- 683.

LOCKNER, D. (1993), The role of acoustic emission in the study of rock fracture, Int.J. Rock Mech. Min. Sci. and Geomech. Abstr. *30,* 7, 883–899.

LOCKNER, D. E. (1996), *Brittle fracture as an analog to earthquakes: can acoustic emission be used to develop a viable prediction strategy,* J. Acous. Emission *14,* 88–101.

LOCKNER, D.E., BYERLEE, J.D., KUKSENKO, V., PONOMAREV, A. and SIDORIN, A. (1991), *Quasi-static fault growth and shear fracture energy in granite,* Nature *350,* 39–42.

LOCKNER, D.A. and MADDEN, T. R. (1991), *A Multiple-crack model of brittle fracture. Non-time- dependent simulation,* J Geophys. Res. *96,* 19623–19643.

MADDEN, T.H. (1983), Microcrack Connectivity in Rocks: A renormalization group approach to the critical phenomena of conduction and failure in crystalline rocks, J. Geophys. Res. *88,* 585–592.

MOORE, D.E. and LOCKNER D.A. (1995), *The role of microcracking in shear-fracture propagation in granite,* J Struct. Geol. *17,* 95–114.

NASSERI, M.H.B., MOHANTY, B. and PRASAD, U. (20002), *Investigation of micro-structural properties of selected rocks and their effect on fracture toughness,* NARMS, Toronto, Canada, 961–968.

NASSERI, M.H.B. MOHANTY, B. and ROBIN, P.-Y.F. (2005a), *Characterization of microstructures and fracture toughness in five granitic rocks,* Int. J. Mech. Min. Sci. *42,* 450–560.

NASSERI, M.H.B., THOMPSON, B., SCHUBNEL, A. and YOUNG R.P. (2005b), *Acoustic emission monitoring of Mode I fracture toughness (CCNBD) test in Lac du Bonnet granite,* ARMA/USRMS, Alaska Rocks, (2005), Anchorage, Alaska, 05-741.

NELDER, J. and MEAD, R. (1965), *A simplex method for function minimization,* Computer J. *7:* 308–312.

PRIKRYL, R. (2001), *Some micromechanical aspects of strength variation in rocks,* Int. J. Rock Mech. Min. Sci. *38,* 671–682.

SANO, O., KUDO, Y. and MIZUTA, Y. (1992), *Experimental determination of elastic constants of Oshima granite, Barre granite, and Chelmsford granite,* J. Geophys. Res. *97,* B3, 3367–3379.

SCHEDL, A., KRONENBERG, A.K. and TULLIS, J. (1986), *Deformation microstructures of Bare granite: An Optical SEM and TEM Study,* Tectonophysics *122,* 149–164.

SCHOLZ, C.H., DAWERS, N.H., YU, J.-Z. and ANDERS, M.H. (1993), *Fault growth and fault scaling laws: preliminary results,* J. Geophys. Res. *98,* 21952–21961.

TAKEMURA, T., GOLSHANI, A., ODA, M. and SUSUKI, K. (2003), *Preferreed orientation of open microcracks in granite and their relation with anisotropic elasticity,* Int. J. Mech. Min. Sci. *40,* 443–454.

VERMILYE, J.M. and SCHOLZ, C.H. (1999), *Fault propagation and segmentation: insight from the microstructural examination of a small fault,* J. Struct. Geology *21,* 1623–1636.

WILSON, J.E., CHESTER, J.S. and CHESTER, F.M. (2003), *Microcrack analysis of fault growth and wear processes, punchbowl fault, Sand Andreas System, California,* J. Struct. Geology *25,* 1855–1873.

YOUNG, R.P. and MARTIN, C.D. (1993), *Potential role of acoustic emission/microseismicity investigations in the site characterization and performance monitoring of nuclear waste repositories,* Int. J. Rock Mech. Min. Sci. *30,* 797–803

YOUNG, R.P., COLLINS, D.S., REYES-MONTES, J.M. and BAKER, C. (2004), *Quantification and interpretation of seismicity,* Int. J. Rock Mech. *41*(8), 1317–1328.

ZANG, A., WAGNER, C., STANCHITS, S., DRESEN, G., ANDRESEN, R. and HAIDEKKER, M.A. (1998), *Source analysis of acoustic emissions in aue granite cored under symmetric and asymmetric compressive loads,* J. Geophys. Res. *135,* 1113–1130.

ZANG, A. WAGNER, C.F., STANCHITS, S., JANSSEN, C. and DRESEN, G. (2000), *Fracture process zone in granite,* J Geophys. R. *105,* B10, 23651–23661.
ZIETLOW, W.K. and LABUZ, J.F. (1998*), Measurement of the intrinsic process zone in rock using acoustic emission,* Int. J. R. Mech. Min. Sci. *3,* 291–299.

(Received May 22, 2005, revised September 28, 2005, accepted September 29, 2005)

 To access this journal online:
http://www.birkhauser.ch

Pure appl. geophys. 163 (2006) 947–973
0033–4553/06/060947–27
DOI 10.1007/s00024-006-0061-y

© Birkhäuser Verlag, Basel, 2006

Pure and Applied Geophysics

Quantifying Damage, Saturation and Anisotropy in Cracked Rocks by Inverting Elastic Wave Velocities

ALEXANDRE SCHUBNEL,[1] PHILIP M. BENSON,[1,2] BEN D. THOMPSON,[1,3]
JIM F. HAZZARD,[1,4] and R. PAUL YOUNG[1]

Abstract—Crack damage results in a decrease of elastic wave velocities and in the development of anisotropy. Using non-interactive crack effective medium theory as a fundamental tool, we calculate dry and wet elastic properties of cracked rocks in terms of a crack density tensor, average crack aspect ratio and mean crack fabric orientation from the solid grains and fluid elastic properties. Using this same tool, we show that both the anisotropy and shear-wave splitting of elastic waves can be derived. Two simple crack distributions are considered for which the predicted anisotropy depends strongly on the saturation, reaching up to 60% in the dry case. Comparison with experimental data on two granites, a basalt and a marble, shows that the range of validity of the non-interactive effective medium theory model extends to a total crack density of approximately 0.5, considering symmetries up to orthorhombic. In the isotropic case, KACHANOV's (1994) non-interactive effective medium model was used in order to invert elastic wave velocities and infer both crack density and aspect ratio evolutions. Inversions are stable and give coherent results in terms of crack density and aperture evolution. Crack density variations can be interpreted in terms of crack growth and/or changes of the crack surface contact areas as cracks are being closed or opened respectively. More importantly, the recovered evolution of aspect ratio shows an exponentially decreasing aspect ratio (and therefore aperture) with pressure, which has broader geophysical implications, in particular on fluid flow. The recovered evolution of aspect ratio is also consistent with current mechanical theories of crack closure. In the anisotropic cases—both transverse isotropic and orthorhombic symmetries were considered—anisotropy and saturation patterns were well reproduced by the modelling, and mean crack fabric orientations we recovered are consistent with *in situ* geophysical imaging.

Our results point out that: (1) It is possible to predict damage, anisotropy and saturation in terms of a crack density tensor and mean crack aspect ratio and orientation; (2) using well constrained wave velocity data, it is possible to extrapolate the contemporaneous evolution of crack density, anisotropy and saturation using wave velocity inversion as a tool; 3) using such an inversion tool opens the door in linking elastic properties, variations to permeability.

Key words: Elastic wave velocities, anisotropy, crack density, saturation, effective medium, attenuation, aspect ratio, Vp/Vs ratio, shear stress, effective pressure.

[1]Lassonde Institute, University of Toronto, 170 College Street, Toronto, ON M5S 3E3, Canada
[2]Mineral, Ice and Rock Physics Laboaratory, University College London, Gower Street, London, WC1E 6BT, UK
[3]Applied Seismology Laboratory, Department of Earth Sciences, Liverpool University, 4 Brownlow St., Liverpool, L69 3GP, UK
[4]Rocksciences Inc., 439 University Ave., Toronto, ON M5G 1Y8, Canada

1. Introduction

In upper-crustal conditions, fractures are ubiquitous (from major fault zones to continental shields) and present at all scales in rocks (from macroscopic fractures to microcracks). Despite the fact that these fractures generally represent only small amounts of porosity, they can exert considerable influence upon the rock physical properties (BRACE *et al.*, 1968) or even the fracture toughness (NASSERI *et al.*, this issue). In particular, the existence of embedded microcrack fabrics in rocks significantly influences the elastic properties (SIMMONS and BRACE, 1965; WALSH, 1965), contributes to the difference between static and dynamic elastic moduli (SIMMONS *et al.*, 1975; MAVKO and JIZBA, 1991) and aligned crack fabrics may produce elastic anisotropy (KERN, 1978; KERN *et al.*, 1997). When fluids are present, cracks also play a key role in making the rock permeable on a macroscopic scale. For such reasons, seismic data and velocity field analysis are used on a daily basis in the oil industry to quantify the oil content and the identify fluid properties, i.e., oil or gas. In seismotectonics, a key unanswered question is whether earthquakes and volcanic eruptions can be predicted by quantifying time dependent precursory damage accumulation using elastic wave velocity variations (CHUN *et al.*, 2004; VOLTI and CRAMPIN, 2003; GAO and CRAMPIN, 2004). Thus the understanding and quantification of elastic wave velocity variations is critical to extract information on the physical state of rocks from seismic and seismological data. This has major implications when considering the possible hydro-mechanical coupling taking place during the seismic cycle (MILLER, 2002), forecasting the life time of oil reservoirs, or the integrity of underground storage of hazardous wastes. In most cases, geophysicists have few means to retrieve information on the physical state of field rock masses, other than to invert elastic wave velocities into damage parameters.

For all those reasons, numerous models which predict properties of materials as a function of damage have been developed in the framework of Effective Medium Theories (EMT) in the last thirty years (ESHELBY, 1957; WALSH; 1965; O'CONNELL and BUDIANSKY, 1974, 1977; ANDERSON *et al.*, 1974; SOGA *et al.*, 1978; CHENG and TOKSÖZ, 1979; HUDSON, 1981, 1982, 1986; NISHIZAWA, 1982; KACHANOV, 1994; SAYERS and KACHANOV, 1991, 1995; LE RAVALEC and GUÉGUEN, 1996; SCHUBNEL and GUÉGUEN, 2003). Here, we report an approach based on KACHANOV's (1994) non-interactive EMT scheme which allows the straightforward inversion of elastic wave-velocity measurements made on several rock types in the laboratory into crack density, mean crack aspect ratio and mean crack fabric (or alignment). As the extensive laboratory input data compiled in this study were measured independently, our study provides an ideal opportunity to test the model applicability and investigate directly the elastic wave inversion results, in particular the evolution of crack density and mean crack aspect ratio with pressure and shear stress.

2. Effective Elastic Medium Containing Cracks

Statistically, the effective elastic properties of an initially isotropic medium depend on a few intrinsic parameters, including:

- the solid matrix elastic properties (Young's modulus E_o and Poisson ratio v_o).
- the fluid bulk modulus K_f and the level of fluid saturation.
- the crack density ρ defined as $\rho = \frac{1}{V}\sum^N c_i^3$ where c_i is the radius of the i-th crack, N being the total number of cracks embedded in the Representative Elementary Volume (REV) V.
- the crack geometry (in our case, we consider penny-shaped geometry (Figure 1a), their average aspect ratio $\zeta = \langle w/c \rangle$ and their spatial distributions.

The concept of crack density and mean aspect ratio is a statistical generalization of the concept of porosity for non-spherical inclusions since the crack porosity ϕ is defined as $\phi = \pi\rho\zeta$. In an isotropic matrix containing a uniform (isotropic) distribution of crack centers, the effective elastic modulus of a rock M^* is a linear function of the crack density that can be written in the form (first perturbation order):

$$\frac{M_o}{M^*} = 1 + h.\rho, \tag{1}$$

where M_o is the solid matrix elastic modulus and h is a positive scaling parameter that depends on the matrix and fluid properties, the geometry of the cracks and the interactions between them. Despite the facts that i) real fractures are not generally uniformly distributed spatially — a fractal-type description is often more realistic

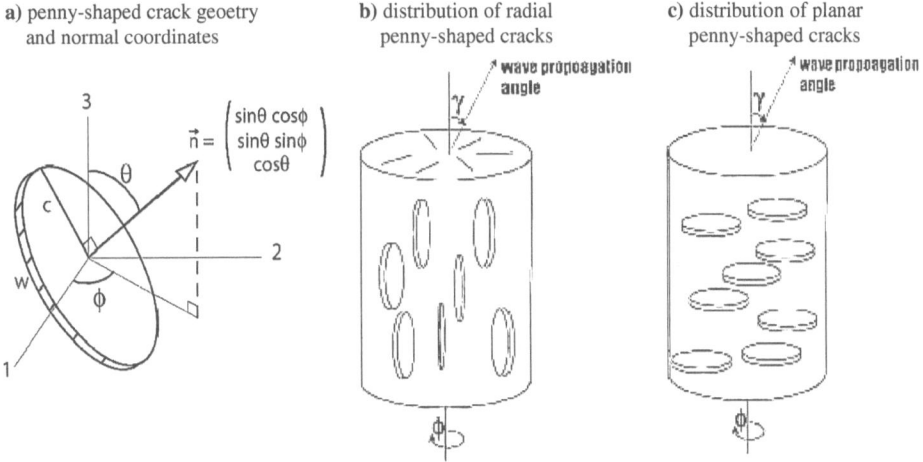

a) penny-shaped crack geoetry and normal coordinates

b) distribution of radial penny-shaped cracks

c) distribution of planar penny-shaped cracks

$$\vec{n} = \begin{pmatrix} \sin\theta\ \cos\phi \\ \sin\theta\ \sin\phi \\ \cos\theta \end{pmatrix}$$

Figure 1 a)
Geometry, conventions and coordinates of a 3-D penny-shaped crack. The two crack distributions considered here: b) the case of radial penny-shaped cracks. c) the case of coplanar penny-shaped cracks.

(e.g., in some crystalline rocks, cracks can be very highly spatially concentrated, with intact rock between these zones), ii) real fractures have intersections and iii) real fractures also have shapes which are complex in detail, most EMT have been derived for non-intersecting cracks, uniform distributions of crack centers and simple crack geometries. Although those three important simplifications can be discussed from a theoretical point of view, considerable insight has been gained from them and the scalar h has been calculated by various authors for all kind of inclusion geometries (spherical, elliptical, penny-shaped discs, rectangular plates or linear cracks — see MAVKO *et al.,* 1998 for a good compilation) and fluid properties. These simplifications (non-intersecting 'penny-shaped' cracks and uniform distributions of crack centers) are used throughout this paper.

Recently, ORLOWSKY *et al.* (2003) and SAENGER *et al.* (2004) showed numerically that the best EMT scheme was the Differential Self-Consistent (DSC) theory (CLEARY *et al.,* 1980; BERRYMAN, 1992; LE RAVALEC and GUÉGUEN, 1996). However the accuracy of the DSC calculation has two main drawbacks: 1) it is limited to isotropic formulations and 2) the way the effective elastic moduli are calculated makes its use for elastic wave inversion complicated.

The most simple of all EMT is certainly the non-interactive theory because it neglects the problem of stress interactions between cracks and is therefore independent of crack centers distribution. It also allows easy computation of anisotropic fluid-filled crack distributions. Non-interactive EMT was shown to be valid when cracks are distributed randomly or when aligned (KACHANOV, 1994; SAYERS and KACHANOV, 1995; SCHUBNEL and GUÉGUEN, 2003) although ORLOWSKY *et al.* (2003) and SAENGER *et al.* (2004) have shown numerically that it would systematically overestimate the crack density. However, it has several advantages over other EMTs: i) it overcomes the divergence problems encountered in HUDSON's theory (HUDSON, 1981, 1982, 1986) or the non-physical drop of an elastic modulus to zero after a critical crack density encountered in the Self Consistent (SC) theory (O'CONNELL and BUDIANSKY, 1974, 1977, ANDERSON *et al.,* 1974; SOGA *et al.,* 1978); ii) any crack distribution can be computed easily, keeping in mind that results are best for aligned or randomly oriented cracks; iii) it depends both on the crack density and the mean crack aspect ratio; iv) and, most importantly, its easy formulation enables one to perform a direct inversion of elastic wave velocities into crack density and aspect ratio, as we will see in the next sections.

2.1 Non-interactive Formulation for Isotropically Distributed Cracks

When neglecting stress interactions between cracks, the effective elastic modulii of a cracked solid can be calculated exactly and rigorously in a unique manner that depends solely upon the crack orientations and distribution. Such a hypothesis is often wrongly confused with the low crack density approximation. Clearly, a low crack density means that cracks are (on average) distant from each other and the non-interactive assumption is likely to be verified. At higher densities however, it is clear that cracks will interact with each other, both in terms of direct stress transfer and fluid flow between fractures (which will become more common with increasing crack densities). Nonetheless one should keep in mind (and this has been forgotten for example in both HUDSON's and the Self Consistent theory) that the stress field is is not only amplified at crack tips, but also shielded on the crack flanks. For such reasons, we can assume that stress interactions are partially compensating geometrically for certain distributions such as random (isotropic) or aligned crack distribution. KACHANOV (1994) showed numerically that the stress interactions in the case of intersecting cracks are small and that stress amplifications are shorter range than shielding. Finally and because of the non-intersecting assumption, EMTs are high-frequency (>KHz) theories for which squirt-flow mechanisms (MAVKO and JIZBA, 1975) can be neglected. It has been shown that, using all those assumptions, the non-interactive approximation gave reasonable and interpretable results at least up to crack densities of 0.5 (KACHANOV, 1994; SAYERS and KACHANOV, 1995; SCHUBNEL and GUÉGUEN, 2003; BENSON et al., 2006).

In the non-interactive approximation, each crack is considered to be isolated and the elastic perturbation ΔS_{ijkl} compliance due to cracks is simply the sum of each crack contribution. Since the average bulk elastic strain is the sum of the matrix elastic strain and the superposition of elastic strain of each individual crack, effective elastic modulii can simply be calculated using the elastic potential f (KACHANOV, 1994; SAYERS and KACHANOV, 1995). In the isotropic case, KACHANOV (1994) showed that the effective Young E^* and shear modulii μ^* of a rock could be written[1] as:

$$\frac{E_o}{E^*} = 1 + \left(1 + \frac{3}{5}\left[\left(1 - \frac{v_o}{2}\right)\left(\frac{\delta}{1+\delta} - 1\right)\right]\right)h\rho \tag{2}$$

and:

$$\frac{\mu_o}{\mu^*} = 1 + \left(1 + \frac{2}{5}\left[\left(1 - \frac{v_o}{2}\right)\left(\frac{\delta}{1+\delta} - 1\right)\right]\right)\frac{h\rho}{(1+v_o)}, \tag{3}$$

where h is a geometrical factor linked to the penny-shaped geometry (see Appendix I) and given by,

$$h = \frac{16(1 - v_o^2)}{9(1 - v_o/2)}. \tag{4}$$

δ is a non-dimensional number which characterizes the coupling between the stress and the fluid pressure, and is equal to:

---

[1] Note that equation (3) has been corrected from KACHANOV's (1994) original manuscript by FORTIN (2005) and BENSON et al. (2006).

$$\delta = (1 - v_o/2) \frac{E_0 \zeta}{K_f} h. \tag{5}$$

δ compares the fluid bulk modulus K_f to the crack bulk modulus $\{(1 - v_o/2)E_o\zeta h\}$ assuming that all change in the crack volume is due to aperture variations (see Appendix I). As a lower bound, i.e., for an incompressible fluid such as water, and using $E_0 \sim 5.10^{10}$, $\delta \to 0$ and is negligible if ζ is small ($<10^{-3}$). As an upper bound, i.e., for a compressible fluid such as dry air, δ considerably exceeds 1 and $\delta/1 + \delta \to 0$.

Figures 2a and 2b present the normalized P- and S- wave velocity isocontours respectively as a function of both crack density and the aspect ratio. The velocities were calculated from equations (2)–(3) using solid matrix and fluid elastic parameters typical of an undamaged water saturated granite and equal to: $E_o = 85$ GPa, $v_o = 0.25$ and $K_f = 2$ GPa. Velocities were normalized to the crack-free velocities (i.e., $\rho = 0$) to remove bulk density effects. On Figure 2a, the *P*-wave velocity is shown to decrease with crack density. However, at a given crack density, *P*-wave velocity remains more or less constant with aspect ratio. On the contrary, whilst *S*-wave velocity also decreases with crack density on Figure 2b, they are very sensitive to the crack mean aspect ratio. Figure 2c shows that the evolution of the Vp/Vs ratio is a function of both crack density and aspect ratio when cracks are saturated with fluid. This way, KACHANOV's (1994) non-interactive model permits the quantification of both the crack density ρ and the aspect ratio ζ, through the saturation coefficient δ (equation 5). As a direct consequence, one can expect to be able to extract from Vp and Vs measurements in saturated rocks containing cracks, not only the crack density, but also the aspect ratio.

2.2 Non-interactive Formulation: General Case

In the most general case, when fractures seem to be aligned in one or several directions and wavelengths are considerably larger than the fracture spacing, it is convenient to formulate the equivalent anisotropic medium problem in terms of compliances (KACHANOV, 1994; SAYERS and KACHANOV, 1995; SCHOENBERG and SAYERS, 1995). For an isotropic matrix containing penny-shaped cracks, the additional elastic compliance ΔS_{ijkl} due to those can be expressed as (SAYERS and KACHANOV, 1995):

$$\Delta S_{ijkl} = \frac{1}{4hE_o} (\delta_{ik}\alpha_{jl} + \delta_{il}\alpha_{jk} + \delta_{jk}\alpha_{il} + \delta_{jl}\alpha_{ik}) + \beta_{ijkl} \tag{6}$$

where α_{ij} is the crack density tensor:

$$\alpha_{ij} = \rho.\langle n_i n_j \rangle \tag{7}$$

and β_{ijkl} the saturation tensor:

$$\beta_{ijkl} = \rho \cdot \left[\left(1 - \frac{v_0}{2}\right)\frac{\delta}{1+\delta} - 1\right]\langle n_i n_j n_k n_l \rangle. \tag{8}$$

δ_{ij} is the Kronecker symbol and we recall that the scalar (or total) crack density is equal to $\rho = \mathrm{tr}(\alpha) = Nc^3/V$. In the case of an orthorhombic crack distribution, the crack density tensor is diagonal, and it follows from equations (6)–(8) that the nine independent effective elastic compliances can be expressed:

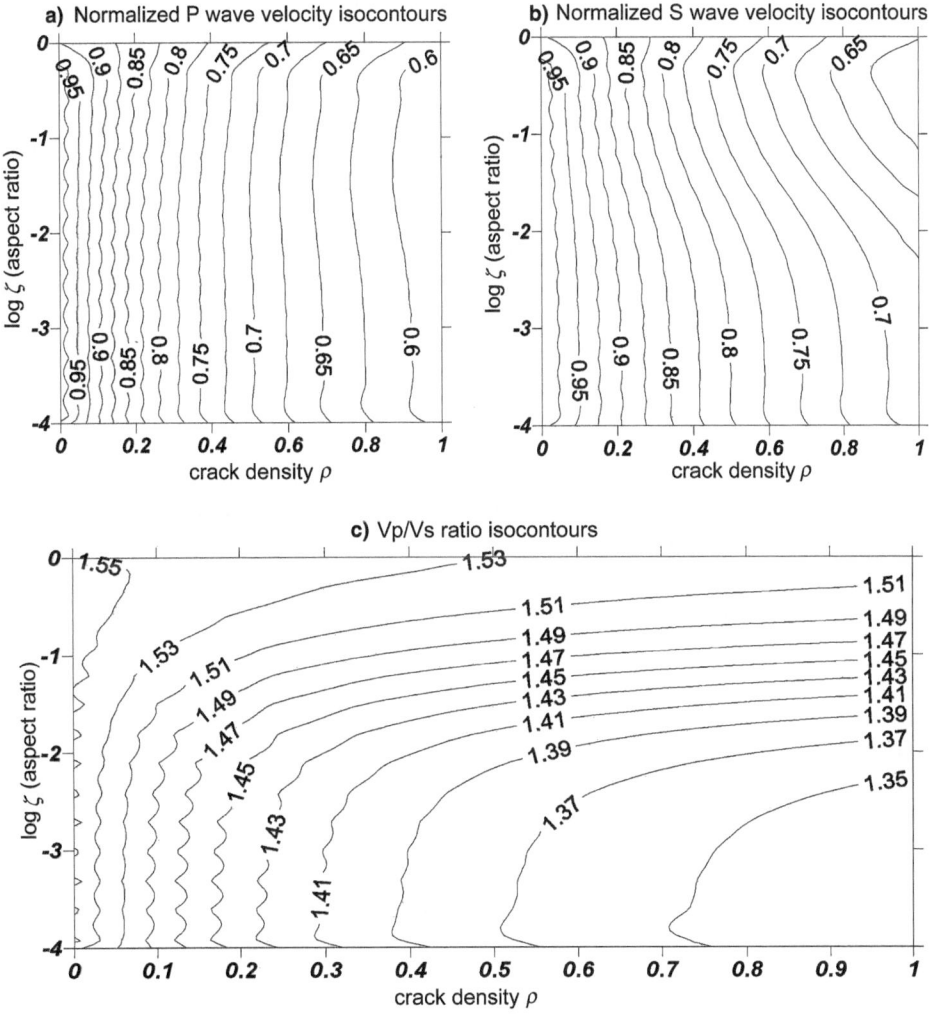

Figure 2

Isotropic case: Elastic wave velocities calculated from equations (2)–(3) using solid matrix and fluid elastic parameters typical of a water saturated granite and equal to: $E_o = 85$ GPa, $v_o = 0.25$ and $K_f = 2$ GPa. a) and b) normalized P- and S-wave velocity isocontours as a function of crack density and aspect ratio. Velocities were normalized to the crack free velocities (i.e., $\rho = 0$) to remove bulk density effects. c) evolution of the Vp/Vs ratio as a function of both crack density and aspect ratio.

$$S_{ijkl} = \begin{cases} = S^o_{ijkl} + \frac{\rho}{E_o h}\left(\langle n_i^2\rangle + \left[(1 - \frac{v_0}{2})\frac{\delta}{1+\delta} - 1\right]\langle n_i^4\rangle\right) \; if \; \{i=j=k=l\} \\[2mm] = S^o_{ijkl} + \frac{\rho}{E_o h}\left(\frac{\langle n_i^2\rangle + \langle n_j^2\rangle}{4} + \left[(1 - \frac{v_0}{2})\frac{\delta}{1+\delta} - 1\right]\langle n_i^2 n_j^2\rangle\right) \; if \; \{i=k \;\; and \;\; j=l\} \quad (9) \\[2mm] = S^o_{ijkl} + \frac{\rho}{E_o h}\left[(1 - \frac{v_0}{2})\frac{\delta}{1+\delta} - 1\right]\langle n_i^2 n_k^2\rangle \; if \; \{i=j \;\; and \;\; k=l\} \end{cases}$$

Recalling the relationship between dependent elastic compliances, the formulation for transverse isotropy can also be obtained from equation (9). Assuming now a continuous distribution function of crack orientations, the average i-th component of the crack fabric can be defined as follows:

$$\langle n_i\rangle = \frac{1}{2\pi}\int_0^{2\pi} d\phi \int_0^{\pi/2} \psi(\theta, \phi) n_i \sin\theta d\theta \;, \tag{10}$$

$\psi(\theta, \phi)$ being the orientation distribution function (cf. fig. 1a). The tensors $\langle n_i n_j\rangle$ and $\langle n_i n_j n_k n_l\rangle$, which represent respectively the second-order and the fourth-order moments of the crack orientation distribution function respectively, can be calculated easily in the same way.

2.3 Wet and Dry Elastic Waves Anisotropy: Numerical Results

For the sake of simplicity, we will consider Dirac distributions, so that $\langle n_i\rangle = n_i$ and subsequently, in such a way that the average crack fabric orientation is described as on Figure 1a for a single crack. In this section, we consider two different types of crack distributions: i) radial cracks ($n_3 = 0, n_1 = n_2 = 1/\sqrt{2}$) and ii) planar cracks ($n_3 = 1, n_1 = n_2 = 0$). The first case corresponds to a distribution of symmetrically distributed radial cracks (symmetry along the vertical axis, cracks normal in the horizontal axis) and is representative of the TI symmetry (Fig. 1b). The second case corresponds to horizontally aligned coplanar penny-shaped cracks and is representative of both TI and orthorhombic symmetry (Fig. 1c). Furthermore, only the lower and upper bound of saturation, i.e., for an incompressible fluid such as water $\delta \to 0$ and for a compressible fluid such as dry air, $\delta/1 + \delta \to 0$, were investigated. In all that follows, the solid matrix Young's modulus and Poisson ratio were taken as equal to the ones of an isotropic and crack-free granite material, i.e., $E_o = 85$ GPa and $v_o = 0.25$, respectively.

2.3.1 Radial cracks

The case of radial cracks is illustrated on Figure 3. Figures 3a and 3b show the normalized P velocities versus the angle of propagation γ of the elastic wave, in wet and dry conditions respectively, and for crack densities equal to 0.1, 0.25, 0.5 and 1. The angle γ is defined as that between the vertical axis Ox_3 and the wave vector. For example, elastic waves propagating at 0 and 90 degrees correspond to propagation along the vertical and horizontal axes, respectively. The normalized wave velocity is

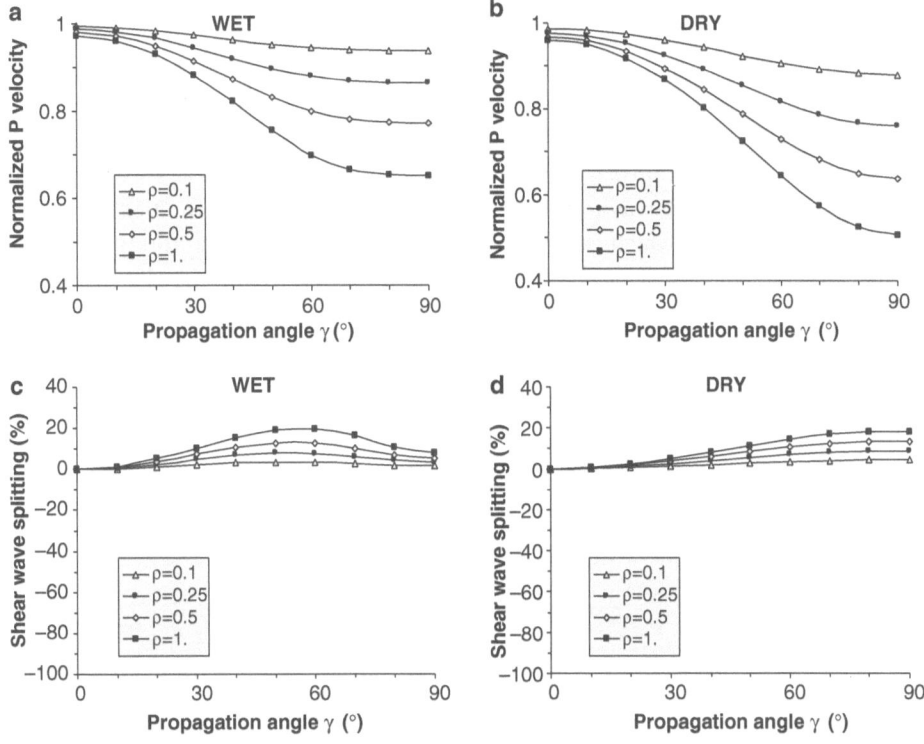

Figure 3

Case of radial cracks (Fig. 1b). Only the lower ($\delta \to 0$) and upper ($\delta/1 + \delta \to 0$) bound of saturation were investigated using equation (9). The solid matrix Young's modulus and Poisson ratio were taken as equal to $E_o = 85$ GPa and $v_o = 0.25$, respectively. Velocities were normalized to the crack-free velocities (i.e., $\rho = 0$) to remove bulk density effects. Results are displayed for crack densities equal to 0.1 (open triangles), 0.25 (plain circles), 0.5 (open diamonds) and 1 (plain squares). a) and b) show the normalized P-wave velocity as a function of the direction of propagation γ in the wet and dry cases respectively. c) and d) show the shear-wave splitting ($\frac{SV(\gamma)-SH(\gamma)}{SV(\gamma)} * 100$ or birefringence) as a function of the direction of propagation γ in the wet and dry cases, respectively.

the ratio of V_γ to the solid grain velocity. In the case of radial cracks, P waves propagating vertically are unaffected and do not "see" the cracks. Anisotropy is larger in the dry case (up to 50% in the dry case, up to $\sim 40\%$ in the wet case) but the pattern is very comparable in both cases. Figures 3c and 3d show the results obtained in terms of the shear-wave splitting $\left(\frac{SV(\gamma)-SH(\gamma)}{SV(\gamma)} * 100 \text{ or birefringence}\right)$ between vertically (SV) and horizontally (SH) polarized S waves in the wet and dry cases, respectively. In the dry case, SV waves always travel faster than SH. In the wet case, maximum splitting is observed at a γ angle of 60 degrees, whereas maximum dry splitting is obtained for horizontally propagating S waves. However, and as one could expect for a set of radial cracks, the effect of fluid is rather small.

2.3.2 Planar cracks

Figure 4 shows numerical results obtained in the case of planar cracks. Figures 4a and 4b present the normalized P velocities as a function of propagation angle γ and crack density in the wet and the dry cases, respectively. As noted earlier, the effect of cracks is considerably stronger in the dry case than in the wet case. This time, cracks are "invisible" to waves propagating in the horizontal direction ($\gamma = 90^{\circ}$); and in the dry case (Fig. 4b), the P-wave velocity decrease is very large (up to 60%) for waves propagating vertically, which is expected from geometrical considerations. However, and because the fluid is assumed to be fully incompressible, cracks are also "invisible" in the direction $\gamma = 0^{\circ}$ in the wet case (Fig. 4a). This can be explained theoretically as in KACHANOV's model, the normal crack compliance is assumed to be equal to zero

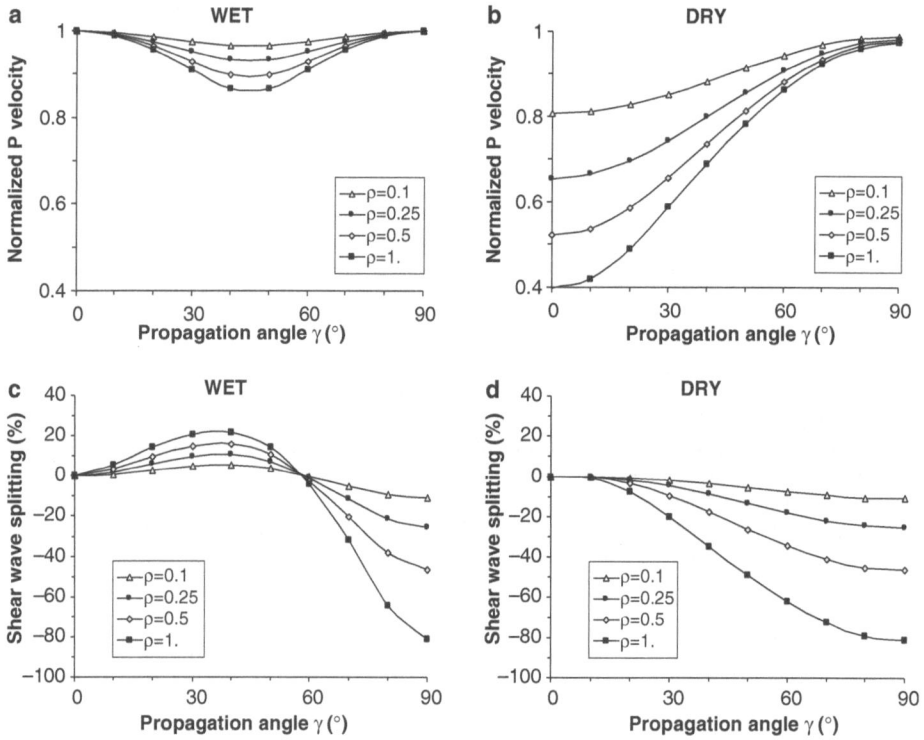

Figure 4

Case of coplanar cracks (Fig. 1c). Only the lower ($\delta \to 0$) and upper ($\delta/1 + \delta \to 0$) bound of saturation were investigated using equation (9). The solid matrix Young's modulus and Poisson ratio were taken as equal to $E_o = 85$ GPa and $v_o = 0.25$, respectively. Velocities were normalized to the crack free velocities (i.e., $\rho = 0$) to remove bulk density effects. Results are displayed for crack densities equal to 0.1 (open triangles), 0.25 (plain circles), 0.5 (open diamonds) and 1 (plain squares). a) and b) show the normalized P-wave velocity as a function of the direction of propagation γ in the wet and dry cases, respectively. c) and d) show the shear-wave splitting ($SV(\gamma) - SH(\gamma)/SV(\gamma) * 100$ or birefringence) as a function of the direction of propagation γ in the wet and dry cases, respectively.

(see Appendix I) when the fluid is fully incompressible (or when the crack is fully constrained). Therefore, in the wet case, the maximum anisotropy (up to 20% for $\rho = 1$) is seen for waves propagating at 45 degrees from the vertical, for which the associated shear strain is maximum. Figures 4c and 4d show the predicted results for shear-wave splitting $(SV(\gamma) - SH(\gamma)/SV(\gamma) * 100)$. Again, the pattern is very different in the wet and the dry cases. As expected, in the dry case SH waves always propagate faster than SV waves. In the wet case however, and for propagation angles between 0 and 60 degrees, SV waves are faster than SH waves. Conversely, SH-wave velocity is higher than SV-wave velocity for angles larger than 60 degrees. The shear-wave splitting maximum observed is thus an effect of fluid incompressibility.

In each of the two distributions discussed above, the non-interactive model manifests a clear difference between dry and fully saturated cracks, both for P waves and S waves. Intermediate results for partial saturation could be obtained considering the δ parameter and average crack aspect ratio. It demonstrates that, to a certain extent, the P-wave anisotropy and S-wave birefringence patterns can correspond to a genuine crack distribution and saturation state. Therefore, combining velocity data on P-wave anisotropy and S-wave birefringence should allow one to investigate the "average" crack distribution in a rock, as well as the saturation state. Unfortunately, such extensive wave-velocity data is very rare in the literature. It exists at the field scale, however one would need to take into account the dispersion effects due to frequency and "squirt flow" mechanisms (MAVKO and NUR, 1975). This may be done using BIOT-GASSMANN or BROWN and KORRINGA's equations (1975), as shown by SCHUBNEL and GUÉGUEN (2003).

3. Inversion of Experimental Data

In the laboratory, experimental studies measuring the evolution of dynamic elastic properties under conditions simulating upper crust burial depths can be performed thanks to the use of piezoelectrics transducers (PZT), for which eigen-frequencies are generally well into the dynamic range (>100 KHz). Despite the experimental difficulties, such studies have been undertaken and can provide 'well-constrained' experimental datasets which are perfectly suited to attempt predicting the evolution of rock fabric parameters such as crack density, aspect ratio and alignment.

When modelling the evolution of elastic velocities, we restricted ourselves to the case of an initially isotropic solid rock matrix. The crack distribution function obeyed either an isotropic, transverse isotropic or orthorhombic symmetry. In such conditions, the effective elastic properties predicted by equations (2), (3) and (9) are dependent only upon the matrix Young's modulus E_o and Poisson ratio v_o, and more importantly, the scalar crack density ρ. Additionally, when cracks are non-randomly oriented, elastic wave velocities also depend on the mean orientation (θ, ϕ — Fig. 1a) of the crack fabric with respect to the axis of symmetry. Finally, when saturated, elastic properties also depend on the saturation coefficient δ and therefore the average crack aspect ratio ζ and fluid bulk modulus K_f. In the following, we simply performed least-square inversions of laboratory datasets:

$$\text{RMS}_i = \frac{1}{N_i} \sum_{N_i} (V_\gamma^{\text{data}} - V_\gamma^{\text{model}}(\rho, \zeta, \theta, \phi))^2 \qquad (11)$$

in order to recover the crack density, average crack aspect ratio and orientation. Inversions were performed using initial *P*- and *S*- wave velocities of the "uncracked" material as calculated from experimental and/or petrological considerations, thus reflecting the average bulk solid matrix elastic properties. Model elastic wave-velocity field was calculated from equations (2), (3) and (9) as a function of ρ, ζ when the rock was saturated with water together with θ and ϕ when the rock was anisotropic. Each modelled velocity field was then compared using a simple RMS technique (equation 11) with experimental measurements. In equation (11), N_i represents the number of *P*- and *S*- velocity measurements along several γ directions at each step i of a given experiment. The lowest RMS error between modelled and data velocities was taken as being the best inversion result, which output the quadruplet $(\rho, \zeta, \theta, \phi)$. The agreement between data and best fit velocities is, in general, very good with the average error between model and data points lower than 0.05 km/sec. This is a direct consequence of the well constrained laboratory data itself.

3.1 Isotropic Inversions

Elastic wave velocities have been widely shown to increase with increasing hydrostatic pressure, because of crack closure. Conversely and in the presence of shear stress, elastic wave velocities can decrease due to nucleation and propagation of new microcracks. In the following, we compile experimental results obtained on two different rock types, representative of contrasting isotropic microcrack fabric, a porphyritic alkali basalt from Mount Etna and Carrara marble.

3.1.1 Crack density and aspect ratio evolution as a function of hydrostatic pressure
In the case of the Etnean basalt, the simultaneous evolutions of *P*- and *S*- wave elastic wave velocities were measured during hydrostatic compression of three different rock samples (38.1 mm diameter by 40 mm length) cored in three orthogonal directions (Fig. 5a). The measurements, performed in a high pressure confining cell installed at University College London (BENSON, 2004; BENSON *et al.*, 2005), evidenced no marked elastic anisotropy in the rock. The experimental *P*-wave velocities ranged from 5.35 ± 0.13 km/s at 5 MPa to 5.88 ± 0.12 km/s at 80 MPa; while *S*-wave velocities ranged from 3.30 ± 0.04 km/s to 3.60 ± 0.04 km/s. *P* and *S* elastic wavespeeds were inverted using equations (2) and (3). Crack-free elastic wave

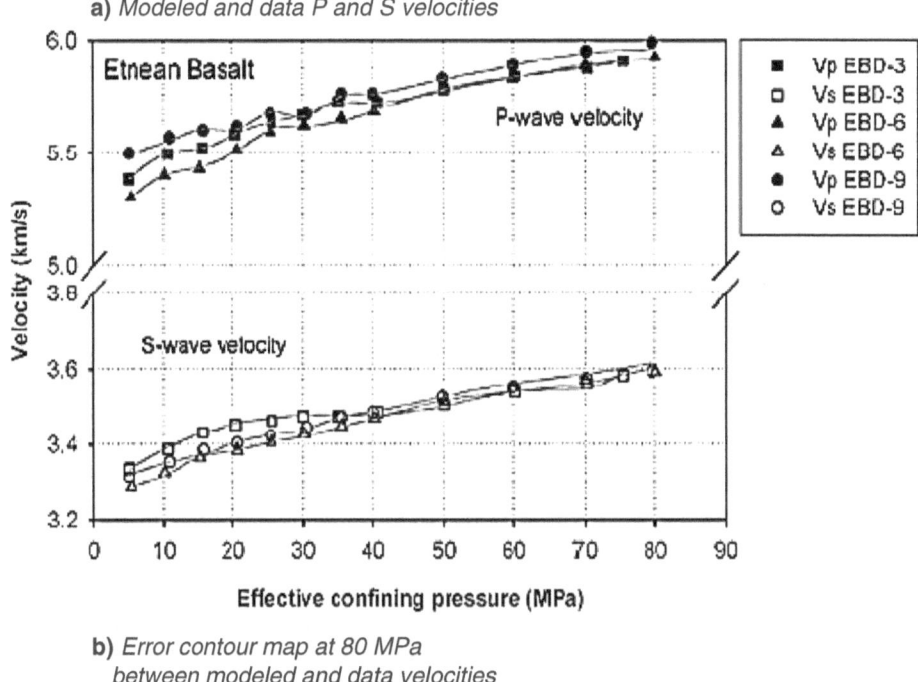

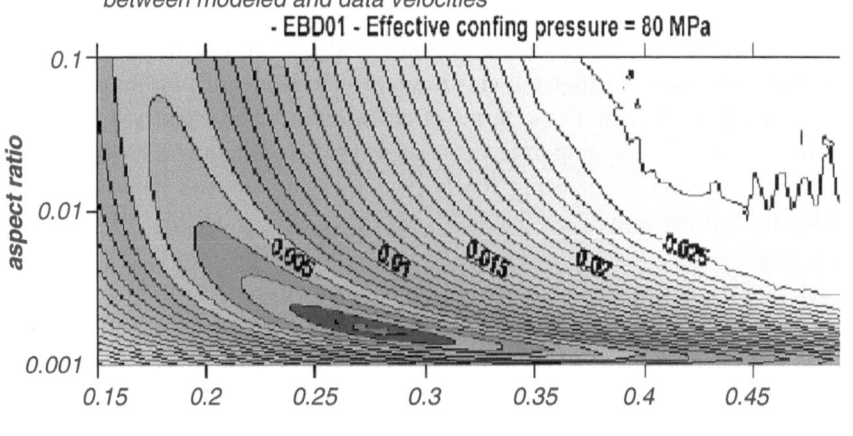

Figure 5

a) Modelled (solid lines) P and S velocities in Etna basalt compared to the laboratory measurements (dots - BENSON et al., 2005). The effective confining pressure was calculated as $P_c - P_p$ where P_c is the confining pressure and P_p the pore pressure. b) Error contour map between modelled and data (P and S) velocities in the case of sample EBD01 at 80 MPa confining pressure. Error contours are displayed as a function of crack density and aspect ratio, respectively. Darker areas represent lower errors.

velocities were taken as equal to $V_p = 6400$ m/s and $V_s = 3750$ m/s (i.e., solid matrix elastic parameter equal to $E = 100$ GPa and $\nu = 0.22$). The fluid bulk modulus was taken as $K_f = 2$ GPa.

Figure 5a compares the model P and S elastic velocities (solid and dashed line) to the experimental results (symbols). Because the degree of freedom of the inversion is zero (i.e., P- and S-wave velocities were modelled using only two distinct parameters: crack density and aspect ratio), the fit appears perfect, well below the experimental error bar. Figure 5b presents the error contour maps between modelled and data velocities as a function of crack density and aspect ratio for sample EBD01 at 80 MPa, respectively. As illustrated, the inversion is stable as only one minima can be distinguished and for a wave-velocity doublet (P,S) corresponds a unique solution for the crack density and aspect ratio doublet (ρ, ζ). As an output, the model gives back the evolution of parameters (ρ, ζ) with confining pressure which are displayed on Figures 6a and 6b. Results from the inversion reveal a decrease in crack density from ~ 0.5 to ~ 0.35 during pressurization, while the average crack aspect ratio decreases by one order of magnitude approximately. The decrease in crack density illustrates a decrease in the cracks apparent radii c as crack density evolves with $\sim c^3$. This is generally attributed to an increase of the crack surfaces contact areas as cracks are being closed. More importantly, the recovered evolution of aspect ratio shows an exponentially decreasing aspect ratio with confining pressure (and thus aperture evolution as the radius is not expected to vary much). The drop of one order of magnitude within the first MPa's is consistent with that expected intuitively, i.e., fast elastic crack closure at low pressures. This is also consistent with current theories on crack closure (KASELOW and SHAPIRO, 2004) where crack closure is generally modelled using an exponentially decreasing law with pressure with varying power exponents. One should note that the elastic wave inversion not only interprets the increase in both P and S velocities in terms of crack density consistently, but also the fact that the ratio P/S is changing in terms of aspect ratio reduction. This is probably one of the key advantages in using KACHANOV's (1994) scheme when modelling elastic properties in isotropy.

3.1.2 Crack density and aspect ratio evolution as a function of shear stress

Carrara marble is a well investigated marble, with coarse grainsize (150 μm) and a very low initial anisotropy ($<1\%$). SCHUBNEL et al. (2005) measured both P- and S- wave velocities during a full tri-axial cycle in wet conditions ($Pc = 260$ MPa and $Pp = 10$ MPa). In this case, the non-interactive scheme becomes particularly relevant as calcite behaves plastically and intragranular plasticity inhibits long-range stress interactions between cracks. Initial P- wave velocity was equal to 5.9 km.s$^{-1} \pm 1\%$ while final P-wave velocity was lower than 3 km.s$^{-1} \pm 1\%$. Figure 7a presents the evolution of both P- and S-wave velocities as a function of effective mean stress $\mathbf{P} = [(\sigma_1 + 2\sigma_3)/3 - P_p]$. In the first phase both velocities increased. When the onset of crack propagation was reached, P- and S-wave velocities began to decrease rapidly due to damage accumulation. During the subsequent relaxation period, wave velocities increased again, before decreasing drastically due to stress relief microcracking as hydrostatic stress was removed. On the figure, the fit between

a) crack density as a function of effective confining pressure

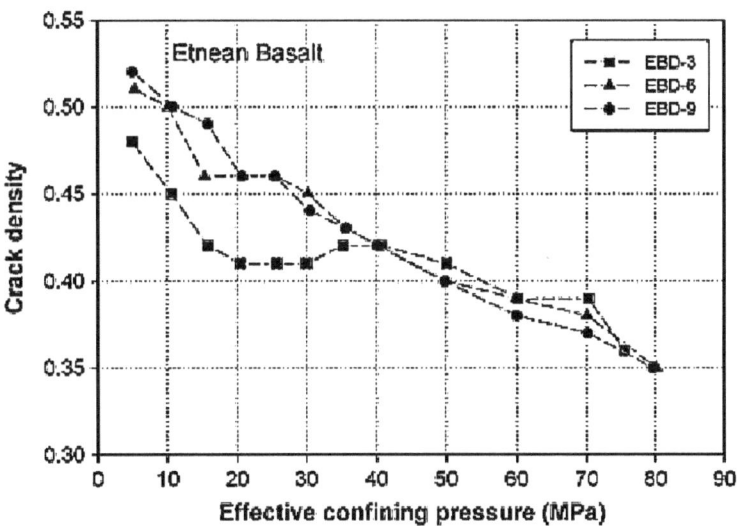

b) aspect ratio as a function of effective confining pressure

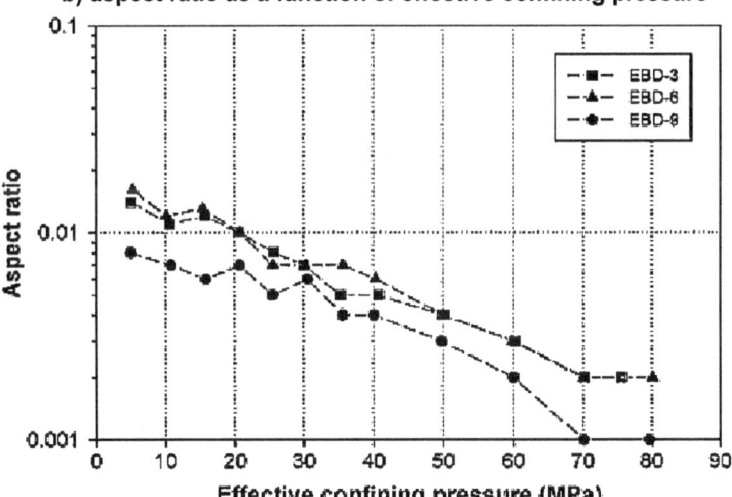

Figure 6

Evolution of crack density (a) and aspect ratio (b) as a function of confining pressure in Etna basalt. Both crack density and aspect ratio were inverted from experimental data presented in Figure 5 and using equations (2) and (3). Matrix elastic parameters were taken as $E = 100$ GPa and $v = 0,22$ (i.e., crack-free elastic wave velocities equal to $V_p = 6400$ m/s and $V_s = 3750$ m/s)

data and modelled velocity appears perfect for the same reasons as on Figure 5a. This time, the model crack-free parameters were taken as that of a non-porous calcite aggregate, i.e., $E = 100$ GPa and $v = 0.32$. Figure 7b shows the inversion results, i.e., the evolution of crack density (solid diamonds) and aspect ratio (open squares) as a

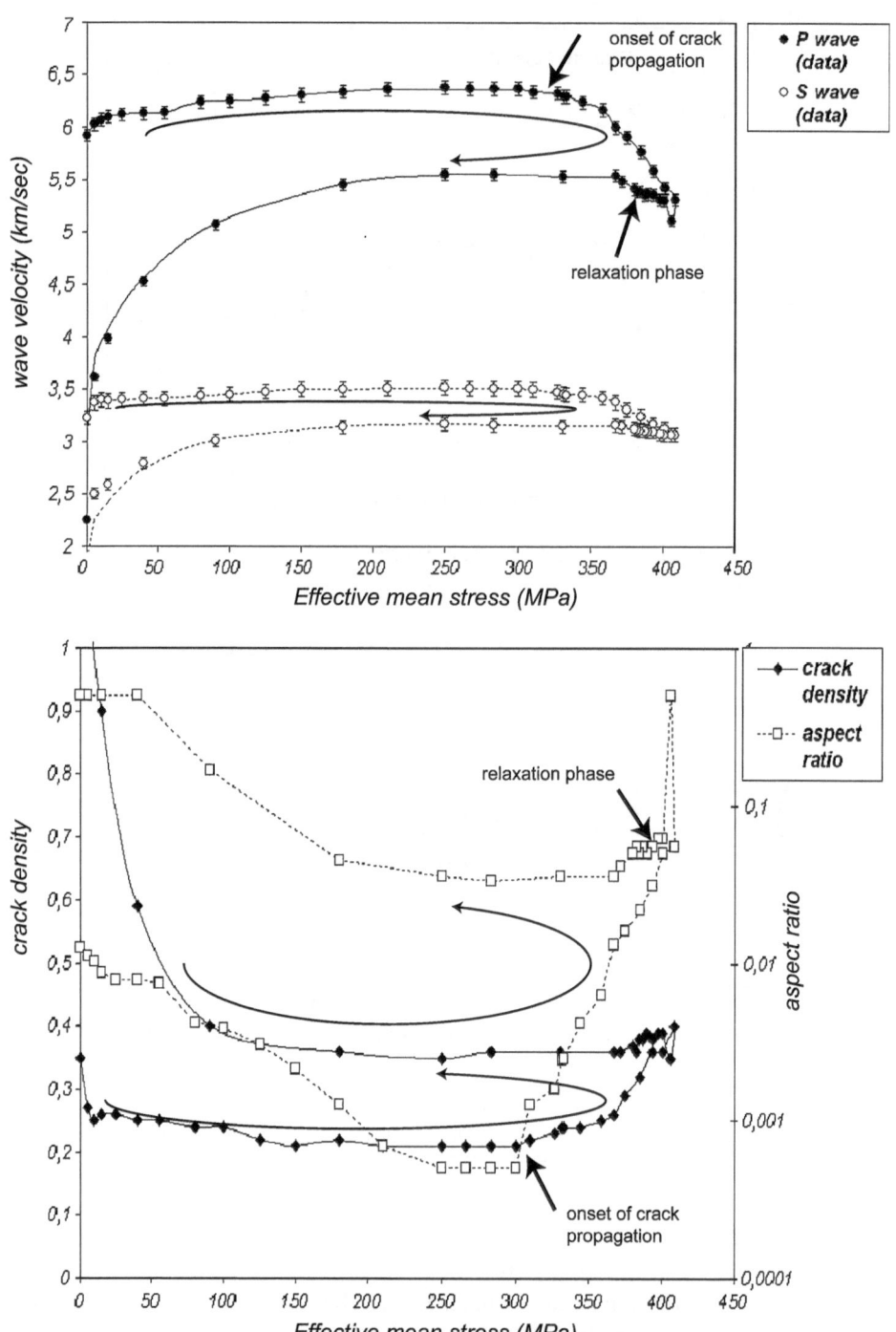

function of effective mean stress. During the first phase, crack density drops from 0.35 to 0.2 while aspect ratio drops again exponentially from 0.01 to lower than 0.001 (more than one order of magnitude) which clearly illustrates crack closure, contact increase in crack surfaces and aperture reduction. During the second phase, crack density increases slightly to 0.5 while aspect ratio increases rapidly to 0.5, revealing that high aspect ratio voids are being opened, which is consistent with microstructural analysis of cataclastic deformation in calcite performed by FREDRICH et al. (1989) and SCHUBNEL et al. (2005). The aspect ratio last value is probably an artifact of the inversion due to a bad Vp/Vs dataset at the onset of relaxation, because deformation rates were rapid at that point and Vp and Vs measurements were probably distant in time. During the relaxation phase, crack density remains more or less constant while the aspect ratio decreases exponentially again, which is a clear geophysical signature of visco-elastic crack closure and aperture reduction during restrengthening (BEELER and TULLIS, 1997). The fourth phase corresponds to a rapid increase in both crack density and aspect ratio due to stress relief crack propagation and opening. It is interesting to note that the final crack density is larger than 1, although no macroscopic rupture or strain localization band was observed.

3.2 Anisotropic Inversions

Most rocks are characterized by anisotropic crack patterns, often produced by deviatoric stress fields. For example, as a rock body is being deformed during triaxial compression experiments, cracks grow and propagate along preferential orientations, leading to an overall anisotropic elastic pattern (HADLEY, 1975; SCHUBNEL et al., 2003; STANCHITS et al., this issue). However and in the following, because the upper-limit of a non-interactive model is obviously that of instable crack propagation, we do not pretend to deal with coalescence or rupture propagation, but with phenomena prior to these.

3.2.1 Crack density as a function of depth and hydrostatic pressure

Granodiorite samples retrieved from the Nojima fault core were investigated experimentally at room pressure by ZAMORA et al. (1999). Elastic wave velocities, measured in the laboratory (500 KHz), are in good agreement with the sonic log

◄

Figure 7

a) Modelled (lines) and data (dots) *P*- and *S*-wave velocities as a function of effective mean stress in Carrara marble. Experiment was performed at *Pc* = 260 MPa confining pressure and *Pp* = 10 MPa. Model crack-free parameters were taken as *E* = 100 GPa and *v* = 0.32. b) Evolution of crack density and aspect ratio as a function of effective mean stress. Both crack density and aspect ratio were inverted from experimental data presented on a) and using equations (2) and (3). Arrows indicate the reading and the important phases of the experiment (onset of crack propagation and relaxation phases - cf. SCHUBNEL *et al.*, 2005).

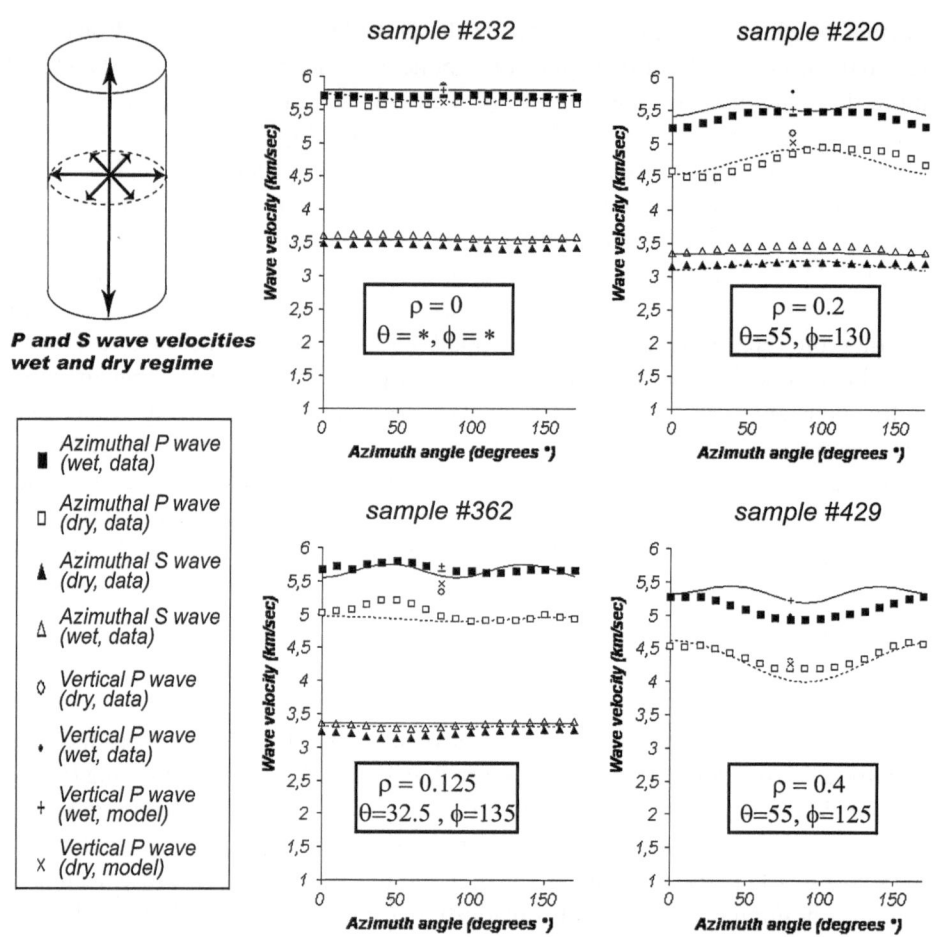

Figure 8

P- and S-wave velocity measurements in dry (empty symbols) and wet conditions (plain symbols) of Nojima fault core granodiorite (ZAMORA *et al.,* 1999). Experimental data obtained on four different samples, retrieved at 220, 232, 362 and 429 meters, respectively is presented. P waves were measured along the vertical axis and along 18 different directions in the horizontal plane; S waves propagating along the vertical axis were measured along 18 different azimuthal polarizations. Modelled velocities for each of these samples are represented by solid and dashed lines. Crack-free elastic parameters were taken as equal to that of sample 232 ($E_o = 85$GPa and $v_o = 0.25$). The inversion outputs both the crack density ρ and the crack fabric orientation (θ, ϕ).

performed during drilling (10 KHz, ZAMORA *et al.,* 1999). In both dry and wet conditions, *P*-wave velocities were measured along the vertical axis and along 18 different azimuthal directions in the horizontal plane; *S*-wave velocities propagating along the vertical axis were measured along 18 different azimuthal polarizations. Figure 8 presents the experimental data (symbols) obtained on four different samples, retrieved at 220, 232, 362 and 429 meters, respectively. Wet velocities are marked by plain symbols, while dry velocities are marked by empty symbols. For each of these samples, a simultaneous inversion of the dry and wet elastic wave velocity field was performed using equation (9). Crack-free elastic parameters were taken as equal to that of sample 232 ($E_o = 85$ GPa and $v_o = 0.25$), which presented the least amount of damage in the column. The inversion was performed simultaneously in both dry (assuming $\delta/1 + \delta \rightarrow 0$) and saturated (assuming $\delta \rightarrow 0$) conditions. Modelled azimuthal velocities are represented on Figure 8 as solid and dashed lines. Figure 8 shows that our inversion is reasonable even when the degree of freedom is drastically increased. Indeed, for a given crack density and crack fabric, our modelling effectively mimics both the observed anisotropy pattern and the saturation effect. The retrieved crack density for samples 220, 362 and 429 was 0.2, 0.125 and 0.4, respectively. Fitting of dry data is generally more accurate than fitting of wet data, which might be due to the fact that the incompressible fluid assumption ($\delta \rightarrow 0$) is not valid at room pressure. The dip of the distribution is not well constrained due to the paucity of vertical measurements (and the absence of diagonal measurements - see Figures 3 and 4), but is nevertheless in overall agreement with the geological setting (SCHUBNEL, 2002). However, the azimuth of the crack fabric is well constrained and was found to be approximately constant in the column, which is what is intuitively expected in a fault zone, where cracks and damage are aligned parallel to the main fracture plane. This gives us additional confidence in our modelling.

3.2.2 Crack density and aspect ratio evolution as a function of shear stress

Here, we present *P*-wave velocity measurements performed along several directions on a sample of dry Westerly granite during a tri-axial compression (Pc = 50 MPa). The experiment was performed at the USGS at Menlo Park (THOMPSON et al., this issue). The initial *P*-wave velocity field was more or less isotropic and equal to 5.9 km.s^{-1} while final *P*-wave anisotropy exceeded 30%. The non-interactive model can provide a useful tool to study and quantify the first phase of crack propagation, which is stable in triaxial compression experiments. Figure 9a presents a non-exhaustive compilation of the *P*-wave velocity data (symbols, see also THOMPSON et al., this issue) as a function of raypath angle and shear stress steps. The anisotropic pattern is very similar to that of the modelled velocity field for radial cracks presented on Figure 3b, which is as you would expect during the primary phases of a triaxial compression experiment. Modelled velocities are represented by solid and dashed lines. Model crack-free parameters were taken as $E_o = 85$ GPa and $v_o = 0.25$. Data and model velocities fit well. The orientation of the crack fabric is first vertical and then appears to continuously diminish: at 617 MPa, the mean crack fabric dip is only 78° with respect to horizontal, while the crack density increased to 0.44. The inversion, although poorly constrained (only *P*-waves), is stable and coherent. The *P*-wave anisotropy pattern is well reproduced. Figure 9b delineates the evolution

a) *P wave anisotropy pattern*

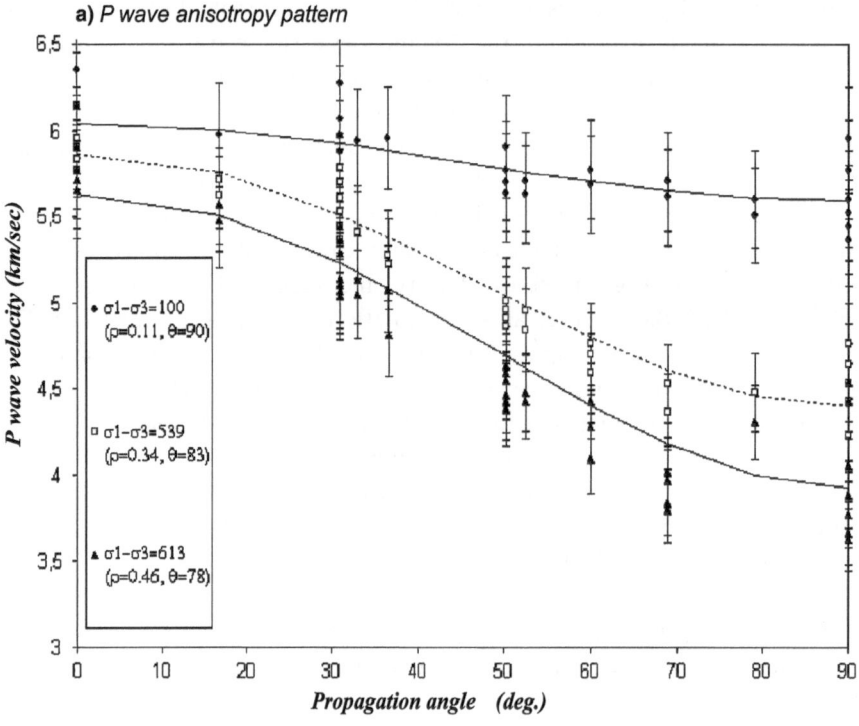

b) *shear stress and crack density evolutions with time*

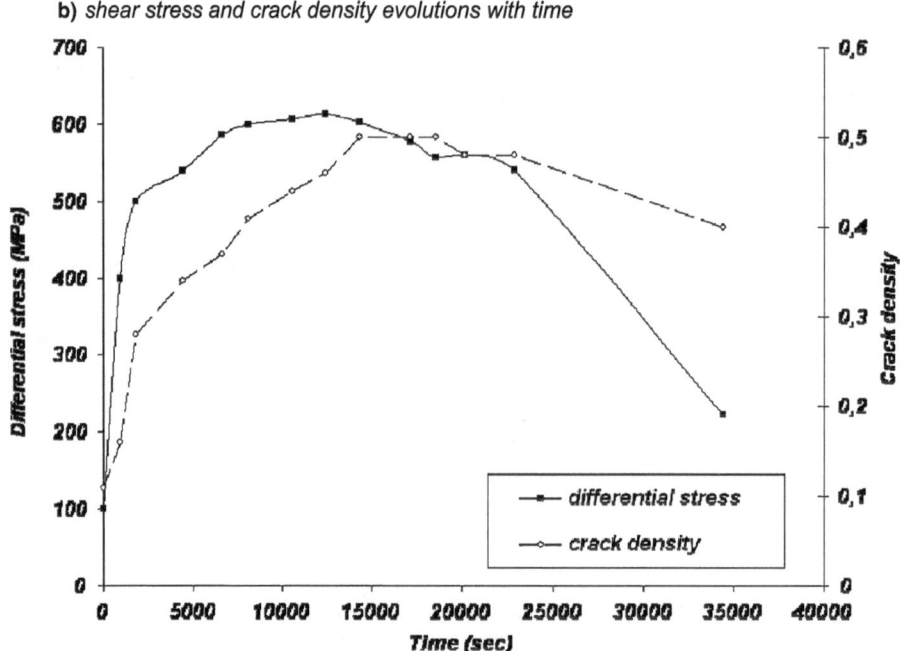

of shear stress and crack density with time. The system was loaded using Acoustic Emission (AE) feedback (see THOMPSON *et al.*, this issue) and one can see that, at first, crack density increased linearly with time. As the system was unloaded, crack density started decreasing, indicating that vertical cracks were closing due to diminishing shear stress. One can see the clear correlation between the two curves. The overall final crack fabric orientation (78°) is in agreement with AE locations, which enabled geophysical imaging of the fracture as it slowly propagated. Although it seems clear that the assumption of homogeneous damage was no longer valid at the end of the test, the inversion continuously output reasonable and physically interpretable results.

4. Discussion and Conclusions

We finally compare our model to that of a numerical simulation of damage evolution using the Particle Flow Code in 3 Dimensions (Itasca Consulting Group). First, a laboratory study was conducted in which a 50 mm cubic sample of Crossland Hill sandstone was subjected to true triaxial loading with velocity measurements taken parallel to each of the principal stress directions. The experiment was performed at Imperial College London (KING, 2002) and then simulated with a distinct element modelling approach using PFC3D. In the numerical experiment, the sample of Crossland Hill sandstone was simulated by an assemblage of 20,000 spherical particles closely packed and bonded together at points of contact. Particle stiffnesses and bond strengths were set such that the macro stiffness and strength of the model matched that of the actual rock. The numerical model and the actual rock sample were both subjected to two episodes of hydrostatic loading up to 100 MPa and then deviatoric loadings. Results of the numerical and laboratory study can be found in HAZZARD and YOUNG (2004). The model was fully dynamic so that changes in wave velocities can be measured with changes in stress. The number of 'cracks' in the model was then directly counted. Figure 10 compares the crack densities calculated from measured velocities and using equations (2)–(3), compared with the crack densities calculated by directly counting the number of cracks (or broken bonds) in the numerical assemblage. To arrive at the curve for 'direct counting', all cracks were assumed to be closed at 100 MPa hydrostatic stress and the number of particle-particle contacts and bonds

◄

Figure 9

a) Modelled (lines) and data (symbols) *P*-wave evolution as a function of shear stress and raypath angle in Westerly granite (THOMPSON *et al.*, 2006). Experiment was performed at $Pc = 50$ MPa confining pressure in dry conditions. Model crack-free parameters were taken as $E = 85$ GPa and $v = 0.25$. Modelled crack density and average crack fabric dip are marked on the legend. b) Shear stress and crack density evolutions as a function of time.

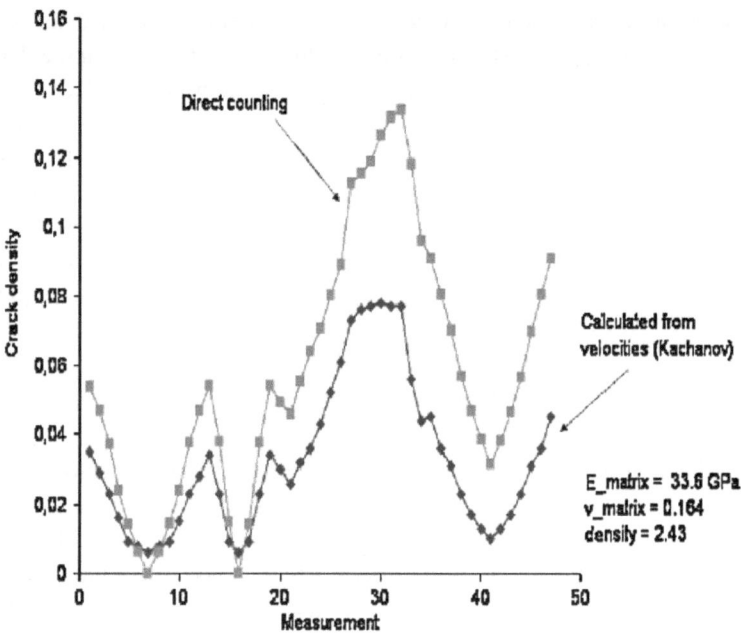

Figure 10

Comparison between PFC model directly observed crack density (crack radii were taken as equal to particles) and Kachanov's model inversion results for Crossland sandstone (KING, 2002; HAZZARD and YOUNG, 2004).

broken relative to this stress state were counted as cracks. Cracks were assumed to have a radius equal to that of the particles. It is clear that there is a fairly good match between the directly counted and calculated crack densities, which validates the modelling. However, the direct counting seems to systematically overestimate the crack density and the deviation between theoretical and numerical modelling increases with increasing crack density. This is probably due to the fact the crack radii are overestimated when taken as equal to the particle's radius. Real fractures also have shapes which are complex in detail, and this may be an additional reason for the differences shown in Figure 10. This may also reflect the fact that the material is no longer isotropic and that a Transversely Isotropic formulation might have produced better results.

The main question that remains is that of the significance of the fit between the laboratory or numerical and theoretical data. In general, this fit (e.g., in Fig. 5) is excellent and clearly, *P*- and *S*-wave velocities will in general be smoothly varying functions of the various parameters involved, in Figure 5a the effective confining pressure for example. If this is the case, then we can expect to describe any such curve with few parameters (i.e., the true empirical curve contains relatively few degrees of freedom). If the true number of degrees of freedom in the model is comparable to the true number of degrees of freedom in the data, then an excellent fit will always be found - but this is not necessarily significant. However, only a relatively good model which relies on good parameters will systematically output, using an inverse method, a physically understandable and expected evolution of the fitting parameters. In our case, the systematic decrease in crack density with confining pressure illustrates a decrease in the crack's apparent radii c as crack density evolves with $\sim c^3$. An increase in crack density also helps in quantifying crack growth. This can also be interpreted as a variation of the crack surfaces contact areas as cracks are being closed or opened, respectively. Even more so, the recovered evolution of aspect ratio shows an exponentially decreasing aspect ratio (and therefore aperture) with confining pressure which is also consistent with current mechanical theories of crack closure (KASELOW and SHAPIRO, 2004). The crack fabric orientations we recovered, when the rock was anisotropic, are also consistent with *in situ* geophysical imaging.

In conclusion, by using a non-interactive crack effective medium theory as a fundamental tool, it is possible to calculate the cracked rock dry and wet elastic properties in terms of a crack density tensor, average crack aspect ratio and mean crack fabric orientation using the solid grains and fluid elastic properties solely. Using the same method, both the anisotropy and shear-wave splitting of elastic waves can be derived. Two simple crack distributions have been considered for which the predicted anisotropy depends strongly on the saturation, reaching 60% in the dry case.

In the isotropic case, KACHANOV's (1994) model was used to invert elastic wave velocities and infer both crack density and aspect ratio evolutions. Inversion results were coherent in terms of crack density and aperture evolutions. A systematic decrease in crack density with confining pressure illustrated a decrease in the crack's apparent radii c. An increase in crack density can facilitate the precise quantification of crack growth (FORTIN *et al.*, 2006). Using such an inversion tool opens the door in linking elastic propertie's variations to permeability, as pointed out GUÉGUEN and SCHUBNEL (2003) or BENSON *et al.* 2006.

Inversion results agreed very well with the data and were consistent with the microstructure of the different rocks investigated here. That the theoretical curves in Figures 5a and 7a so closely follow the minor fluctuations in the empirical data curves might also suggest that the data are being overfitted. Obviously, if the first step in verifying a theoretical model is to compare it directly to data, a logical next step would be to perform a variance, covariance analysis in order to more reliably assess what the fit between data and model really means. It needs to be pointed out that for crack densities larger than 0.5, the predictions of the model are less accurate but would nevertheless remain within the same physical trend, which is a major difference with the Self-Consistent method (O'CONNELL and BUDIANSKY, 1975). A natural extension of this work would be to incorporate inversion of porosity into the model. Unfortunately, such extensive laboratory data is very rare in the literature.

Acknowledgements

The authors would like to acknowledge MARIA ZAMORA, ANNE-MARIE BOULIER and PHILIPPE PEZARD for providing the data on the Nojima fault core. SERGIO VINCIGUERRA, CONCETTA TROVATO and PHILIP MEREDITH provided the UCL data and are duly thanked. DAVID LOCKNER, LUIGI BURLINI and JÉRÔME FORTIN assisted in acquiring the data on Westerly granite and Carrara marble, respectively. This work was developed thanks to many useful discussions with MARK KACHANOV, IAN JACKSON and YVES GUÉGUEN. Additional and useful comments were provided by DAVID COLLINS and FARZINE NASSERI. We would also like to thank our two reviewers, ERIK SAENGER and ROLAND ROBERTS, for their very constructive comments which helped to greatly improve the quality of this manuscript. This study was partially supported by an NSERC discovery grant obtained by Prof. R. PAUL YOUNG.

Appendix I

Elastic Deformation of a Single Penny-shaped Crack

For a single penny-shaped crack under a uniform normal load p or shear stress τ, the elastic normal and shear displacements of the crack surfaces are not equal (and thus there can be a non-collinearity between the stress and displacement vectors). Such elastic displacements $\langle \mathbf{b} \rangle$ can be written respectively as (KACHANOV, 1994):

$$\langle \mathbf{b} \rangle = \begin{cases} \langle b_n \rangle = \frac{16(1-v^2)c}{3\pi E}p \\ \\ \langle b_\tau \rangle = \frac{16(1-v^2)c}{3\pi(1-v/2)E}\tau \end{cases} \tag{12}$$

where c is the crack radius, E and v the matrix YOUNG's modulus and Poisson ratio respectively. In the case of a crack filled up with fluid, an applied stress on the crack surface will generate a pore pressure variation Δp_f. Considering the crack aspect ratio $\zeta = w/c$, and assuming that $w \ll c$, the volume variation of the crack is mainly related to a change in its aperture. Thus the total traction applied on the crack surface is $\mathbf{n}.\sigma.\mathbf{n} + \Delta p_f$. From equation (12), one obtains:

$$\langle \Delta b_n \rangle = \frac{16(1-v^2)c}{3\pi E}(\mathbf{n} \cdot \sigma \cdot \mathbf{n} + \Delta p_f). \tag{13}$$

Considering that the fluid mass is constant, the variation in fluid density, volume and aperture are linked:

$$\frac{\Delta\varrho}{\varrho_o} = -\frac{\Delta V}{V_o} = \frac{-\langle \Delta w \rangle}{\langle w \rangle} = -\frac{\langle \Delta b_n \rangle}{\langle w \rangle}. \tag{14}$$

Remembering that $\Delta p_f = -K_f \frac{\Delta\varrho}{\varrho_o}$, the crack normal compliance B_N is such that (KACHANOV, 1994):

$$\frac{B_N}{B_T} = \left(1 - \frac{v_0}{2}\right)\frac{\delta}{1 + \delta}, \tag{15}$$

where δ is given by equation (5) of the main text.

REFERENCES

ANDERSON, D.L., MINSTER, B., and COLE, D. (1974), *The effect of oriented cracks on seismic velocities*, JGR *79*, 4011–4015.

BEELER, N.M. and TULLIS, T.E. (1997), *The roles of time and displacement in velocity dependent volumetric strain of fault zones*, JGR *102*, 22595–22609.

BENSON, P. M. (2004), *Experimental study of void space, permeability and elastic anisotropy in crustal rock under ambient and hydrostatic pressure*, Ph.D. Thesis, 272 pp., University of London, London.

BENSON, P., SCHUBNEL, A., VINCIGUERRA, S., TROVATO, C., MEREDITH, P., and YOUNG, R.P. (2006), *Modelling the permeability evolution of micro-cracked rocks from elastic wave velocity inversion at elevated hydrostatic pressure*, JGR *111*, in press.

BRACE, W.F., WALSH, J.B., and FRANGOS, W.T. (1968), *Permeability of granite under high pressure*, JGR *73*, 2225–2236.

BERRYMAN, J.G. (1992), *Single-scattering approxmations for coefficient in Biot equations of poroelasticity*, J. acoustical Soc. Am. *92*, 551–571.

BROWN, R. and KORRINGA, J. (1975), *On the dependence of the elastic properties of a porous rock on the compressibility of the pore fluid*, Geophysics 40, 608–616.

CHENG, C.H. and TOKSÖZ, M.N. (1979), *Inversion of seismic velocities for the pore aspect ratio spectrum of a rock*, JGR *84*, 7533–7543.

CHUN, K.-Y., HENDERSON, G. A., and LIU, J. (2004), *Temporal changes in P-wave attenuation in the Loma Prieta rupture zone*, JGR *109*, B02317, doi:10.1029/2003JB002498.

CLEARY, M.P., CHEN, I.W. and LEE S.M. (1980), *Self-consistent techniques for heterogeneous media*, J. Engin. Div. *106*, 861–887.

ESHELBY, J.D. (1957), *The determination of the elastic field for an elliptical inclusion and related problems*, Proc. R. Soc. London *241*, 376–396.

FORTIN, J. (2005), Compaction homogène et compaction localisée des roches poreuses, PhD. thesis, Université Pierre et Marie Curie, Paris, France.

FORTIN J., GUÉGUEN, Y., and SCHUBNEL, A. (2006), *Consequences of pore collapse and grain crushing on ultrasonic velocities and Vp/Vs*, JGR, in press.

FREDRICH, J.T., EVANS, B., and WONG, T.-F. (1989), *Micromechanics of the brittle to plastic transition in Carrara marble*, JGR *94*, 4129–4145.

GAO, Y. and CRAMPIN, S. (2004), *Observations of stress relaxation before earthquakes*, Geophys. J. Int. *157*, 578–582.

GUÉGUEN, Y. and SCHUBNEL, A. (2003), *Elastic wave velocities and permeability of cracked rocks*, Tectonophysics *370*, 163–176.

HADLEY, K. (1975), *Azimuthal variation of dilatancy*, JGR *80*, 4845–4850.

HAZZARD, J.F. and YOUNG, R.P. (2004), *Numerical investigation of induced cracking and seismic velocity changes in brittle rock*, GRL *31*, Art. No. L01604.

HUDSON, J.A. (1981), *Wave speeds and attenuation of elastic waves in material containing cracks*, Geophys. J. R. Astr. Soc. *64*, 133–150.

HUDSON, J.A. (1982), *Overall properties of a cracked solid*, Math. Proc. Cambridge Phil. Soc. *88*, 371–384.

HUDSON, J.A. (1986), A higher order approximation to the wave propagation constants for a cracked solids, Geophys. J. R. Astr. Soc. *87*, 265–274.

KACHANOV, M. (1994), *Elastic solids with many cracks and related problems*, Adv. Appl.Mech. *30* 259–445.

KASELOW, A. and SHAPIRO, S.A. (2004), *Stress sensitivity of elastic moduli and electrical resistivity in porous rocks*, J. of Geophy. Engin *1*, 1–11.

KERN, H. (1978), *The effect of high temperature and high confining pressure on compressional wave velocities in quartz bearing and quartz free igneous and metamorphic rocks*, Tectonophysics *44*, 185–203.

KERN, H., LIU, B., and POPP, T. (1997), *Relationship between anisotropy of P- and S-wave velocities and anisotropy of attenuation in serpentinite and amphibolite*, JGR *102*, 3051–3065.

KING, M.S. (2002), *Elastic wave propagation in and permeability for rocks with multiple parallel fractures*, Int. J. Rock Mech. Min. Sci. *39*, 1033–1043.

LE RAVALEC, M. and GUÉGUEN, Y. (1996), *High and low frequency elastic moduli for a saturated porous/cracked rock - Differential self-consistent and poroelastic theories*, Geophys. *61*, 1080–1094.

MAVKO, G. and NUR, A. (1975), *Melt squirt in the asthenosphere*, JGR *80*, 1444–1448.

MAVKO, G. and JIZBA, D. (1991), *Estimating grain-scale fluid effects on velocity dispersion in rocks*, Geophysics *56*, 1940–1949.

MAVKO, G., MUKERJI, T., and DVORKIN, J. In *The Rock Physics Handbook* (Cambridge University Press 1998), 329 pp.

MILLER, S. A. (2002), *Properties of large ruptures and the dynamical influence of fluids on earthquakes and faulting*, JGR *107(B9)*, 2182, doi:10.1029/2000JB000032.

NASSERI, M.H.B., MOHANTY, B., and YOUNG, R.P. (2006), *Fracture toughness measurements and acoustic emission activity in brittle rocks*, Pure Appl. Geophys. this issue.

NISHIZAWA, O. (1982), *Seismic velocity anisotropy in a medium containing oriented cracks: Transverse isotropy case*, J. Phys. Earth *30*, 331–347.

O'CONNELL, R. and BUDIANSKY, B. (1974), *Seismic velocities in dry and saturated rocks*, JGR *79*, 5412–5426.

O'CONNELL, R. and BUDIANSKY, B. (1977), *Viscoelastic properties of fluid saturated cracked solids*, JGR *82*, 5719–5736.

ORLOWSKY, B., SAEGER, E.H., GUÉGUEN, Y., and SHAPIRO, S.A. (2003), *Effects of parallel crack distributions on effective elastic properties — a numerical study*, Internat. J. Fract. *124*, 171–178.

SAENGER, E.H., KRUGER, O.S., and SHAPIRO, S.A. (2004), *Effective elastic properties of randomly fractured soils: 3-D numerical experiments*, Geophys. Prosp. *52*, 183–195.

SAYERS, C.M. and KACHANOV, M. (1991), *A simple technique for finding effective elastic constants of cracked solids for arbitrary crack orientation statistics*, Int.J Sol. Struct. *12*, 81–97.

SAYERS, C.M. and KACHANOV, M. (1995), *Microcrack induced elastic wave anisotropy of brittle rocks*, JGR *100*, 4149–4156.

SCHOENBERG, M. and SAYERS, C.M. (1995), *Seismic anisotropy of fractured rock*, Geophysics *60*, 204–211.

SCHUBNEL, A. (2002), *Mécanique de la dilatance et de la compaction des roches de la croûte*, Ph.D. Thesis of Institut de Physique du Globe de Paris, 229pp.

SCHUBNEL, A., NISHIZAWA, O., MASUDA, K., LEI, X.J., XUE, Z. and GUÉGUEN, Y. (2003), *Velocity measurements and crack density determination during Wet Triaxial Experiments on Oshima and Toki Granites*, Pure Appl. Geophy. *160*, 869–887.

SCHUBNEL, A. and GUÉGUEN, Y. (2003), *Dispersion and anisotropy in cracked rocks*, JGR *108*, 2101, doi:10.1029/2002JB001824.

SCHUBNEL A., FORTIN, J., BURLINI L., and GUÉGUEN, Y. (2005), *Damage and recovery of calcite rocks deformed in the cataclastic regime*. In Geological Society of London special publications on *High Strain Zones* (eds. by Bruhn, D. and Burlini, L.) *245*, 203–221.

SIMMONS, G. and BRACE, W.F. (1965), *Comparison of static and dynamic measurements of compressibility of rocks*, JGR *70*, 5649–5656.

SIMMONS, G., TODD, T., and BALRIDGE, W.S. (1975), *Toward a quantitative relationship between elastic properties and cracks in low porosity rocks*, AM. J. Sci. *275*, 318–345.

SOGA, N., MIZUTANI, H., SPETZLER, H. and MARTIN, R. J. III. (1978),*The effect of dilatancy on velocity anisotropy in Westerly Granite*, JGR *83*, 4451–4458.

STANCHITS, S., VINCIGUERRA, S., and DRESEN, G. (2006), *Ultrasonic velocities, acoustic emission characteristics and crack damage of basalt and granite*, Pure Appl. Geophys. this issue.

THOMPSON, B.D., YOUNG, R.P., and LOCKNER D.A. (2006), *Observations of fracture in Westerly granite under AE feedback and constant strain rate loading: Nucleation, quasi-static propagation, and the transition to unstable fracture propagation*, Pure Appl. Geophys. this issue.

VOLTI, T. and CRAMPIN, S. (2003), *A four-year study of shear-wave splitting in Iceland: 2. Temporal changes before earthquakes and volcanic eruptions*. In Geological Society of London special publications on *New Insights into Structural Interpretation and Modelling* (ed. Nieuwland, D.A.) *212*, 135–149.

WALSH, J.B. (1965), *The effect of cracks on the compressibility of rock*, JGR *70*, 381–389.

ZAMORA, M., PEZZARD, P.A., and ITO, H. (1999), *Anisotropy of elastic properties of granites from the Hirabayashi borehole, Japan*, internal note, GSJ/USGS.

(Received May 4, 2005, revised October 28, 2005, accepted October 29, 2005)

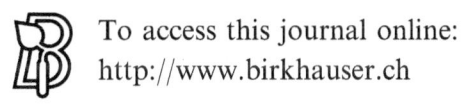

To access this journal online:
http://www.birkhauser.ch

Pure appl. geophys. 163 (2006) 975–994
0033–4553/06/060975–20
DOI 10.1007/s00024-006-0059-5

❙ Pure and Applied Geophysics

Ultrasonic Velocities, Acoustic Emission Characteristics and Crack Damage of Basalt and Granite

SERGEI STANCHITS,[1] SERGIO VINCIGUERRA,[2] and GEORG DRESEN[1]

Abstract—Acoustic emissions (AE), compressional (P), shear (S) wave velocities, and volumetric strain of Etna basalt and Aue granite were measured simultaneously during triaxial compression tests. Deformation-induced AE activity and velocity changes were monitored using twelve P-wave sensors and eight orthogonally polarized S-wave piezoelectric sensors; volumetric strain was measured using two pairs of orthogonal strain gages glued directly to the rock surface. P-wave velocity in basalt is about 3 km/s at atmospheric pressure, but increases by > 50% when the hydrostatic pressure is increased to 120 MPa. In granite samples initial P-wave velocity is 5 km/s and increases with pressure by < 20%. The pressure-induced changes of elastic wave speed indicate dominantly compliant low-aspect ratio pores in both materials, in addition Etna basalt also contains high-aspect ratio voids. In triaxial loading, stress-induced anisotropy of P-wave velocities was significantly higher for basalt than for granite, with vertical velocity components being faster than horizontal velocities. However, with increasing axial load, horizontal velocities show a small increase for basalt but a significant decrease for granite. Using first motion polarity we determined AE source types generated during triaxial loading of the samples. With increasing differential stress AE activity in granite and basalt increased with a significant contribution of tensile events. Close to failure the relative contribution of tensile events and horizontal wave velocities decreased significantly. A concomitant increase of double-couple events indicating shear, suggests shear cracks linking previously formed tensile cracks.

Key words: Acoustic emission, ultrasonic velocity, fracture, rock.

1. Introduction

The physical properties of rocks such as elastic constants and the speed of elastic waves are affected significantly by the volume, distribution and shape of the rock pore space (WALSH, 1965a; O'CONNELL and BUDIANSKY, 1974; PATERSON and WONG, 2005). In particular, it is well known that elastic wave velocities P and S may be reduced substantially in the presence of thin cracks (e.g., HADLEY, 1976). In a stressed rock volume, tensile cracks open preferentially normal to the

[1]Department 3.2, Geo Forschungs Zentrum Potsdam, Telegrafenberg D420, 14473, Potsdam, Germany. E-mail: stanch@gfz-potsdam.de
[2]Istituto Nazionale di Geofisica e Vulcanologia, Via di Vigna Murata 605, 00143, Rome, Italy. E-mail: vinciguerra@ingv.it

maximum compressive stress direction (TAPPONNIER and BRACE, 1976; RECHES and LOCKNER, 1994) resulting in an anisotropy of the elastic wave velocities. The speed of waves propagating in the direction normal to the planes of preferred crack orientation is more significantly decreased than the speed of waves travelling parallel to the crack planes (NUR, 1971; BONNER, 1974; LOCKNER *et al.*, 1977; SCHUBNEL *et al.*, 2003). Assuming transverse isotropy observed elastic wave speeds may be used to estimate the respective anisotropic elastic constants and crack damage (MAVKO *et al.*, 1998; SOGA *et al.*, 1978; AYLING *et al.*, 1995; SCHUBNEL *et al.*, 2003). HADLEY (1976) observed a significant decrease of P-wave and S-wave speeds and comparable changes in velocity ratios with increasing dilatancy in both Westerly granite and San Marcos gabbro. Upon unloading, plots of velocities versus volumetric strain showed a hysteresis with lower velocities during unloading. In general, the experimental studies show that wave velocities increase with increasing confining pressure and decrease with increasing deviatoric stresses; both observations have been related to the pressure-induced closure and stress-induced opening of narrow cracks (e.g., GUEGUEN and PALCIAUSKAS, 1994; PATERSON and WONG, 2005).

Mount Etna with its diameter of 40 km and its height of ~ 3300 m is the largest volcano in Europe, characterized by mostly effusive eruptions constituting a major natural hazard for human settlements in the Etna region. Overpressured magma stored in shallow reservoirs and time-dependent failure of the country rock, induce bursts of seismic activity and episodes of ground deformations that occur from years to months before a new major eruption. Recent pre-eruptive stages have been closely monitored by using both ground deformation and seismic arrays (CASTELLANO *et al.*, 1993; FERRUCCI *et al.*, 1993; BRIOLE *et al.*, 1990; BONACCORSO *et al.*, 1990). The observations have been related to increasing damage and a changing stress field within the volcanic edifice largely consisting of basalt. However, the mechanical behavior of Etna basalt and the effect of deviatoric stresses on crack damage and elastic wave velocities are not well understood. Previous laboratory measurements of P-wave and S-wave velocities were solely performed at hydrostatic pressures up to 80 MPa and indicate that Etna basalt contains substantial crack damage (VINCIG-UERRA *et al.*, 2005). Crack densities inverted from velocity measurements at increasing pressure (BENSON *et al.*, 2006; SCHUBNEL *et al.*, this issue), manifest a progressive closure of the cracks.

Here we combine laboratory measurements of elastic wave velocities (P and S) at varying confining pressures up to 120 MPa and under deviatoric stresses increased to sample failure with an analysis of acoustic emissions (AE) recorded during the experiments. We determined the location and AE source types in experiments performed on Etna basalt and Aue granite at identical loading conditions. Aue granite was used as a reference material containing low initial crack damage and low porosity. Anisotropic elastic constants, crack damage and crack types are estimated from the data.

2. Experimental Techniques

2.1 Sample Material and Testing Procedure

Experiments were performed on cores drilled from typical Etna basalt and on granite samples from Aue, Germany. Etna basalt (EB) is a porphyritic alkali basalt from Mount Etna, Italy, containing mm-sized phenocrysts of pyroxene, olivine and feldspar in a fine-grained groundmass. Initial density is 2.86 ± 0.01 g/cm^3 and porosity is $2.1 \pm 0.1\%$. The granite has an initial density of 2.62 g/cm^3 and an initial porosity of 1.3%. Mineral composition of granite contains 30% quartz, 40% plagioclase, 20% potassium feldspar and 10% mica (ZANG et al., 2000). The grain size ranges from 0.9 mm to 1.8 mm with an average value of 1.3 mm (Fig. 1). Microstructure analysis of undeformed Aue granite yielded crack densities in the range of 0.3–0.6 mm^{-1} (JANSSEN et al., 2001). No data exist for the Etna basalt, however, the block from which samples were cored contains abundant mm- to cm-scale cracks that possibly result from rapid cooling (VINCIGUERRA et al., 2005).

Cylindrical samples of Etna basalt and Aue granite with a diameter of 50 mm and 100 mm in length were subjected to two types of loading: All samples were first subjected to hydrostatic pressure $(\sigma_1 = \sigma_2 = \sigma_3)$ increasing from 5 MPa up to 120 MPa and subsequently decreasing to 60 MPa. Pressure changes were performed at a rate of 1.5 MPa per minute. Subsequently, an axial stress was applied to the specimens. The samples were subjected to three stress cycles at decreasing confining pressures of 60, 40 and 20 MPa, respectively, in a servo-controlled 4600 kN MTS loading frame (Fig. 2a). At the lower confining pressures, samples were loaded until catastrophic failure occurred.

Volumetric strain was estimated using two pairs of strain gages glued directly onto the sample surface. The strain gages were oriented parallel to the sample axis (ε_1) and in a circumferential direction (ε_3) (Fig. 3). Volumetric strain Δ was calculated using: $\Delta = \varepsilon_1 + 2\varepsilon_3$. Axial strain ε_1 was calculated using the average values of the two vertically oriented strain gages. Compaction is indicated by $\Delta > 0$ and dilatancy by $\Delta < 0$.

Axial loading rate was servo-controlled using the AE rate recorded by a piezoceramic sensor attached to the sample. For AE rates below a defined trigger level, the displacement rate was set to 0.02 mm/min. When the AE rate exceeds the trigger level the displacement rate is decreased significantly to about 10^{-10} mm/min. The threshold level of the rate control sensor allowed variation of the speed of fault propagation by three orders of magnitude, i.e., from mm/s in fixed displacement rate tests to μm/s in AE rate-controlled tests.

2.2. Velocity and Acoustic Emission Monitoring

AE activity and velocity changes were monitored by twelve *P*-wave and eight *S*-wave piezoelectric sensors either embedded in the pistons or glued to the sample

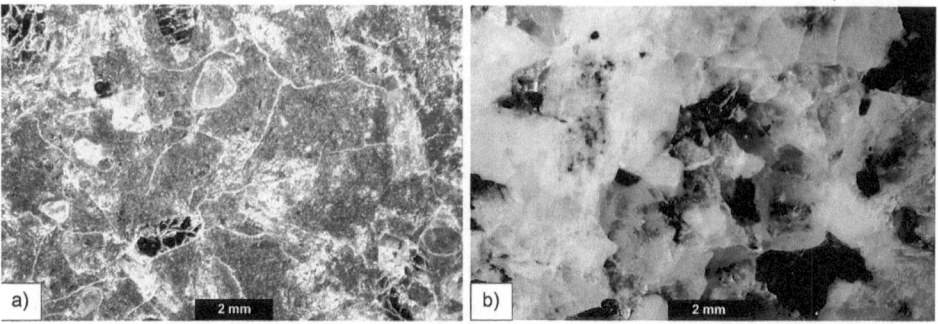

Figure 1
Optical micrograph (crossed polars) of Etna basalt (a) and Aue granite (b). Note array of mm-scale cracks in basalt.

Figure 2
Experimental setup. (a) MTS loading frame with 200 MPa pressure vessel. (b) Cylindrical specimen encapsulated in rubber jacket with P and S sensors glued directly to the sample surface. For velocity measurements parallel to sample axis, two P sensors were embedded in upper and lower endcaps.

surface and sealed in a Neoprene jacket using two-component epoxy (Fig. 2b). P-wave velocities were measured parallel (P_V) and normal (P_H) to the loading direction. To analyze stress-induced shear-wave splitting, we installed four S-wave piezoelectric sensors polarized in the vertical direction (S_{HV}) and horizontal direction

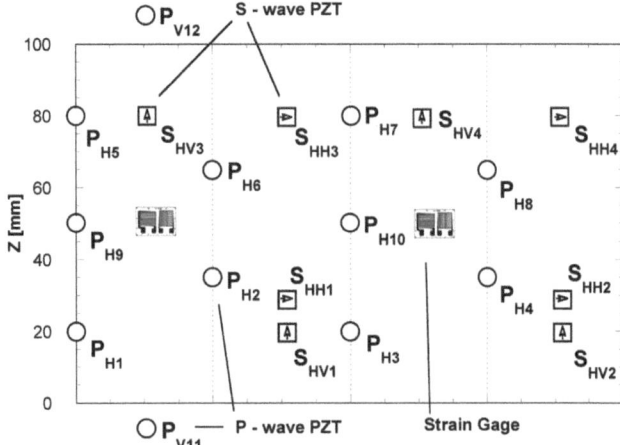

Figure 3

Projected sample surface showing position of 20 ultrasonic sensors and 4 strain gages installation (Z is sample axis). P_H and P_V are piezoelectric sensors measuring P-wave velocity in horizontal and vertical directions. S_{HH} are shear sensors polarized in horizontal direction, S_{HV} are shear sensors polarized in vertical direction.

(S_{HH}), respectively. The first index on P and S indicates wave propagation direction and the second index on S indicates the plane of polarization. The piezoelectric sensors were manufactured at the GFZ and tested at confining pressures up to 200 MPa. P- and S-wave sensors were produced from PZT piezoceramic discs with 5 mm diameter and 1 mm thickness and square shape piezoceramic plates $5 \times 5 \times 1$ mm, respectively. The thickness and diameter-related resonant frequencies of the P-wave sensors are about 2 MHz and 500 KHz, respectively.

Transducer signals are amplified by 40 dB using Physical Acoustic Corporation (PAC) preamplifiers equipped with 100 kHz high-pass filters. Full-waveform AE data and the ultrasonic signals for P-wave velocity measurements are stored in a 12 channel transient recording system (PRÖKEL, Germany) with an amplitude resolution of 16 bit at 10 MHz sampling rate. S-wave sensors were connected to a separate transient recorder (KRENZ, Germany, 10 bit amplitude resolution, 10 MHz sampling rate) (Fig. 4). For periodic elastic wave speed measurements we used six P-sensors and four S-sensors as senders applying 100 V pulses every 30–40 seconds during loading. Ultrasonic transmissions and AE waveforms were discriminated automatically after the experiments.

The AE recording system is equipped with a 6 Gb memory buffer for temporary storage during the experiments that allows recording of digitized waveforms of 256 thousand AE signals with a length of about 100 μs. Continuous AE recording with zero dead time between consecutive signals is possible. For AE hypocenter location, P-wave onset time is picked automatically using different criteria including Akaike's

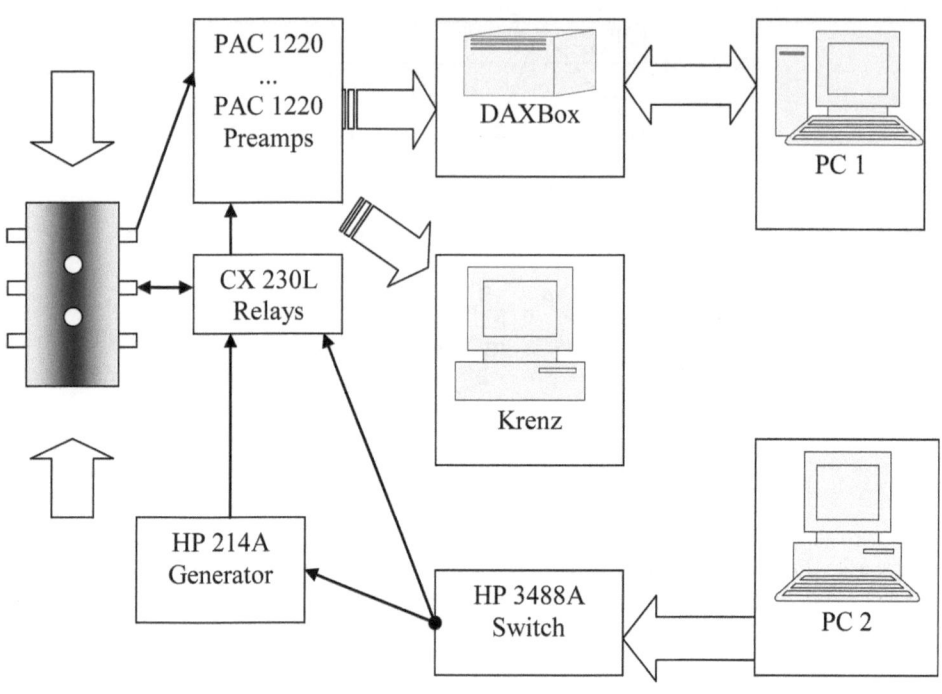

Figure 4
Block diagram of experimental setup to monitor acoustic emission activity and ultrasonic transmission.

information criterion (LEONARD and KINNETT 1999). An automatic minimization procedure of travel-time residuals was used to exclude noisy channels and decide for the preferred onset time criterion. Hypocenter locations were estimated using a downhill simplex algorithm (NELDER and MEAD, 1965) considering time-dependent changes of the anisotropic velocity field. We estimate the AE location accuracy to be 2.5 mm. First motion amplitudes were picked automatically and first motion polarities were used to discriminate AE source types in tensile, shear and collapse events (ZANG et al., 1998).

3. Experimental Results

3.1. Hydrostatic Loading

Compaction of the basalt samples at 120 MPa pressure is $\approx 0.8\%$ and about twice that of the granite samples ($\Delta \approx 0.4\%$). With increasing pressure the elastic wave speeds P and S increased (Figs. 5a-f). The maximum difference in wave speeds in different directions was <2% and close to experimental error. Therefore we

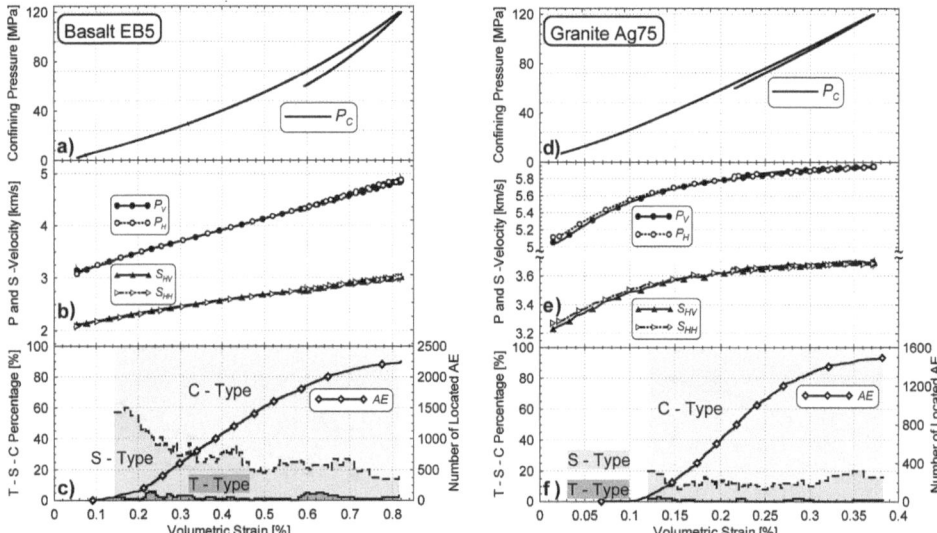

Figure 5
Plot of confining pressure, elastic velocities, cumulative AE number and AE types versus volumetric strain during hydrostatic loading of Etna basalt EB5 and Aue granite Ag75. Volumetric strain was calculated as $\Delta = \varepsilon_1 + 2\varepsilon_3$, where ε_1 and ε_3 are axial and horizontal components of the strain measured during the experiment. Separation of tensile, shear and collapse types of AEs was performed using AE first motion polarity analysis. Estimated error of velocity measurements corresponds approximately to the size of symbols on plots b) and e).

consider the velocity field to be isotropic. Horizontal P-wave velocities (P_H) and the S-wave components (S_{HH} and S_{HV}) were averaged from five and two horizontal traces, respectively. In granite, P- and S-wave velocities approach maximum values of about 6 km/s and 3.7 km/s with increasing pressure and compaction of the sample (Fig. 5). However, in basalt velocities show a continuous increase with pressure up to 120 MPa. With increasingly confining pressure, P-wave velocities in basalt increased by about 50% (from 3.1 to 4.8 km/s), however in granite P-wave speeds increased only by about 18% (from 5.05 to 5.94 km/s).

Significant AE activity was observed during compaction of the basalt and granite samples starting at relatively small volumetric strains of 0.1–0.2%. AE hypocenter distribution is almost random, suggesting that compaction is relatively homogeneous. The relative source type distribution reveals a dominance of pore collapse (C-type) sources and about 20% double-couple (S-type) shear events. The contribution of shear type events during compaction is higher in basalt than in granite, in particular at low volumetric strain. AE activity, AE source types and the hysteresis in the plot of confining pressure versus volumetric strain all indicate that crack closure during hydrostatic loading is not purely elastic but produces some irreversible deformation (Fig. 5).

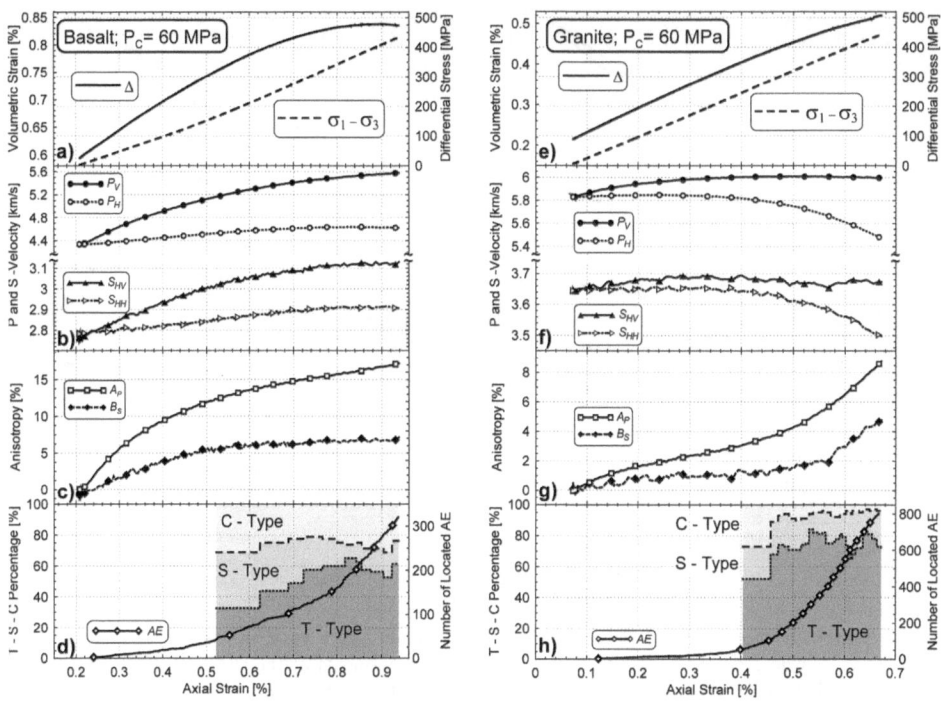

Figure 6

Plot of volumetric strain, elastic velocities, *P*-wave anisotropy, *S*-wave birefringence, cumulative AE number and AE source types versus axial strain during loading of Etna basalt EB5 and granite Ag75 at 60 MPa confining pressure (see text for details).

3.2. Axial Compression Testing

The basalt and granite specimens were subjected to increasing stresses in three consecutive cycles at decreasing confining pressures of 60 MPa, 40 MPa and 20 MPa. The first two cycles were interrupted at differential stresses of about 500 MPa and axial strains between 0.7% – 0.8%. During the third cycle the specimens were loaded to failure. With increasing axial stress and strain, basalt and granite samples show initial overall compaction, but at smaller confining pressures and at axial strains $\varepsilon_1 > 0.7\%$ dilatant cracking becomes increasingly important $\left(\frac{d\Delta}{d\varepsilon_1} < 0\right)$ (Fig. 8a). Maximum compaction of the basalt samples reached during each load cycle decreases with decreasing confining pressure but always remains higher than for granite.

At all confining pressures *P*- and *S*-wave velocities in basalt and granite manifest significant anisotropy (Figs 6–8). In general, P_V -wave speeds are higher than P_H – velocities, and *S*-waves show acoustic birefringence (NUR, 1971; BONNER, 1974; HADLEY, 1975; LOCKNER *et al.*, 1977) with S_{HV} always being faster than S_{HH}. We define anisotropy and birefringence as $A_P = \frac{(P_V - P_H)}{P_V}$ and $B_S = \frac{(S_{HV} - S_{HH})}{S_{HV}}$, respectively. *P*-wave anisotropy and *S*-wave birefringence increase with increasing axial stress and

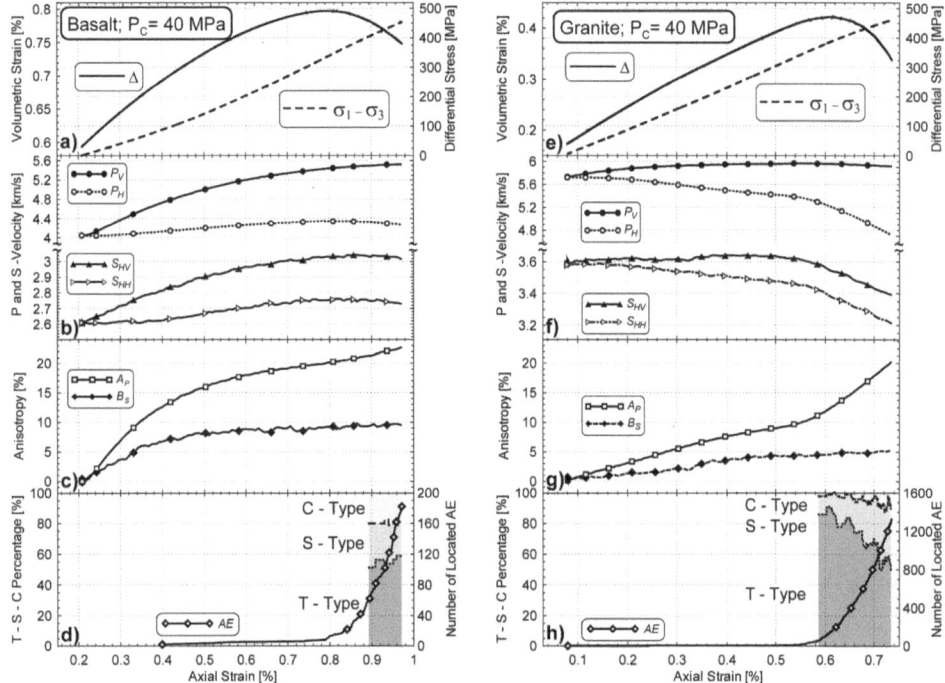

Figure 7

Plot of volumetric strain, elastic velocities, P-wave anisotropy, S-wave birefringence, cumulative AE number and AE types versus axial strain during loading of Etna basalt EB5 and granite Ag75 at 40 MPa confining pressure (see text for details).

axial strain and as samples approach failure. These observations suggest that stress-induced cracks are mainly vertically oriented with crack plane normal directions defining a zone perpendicular to the axial stress direction. However, changes in wave speeds are significantly more pronounced in basalt than in granite. With increasing stress and axial strain, P-wave anisotropy and S-wave birefringence exhibit different behavior in basalt and granite (Figs. 6–8). For example, at all confining pressures and at low stresses, anisotropy and birefringence increase more rapidly in basalt than in granite. However, at elevated stresses and larger axial strain, anisotropy increases more strongly in granite than in basalt.

Acoustic emission activity during loading cycles shows a pronounced Kaiser effect (HOLCOMB, 1993) (Figs. 6–8). For granite, the onset of AE activity is associated with a change in the slope of P_H and S_{HH} velocities versus axial strain (Figs. 6h–8h). The effect is less pronounced for basalt (Figs. 6d–8d). Acoustic emission source types are distinctly different from those found at hydrostatic compaction (Fig. 5). With increasing axial stress and decreasing confining pressure, T-type events are dominant in both basalt and granite specimens. Close to failure, however, double-couple S-type events become increasingly important. The contribution of C-type events related to

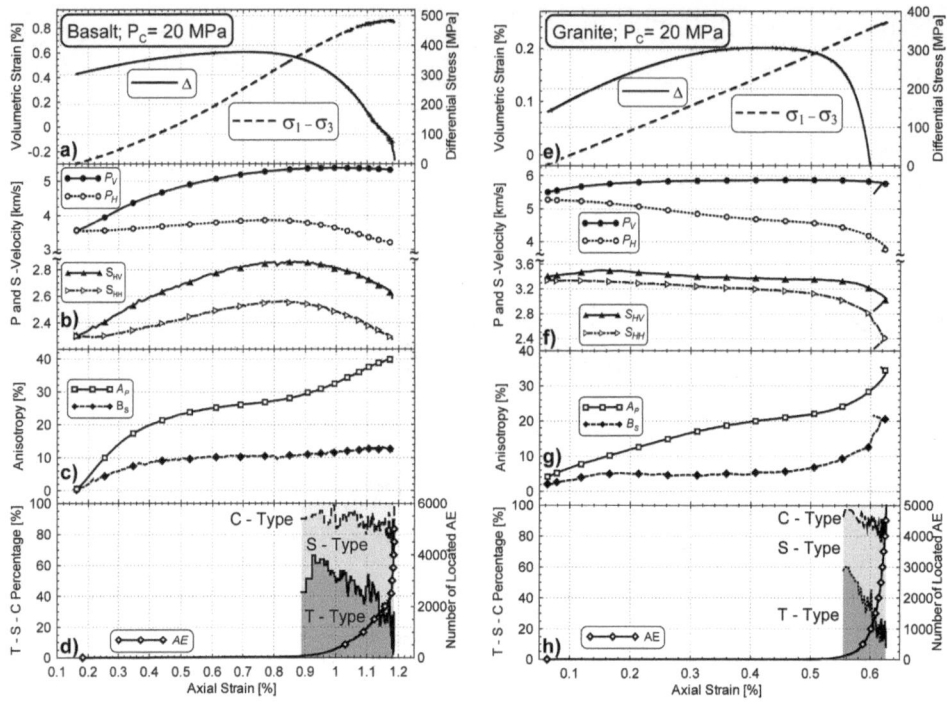

Figure 8

Plot of volumetric strain, elastic velocities, *P*-wave anisotropy, *S*-wave birefringence, cumulative AE number and AE types versus axial strain during loading of Etna basalt EB5 and granite Ag75 at 20 MPa confining pressure. The sample was loaded up to failure (see text for details).

collapsing pore space in basalt during loading at 40 and 60 MPa is 10%–20%, and < 5% in granite.

4. Discussion

4.1. Hydrostatic Compaction of Basalt and Granite

The effective dynamic and static elastic Young's modulus E_{eff}, bulk modulus K_{eff} and the Poisson ratio v_{eff} of basalt and granite increase with increasing pressure (Fig. 9) and show a hysteresis when pressure is released. The dynamic moduli are estimated using the equations:

$$v_{eff} = \left[\left(\frac{P}{S} \right)^2 - 2 \right] \Big/ \left[\left(\frac{P}{S} \right)^2 - 1 \right], \tag{1}$$

$$K_{eff} = \rho \left(P^2 - \frac{4}{3} S^2 \right), \tag{2}$$

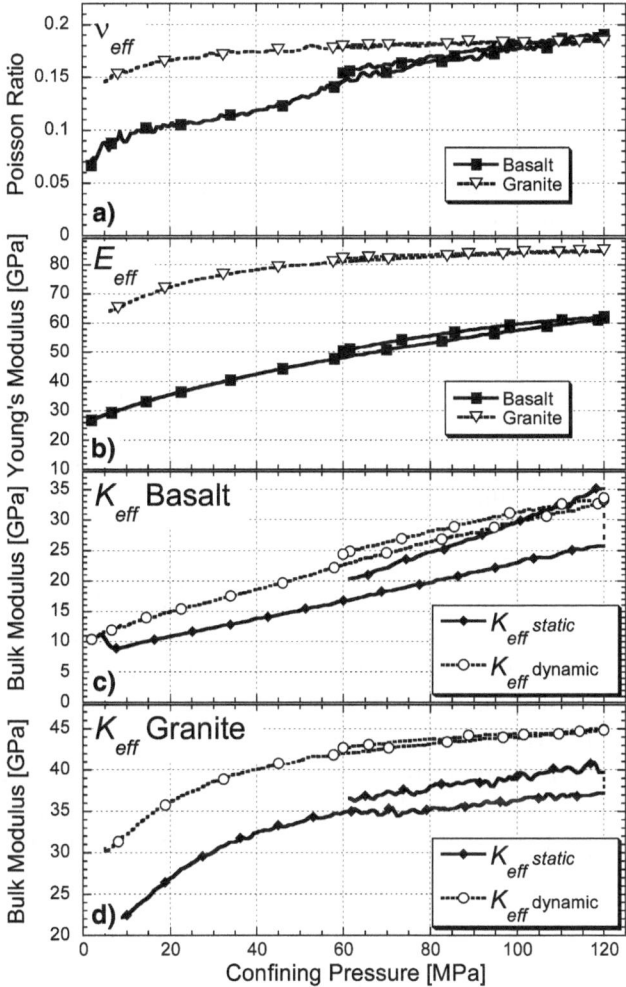

Figure 9

Effective Poisson ratio, Young's modulus and bulk modulus, calculated from elastic wave velocities (dynamic parameters) and mechanical measurements (static parameters) during hydrostatic loading of Basalt Eb5 and Granite Ag75. Samples were first pressurized to 120 MPa and then pressure was reduced to 60 MPa.

$$E_{eff} = 3K_{eff}\left(1 - 2v_{eff}\right), \tag{3}$$

where ρ is density and P, S are average P-wave and S-wave speeds. The static bulk modulus was also estimated from the change of volumetric strain $d\Delta$ with increasing pressure dP_c:

$$K_{eff} = \frac{dP_C}{d\Delta}. \tag{4}$$

The elastic moduli are somewhat higher when estimated during unloading than during loading. This corresponds to a small hysteresis of the wave speed versus volumetric strain relation between loading and unloading. In contrast to loading cycles at differential stress (see below, HADLEY, 1975) velocities remain slightly higher during unloading. Compaction of the basalt and granite samples with increasing hydrostatic pressure is associated with high AE activity and dominantly *C*- and *S*-type AE sources. This suggests that compaction of the samples involves irreversible pore collapse and frictional sliding along pre-existing defects.

For the same sample at similar pressures, static bulk moduli are smaller and show a larger hysteresis with increasing and decreasing confining pressure than the dynamic moduli. Static bulk moduli of dry rocks are often found to be lower than dynamic values although the effect is not well understood (SIMMONS and BRACE, 1965; KING, 1969; CHENG and JOHNSTON, 1981). Often the observed difference between static and dynamic moduli decreases with increasing pressure and is commonly attributed to the presence of cracks. Possibly, larger strain amplitudes for static measurements may produce inelastic effects that are not observed during dynamic measurements (DRESEN and GUEGUEN, 2004). For example, irreversible mechanical compaction and interlocking of crack faces may explain the observed hysteresis of velocities and elastic moduli between loading and unloading cycles (WALSH, 1965b). Deformation of crack asperities and the effect of frictional contacts may differ significantly for strains that are several orders of magnitude larger for static compared to dynamic measurements.

At elevated confining pressure > 80 MPa the bulk modulus for granite remains almost unchanged, but for basalt K_{eff} continues to increase almost linearly with pressure, suggesting that low aspect ratio cracks contribute significantly to the basalt total porosity. The effective compressibility of a rock is mostly affected by compliant cracks and less by stiff spherical pores (WALSH, 1965a). To estimate the contribution of crack porosity to the bulk porosity estimated using Archimedes method we use the approach of WALSH (1965a). The initial crack porosity is estimated graphically (Fig. 10). In granite, the crack porosity (0.07%) is approximately 20 times smaller than the total unconnected porosity (1.3%). In basalt the compliant crack porosity (0.46%) is approximately only 5 times smaller than the total porosity (2.1%). The contribution of crack porosity to the total porosity is significantly larger for basalt than for granite, which is in agreement with preliminary microstructure observations. Granite contains predominantly intergranular cracks. Etna basalt contains abundant long and thin cracks that formed during rapid cooling and spheroidal pores possibly produced from lava degassing (Fig. 1). However, the effect of these spheroidal pores on the bulk modulus is probably subordinate; for example, a total porosity of 2.1% contained in spherical pores reduces the bulk modulus by < 5% (WALSH, 1965a).

Background crack densities estimated from the microstructure at atmospheric pressure only exist for undeformed Aue granite (JANSSEN *et al.*, 2001), ranging between 0.3–0.6 mm^{-1}. However, to assess decreasing crack density with increasing

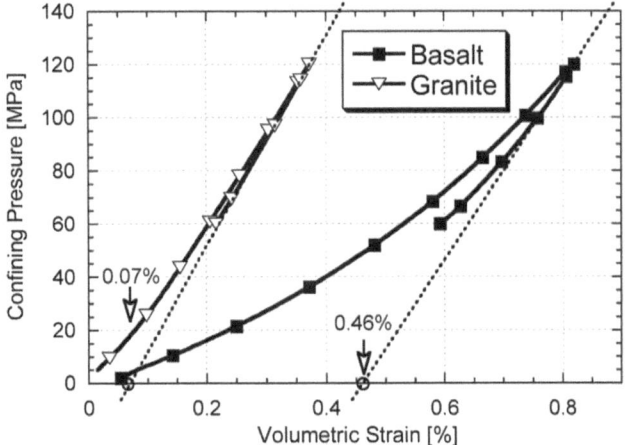

Figure 10
Hydrostats of Basalt Eb5 and Granite Ag75. Crack porosity was estimated graphically (WALSH, 1965a);
for granite the crack porosity is 0.07% compared to 1.3% total porosity, for basalt the crack porosity is
0.46% compared to 2.1% total porosity.

confining pressure we compare the results from Walsh's mechanical crack closure
model with crack densities predicted from seismic velocities using self-consistent
theory. The studies of WALSH (1965a,b) and O'CONNELL and BUDIANSKY (1974)
analyze the effect of cracks on the elastic properties of rocks. A central assumption in
both models is that crack density is low, crack interaction may largely be neglected
and crack distribution in the rock is isotropic. The density of low aspect ratio cracks
is defined as:

$$\Gamma = N \langle c \rangle^3. \tag{5}$$

N is the number of cracks per unit volume and $\langle c \rangle$ is the mean crack radius. Crack
density may be estimated using mechanical closure data or wave speeds from the
equation (WALSH, 1965a):

$$\Gamma = \frac{9}{16} \frac{(1 - 2v_0)}{(1 - v_0^2)} \left(\frac{K_0}{K_{eff}} - 1 \right), \tag{6}$$

where K_0 and v_0 are the bulk modulus and Poisson ratio of the crack-free rock,
respectively. In the self-consistent model of O'CONNELL and BUDIANSKY (1974) it is
assumed that the cracks are contained in a cracked matrix with as-yet-unknown
effective elastic properties. Crack density may then be estimated from the effective
bulk modulus and Poisson ratio:

$$\Gamma = \frac{9}{16} \frac{(1 - 2v_{eff})}{(1 - v_{eff}^2)} \left(1 - \frac{K_{eff}}{K_0} \right). \tag{7}$$

For comparison we also apply the model of SOGA *et al.* (1978) to invert crack densities from the velocity data. This model is largely based on the numerical calculation of effective elastic constants of a cracked solid by ANDERSON *et al.* (1974). For an isotropic material, *P*-wave velocities are related to the crack density by the relation:

$$\Gamma = \left[1 - \left(\frac{P}{P_0}\right)^2\right] \Big/ (a_1 + 2a_2).$$ (8)

P_0 is the *P*-wave velocity of the crack-free rock and the coefficients $a_1 = 1.452$ and $a_2 = 0.192$ are taken from SOGA *et al.* (1978).

Maximum velocities P_0 and S_0 of the crack-free and fully dense materials are not known. We assume P_0 and S_0 exceed by 5% the maximum vertical components of velocities registered in our experiments (P_V and S_{HV}). For basalt $P_{Ob} = 5.9$ km/s; $S_{Ob} = 3.29$ km/s and for granite $P_{Og} = 6.34$ km/s; $S_{Og} = 3.88$ km/s. Equations (1)–(2) give $v_{Ob} = 0.27$ and $K_{Ob} = 58.3$ GPa for basalt and $v_{Og} = 0.2$ and $K_{Og} = 52.7$ GPa for granite.

Crack densities are predicted to decrease with increasing confining pressure. In all models, the estimated crack density of basalt is significantly higher than that of granite. At confining pressures > 20 MPa, there is very good agreement between the model predictions for granite (Fig. 11). For basalt at elevated pressures, the agreement is fair, but at pressures < 80 MPa and crack densities > 0.3 the Walsh model deviates more strongly from other models and predicts considerably higher crack densities than self-consistent and SOGA *et al.* (1978) models. At low pressures, the crack porosity of basalt is > 6× higher than of granite (Fig. 10), rendering the model assumption of a dilute concentration of non-interacting cracks invalid. The self-consistent approach of O'CONNELL and BUDIANSKY (1974) may account for limited interaction between individual cracks, resulting in predictions that are possibly more accurate towards low confining pressures and elevated crack densities. The approach of SOGA *et al.* (1978) predicts crack densities very similar to the self-consistent model.

4.2. *The Effect of Cyclic Stress on Elastic Properties of Basalt and Granite*

A wealth of studies exists investigating the effect of differential stress on the elastic properties of dry rock (e.g., PATERSON and WONG, 2005; SCHUBNEL *et al.*, this issue). In general, stress-induced anisotropy of the crack orientation distribution and the *P*-wave anisotropy and acoustic birefringence of *S*-waves are observed to increase with increasing differential stress extending to rock failure (BONNER, 1974; HADLEY, 1975; WINKLER and MURPHY, 1995). This study illustrates that *P*-wave anisotropy A_P and *S*-wave birefringence B_S increase with increasing axial strain and decreasing confining pressure. Close to failure A_P and B_S attain values of 40% and 12%,

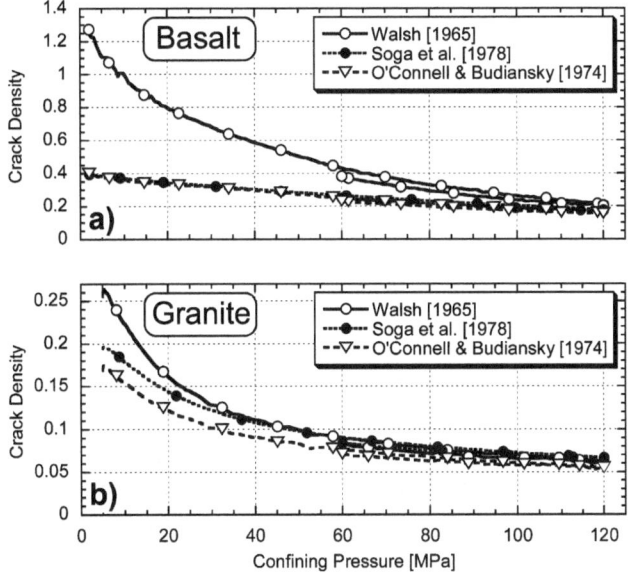

Figure 11
Crack densities of basalt sample Eb5 (a) and granite sample Ag75 (b) inverted from elastic wave velocities and from crack closure data vs. confining pressure. Open circles — predictions of WALSH (1965a); solid circles — model of SOGA *et al.* (1978), open triangles —self-consistent approximation of O'CONNELL and BUDIANSKY (1974). Model predictions for crack density in granite are in good agreement at pressures > 20 MPa; for basalt model predictions deviate significantly at pressures < 80 MPa and crack densities > 0.3.

respectively for basalt, and 35% and 20%, respectively for granite. However, a striking observation of this study is that changes in P-wave anisotropy and S-wave birefringence with increasing strain, differ significantly for basalt and granite (Figs. 6–8). For basalt, A_P and B_S increase much more rapidly at low stresses and axial strains than for granite that shows a different behavior, i.e., a significant increase of A_P and B_S at elevated stresses. At 20 MPa confining pressure and close to failure, P-wave anisotropy in basalt shows a rapid increase.

We attribute this behavior to the different crack microstructures of the starting basalt and granite samples. In particular, basalt contains multiple mm-scale thin cracks often associated with pyroxene inclusions. Closing of these cracks with increasing pressure at hydrostatic conditions (Fig. 5) and with increasing mean and axial stress (Figs. 6–8) produces significantly larger compaction of basalt compared to granite. Development of a strong initial P-wave anisotropy in basalt is related to a significant increase of the axial velocity P_V compared to P_H during initial loading (Fig. 12a). This anisotropy may be due to differential crack closure parallel and normal to the compression direction. However, for granite increasing anisotropy is rather related to a decrease of the horizontal velocity P_H (Fig. 12c). This is due to an increasing density of dilatant cracks, with planes oriented parallel to the compression direction.

The development of an anisotropic crack orientation distribution with increasing stress and axial strain is also shown by the evolution of the crack density parameter. We applied the model of SOGA *et al.* (1978) to invert Γ from the velocity data using the equations:

$$\Gamma_H = \left[(a_1 - a_2) + a_2 \left(\frac{P_V}{P_0} \right)^2 - a_1 \left(\frac{P_H}{P_0} \right)^2 \right] \bigg/ (a_1^2 + a_1 a_2 - 2a_2^2), \qquad (9a)$$

$$\Gamma_V = \left[a_1 - a_2 - (a_1 + a_2) \left(\frac{P_V}{P_0} \right)^2 + 2a_2 \left(\frac{P_H}{P_0} \right)^2 \right] \bigg/ (a_1^2 + a_1 a_2 - 2a_2^2). \qquad (9b)$$

The crack density parameters Γ_H and Γ_V refer to the densities of cracks, with planes oriented respectively parallel and perpendicular to the compression direction z, a_1 and a_2 are as in equation (8).

For basalt, crack densities Γ_H and Γ_V decrease with increasing axial strain except close to failure at 20 MPa confining pressure (Fig. 12b). The density of cracks, with crack planes oriented parallel to the compression direction (Γ_H), starts to increase rapidly at axial strains > 0.8%. In granite, closure of cracks oriented perpendicular to the compression axis is far less than in basalt (Fig. 12d). The SOGA *et al.* (1978) model predicts increasing density of cracks oriented parallel to the compression axis (Γ_H) with increasing stress at 60–20 MPa confining pressure.

The model results are in good qualitative agreement with the observed acoustic emission activity and the distribution of AE source types during stress cycling. At 60 and 40 MPa pressure, AE activity in basalt is substantially less than in granite (Figs. 6d, 6h–8d,8h), suggesting that compaction of basalt largely involves closure of pre-existing cracks. At 40 MPa a significant increase of A_P and Γ_H is observed in granite at an axial strain of about 0.6%, coinciding with a strong increase in AE-activity (compare Figs. 7g, 7h, 12d). A similar observation holds for basalt and granite close to failure. Distribution of AE source types during stress cycling is different compared to the AE distribution during hydrostatic loading. Tensile AE source types dominate in both materials during stress cycling. At elevated confining pressure only basalt samples display C-type sources that may be related to collapsing pore space. Close to failure we found a substantial increase of shear-type events largely at the expense of tensile events.

The strong effect of stress cycling on P-wave anisotropy and S-wave birefringence at low deviatoric stresses may be even more pronounced in the field compared to our laboratory measurements, because effective elastic moduli scale with crack length and aspect ratio (WALSH, 1965a). For example, it is conceivable that loading of the Etna volcanic edifice by increasing magma pressure may result in pronounced

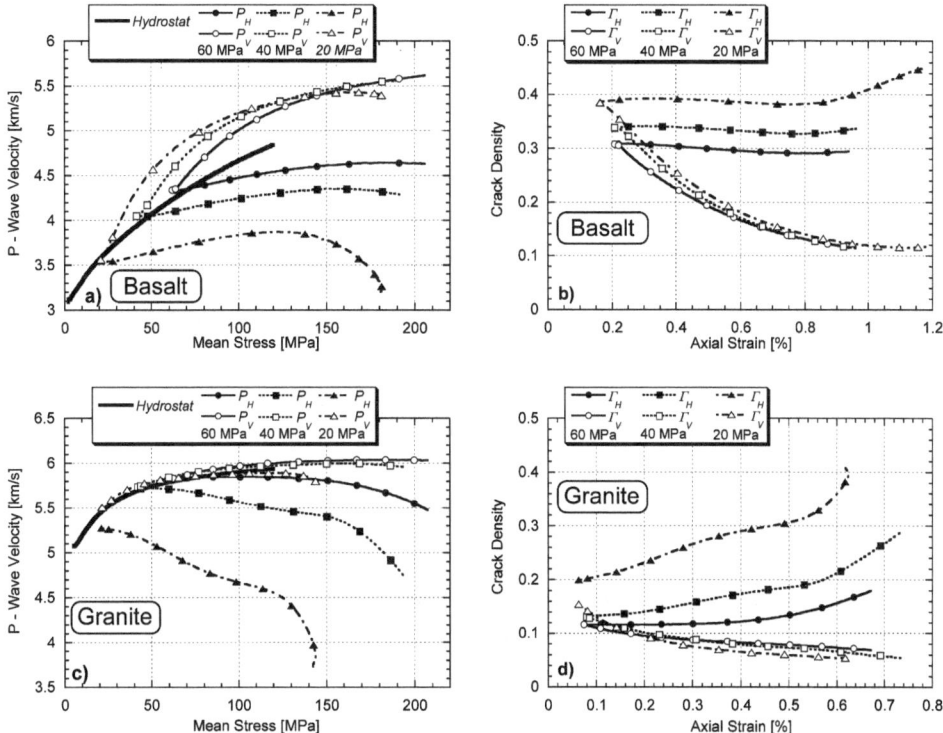

Figure 12

P wave velocity measured in horizontal (P_H) and vertical (P_V) direction during triaxial compression of basalt Eb5 (a) and granite Ag75(c). Solid symbols — horizontal components, open symbols — vertical components of *P* wave velocity, bold curve — *P* wave velocity measured during hydrostatic loading. Mean stress was calculated as $(\sigma_1 + 2\sigma_3)/3$, where σ_1 is axial stress and σ_3 is confining pressure. Crack density Γ was inverted using the model of SOGA *et al.* (1978) for Basalt Eb5 (b) and Granite Ag75 (d). Γ_H (solid symbols) is density of cracks, with planes oriented parallel to the vertical sample axis compression direction, Γ_V (open symbols) is the density of cracks, with planes oriented normal to the sample axis. Crack densities shown in (b) and (d) were estimated by inversion of *P*- wave velocity measurements presented in (a) and (c).

P-wave anisotropy and shear-wave splitting. Our results suggest that during loading anisotropy of elastic wave speeds may evolve before appreciable seismic activity occurs.

5. Conclusions

We investigated the evolution of acoustic emissions activity, compressional and shear-wave velocities, and volumetric strain of Etna basalt and Aue granite in hydrostatic and triaxial compression tests. With increasing hydrostatic pressure to 120 MPa, *P*-wave velocities in basalt increase by > 50% and by < 20% in granite. In both materials compaction is associated with significant AE activity suggesting irreversible closure of pore space. Analysis of source types reveals dominantly *C*-type events indicating pore collapse. Compliant pore space present in basalt is at least six times larger than in granite. Static moduli estimated during hydrostatic compression are substantially lower than dynamic values. Static and dynamic moduli evince a pronounced hysteresis during a pressure cycle. Crack densities inverted from seismic velocities compare favourably with predictions from mechanical closure models at pressures > 80 MPa, but deviate significantly at lower pressures.

In triaxial compression experiments performed in pressure cycles at 20, 40, 60 MPa, pronounced *P*-wave anisotropies and shear-wave birefringence develop in both materials, but are significantly stronger in basalt than in granite. Change in ultrasonic wave speeds and evolution of anisotropy in basalt is more pronounced than in granite, particularly at low confining pressures. With increasing differential stress, AE activity in granite and basalt increased with a significant contribution of tensile events. Close to failure the relative contribution of tensile events and the speed of horizontally oriented elastic waves decreased. A concomitant increase of double-couple events indicating shear, suggests shear cracks connect previously formed tensile cracks.

Acknowledgements

We acknowledge the constructive reviews of Alexandre Schubnel and Phil Benson. Their suggestions and comments, helped to enhance the manuscript.

REFERENCES

ANDERSON, D.L., MINSTER, B. and COLE, D. (1974), *The effect of oriented cracks on seismic velocities*, J. Geophys. Res. *79*, 4011–4015.

AYLING, M.R., MEREDITH, P.G., and MURRELL, A.F. (1995), *Microcracking during triaxial deformation of porous rocks monitored by changes in rock physical properties, I. Elastic-wave propagation measurements on dry rocks*, Tectonophys. *245*, 205–221.

BENSON, P., SCHUBNEL, A., VINCIGUERRA, S., TROVATO, C., HAZZARD, J., MEREDITH, P.G., and YOUNG, R.P. (2006), *Modelling the permeability evolution of micro-cracked rocks from elastic wave velocity inversion at elevated hydrostatic pressure*, J. Geophys. Res. *111*, BO4202, doi: 10.1029/2005JB03710.

BONACCORSO, A., CAMPISI, O., FALZONE, G., PUGLISI, G., VELARDITA, R., and VILLARI, L. (1990), *Ground deformation: Geodimeter trilateration and borehole tiltmetry*. In (Barberi, F., Bertagnini, A., and Landi, P., eds.), Mt. Etna 1989 Eruption, Consiglio Nazionale delle Ricerche-Gruppo Nazionale per la Vulcanologia (Giardini, Pisa, Italy 1990) pp. 44–47.

BONNER, B. P. (1974), *Shear-wave birefringence in dilating granite*, Geophys. Res. Lett. *1*, 217–220.

BRIOLE, P., NUNNARI, G., PUGLISI, G., and MURRAY, J.B. (1990), *The 1989 September-October eruption of Mt. Etna (Italy): some quantitative information obtained by geodesy and tiltmetry.*, C.R. Acad. Sci. Paris *310*, II, 1747–1754.

CASTELLANO, M., FERRUCCI, F., GODANO C., IMPOSA S. and MILANO G. (1993), *Upwards migration of seismic focii: A forerunner of the 1989 eruption of Mt Etna (Italy)*, Bull. Volcanol., *55*, 357–361.

CHENG, C.H. and JOHNSTON, D. (1981), *Dynamic and static moduli*, Geophys. Res. Lett. *8*, 39–42.

DRESEN, G. and GUEGUEN, Y. (2004), *Damage and rock physical properties*. In *Mechanics of fluid-saturated Rocks* (eds. Y. GUEGUEN and M. Bouteca) pp. 169–217 (Elsevier Academic Press, Amsterdam 2004).

FERRUCCI, F., RASA, R., GAUDIOSI, G., AZZARO, R., and IMPOSA, S. (1993), *Mt. Etna: A model for the 1989 eruption*, J. Volc. Geoth. Res. *56*, 35–55.

GUEGUEN, Y., and PALCIAUSKAS, V. (1994), *Introduction to the Physics of Rocks.*, (Princeton University Press, Princeton, ISBN 0-691-03452-4). 294 pp.

HADLEY, K. (1975), *Dilatancy: Further Studies in Crystalline Rocks*, Ph.D. Thesis, 202 pp. Massachusetts Institute of Technology, Cambridge.

HADLEY, K. (1976), *Comparison of calculated and observed crack densities and seismic velocities in Westerly granite*, J. Geophys. Res. *81*, 3484–3493.

HOLCOMB, D. (1993), *General theory of the Kaiser effect*, J. Rock Mech. Min. Sci. and Geomech. Abstr. *30*, 929–935.

JANSSEN, C., WAGNER, C.F., ZANG, A., and DRESEN, G. (2001), *Fracture Process Zone in Granite: A Microstructural Analysis*, Int. J. Earth Sci. *90*, 46–59.

KING, M. S. (1969), *Static and dynamic elastic moduli of rocks under pressure*, Paper presented at Rock Mechanics-Theory and Practice, University of California, June 16–19.

LEONARD, M. and KENNETT, B.L.N. (1999), *Multi-component autoregressive techniques for the analysis of seismograms,* Phys. Earth Planet. Int. *113*(1–4), 247.

LOCKNER, D.A., WALSH, J.B., and BYERLEE, J.D. (1977), *Changes in seismic velocity and attenuation during deformation of granite*, J.Geophys. Res. *82*, 5374–5378.

MAVKO, G., MUKERJI, T., and DVORKIN, J. (1998), *The Rock Physics Handbook-Tools for Seismic Analysis in Porous Media*, 329 pp. (Cambridge University Press, Cambridge 1998).

NELDER, J. and MEAD, R. (1965). *A Simplex method for function minimization*, Computer J. *7*, 308–312.

NUR, A. (1971), *Effects of stress on velocity anisotropy in rocks with cracks*, J. Geophys. Res. *76*, 2021–2034.

O'CONNELL, R.J. and BUDIANSKY, B. (1974), *Seismic velocities in dry and saturated cracked solids*, J. Geophys. Res. *79*, 5412–5426.

PATERSON, M.S. and WONG, T.F. (2005), *Experimental Rock Deformation–The Brittle Field*, 347 pp. (Springer, Berlin).

RECHES, Z. and LOCKNER, D. A. (1994), *Nucleation and growth of faults in brittle rocks*, J. Geophys. Res. *99*, 18,159–18,173.

SCHUBNEL, A., NISHIZAWA, O., MASUDA, K., LEI, X., XUE, Z., and GUEGUEN, Y. (2003), *Velocity Measurements and crack density determination during wet triaxial experiments on Oshima and Toki granites*, Pure Appl. Geophys. *160*, 869–887.

SCHUBNEL, A., BENSON, P., THOMPSON, B.D., HAZZARD, J., and YOUNG, R.P. (2006), *Quantifying damage, saturation and anisotropy in cracked rocks by inverting elastic wave velocities*, Pure Appl. Geophys., this issue.

SIMMONS, G. and BRACE, W.F. (1965), *Comparison of static and dynamic measurements of compressibility of rocks*, J. Geophys. Res. *70*, 5649–5656.

SOGA, N., MIZUTANI, H., SPETZLER, H., and MARTIN, R. J. III. (1978), *The effect of diltancy on velocity anisotropy in Westerly Granite*, J. Geophys. Res. *83*, 4451–4458.

TAPPONNIER, P. and BRACE, W.F. (1976), *Development of stress-induced microcracks in Westerly Granite*, Int. J. Rock Mech. Min. Sci. and Geomech. *13*, 103–112.

VINCIGUERRA, S., TROVATO, C., MEREDITH, P.G. and BENSON, P.M. (2005), *Relating seismic velocities, thermal cracking and permeability in Mt. Etna and Iceland basalts*, Int. J. Rock Mech. Min. Sci. *42*/7-8, 900–910.

WALSH, J.B. (1965a), *The effect of cracks on the compressibility of rock*, J. Geophys. Res. *70*, 381–389.

WALSH, J.B. (1965b), *The effect of cracks on the uniaxial elastic compression of rocks*, J. Geophys. Res. *70*, 399–411.

WINKLER, K. W. and MURPHY III, W. F. *Acoustic velocity and attenuation in porous rocks*, In *Rock Physics and Phase Relations*, AGU Reference Shelf (ed. T. J. Ahrens) pp. 20–34 (AGU, Washington (1995)

ZANG, A., WAGNER, F.C., STANCHITS, S., DRESEN, G., ANDRESEN, R. and HAIDEKKER, M.A. (1998), Source analysis of acoustic emissions in Aue granite cores under symmetric and asymmetric compressive loads, Geophys. J. Int. *135*, 1113–1130.

ZANG, A., WAGNER, F.C., STANCHITS, S., JANSSEN, C., and DRESEN, G. (2000), *Fracture process zone in granite*, J. Geophys. Res. *105*, 23651–23661.

(Received May 20, 2005, revised November 3, 2005, accepted November 4, 2005)

To access this journal online:
http://www.birkhauser.ch

Pure appl. geophys. 163 (2006) 995–1019
0033–4553/06/060995–25
DOI 10.1007/s00024-006-0054-x

© Birkhäuser Verlag, Basel, 2006

Pure and Applied Geophysics

Fracture in Westerly Granite under AE Feedback and Constant Strain Rate Loading: Nucleation, Quasi-static Propagation, and the Transition to Unstable Fracture Propagation

BEN D. THOMPSON,[1,2] R. PAUL YOUNG,[2] and DAVID A. LOCKNER[3]

Abstract—New observations of fracture nucleation are presented from three triaxial compression experiments on intact samples of Westerly granite, using Acoustic Emission (AE) monitoring. By conducting the tests under different loading conditions, the fracture process is demonstrated for quasi-static fracture (under AE Feedback load), a slowly developing unstable fracture (loaded at a 'slow' constant strain rate of 2.5×10^{-6} /s) and an unstable fracture that develops near instantaneously (loaded at a 'fast' constant strain rate of 5×10^{-5} /s). By recording a continuous ultrasonic waveform during the critical period of fracture, the entire AE catalogue can be captured and the exact time of fracture defined. Under constant strain loading, three stages are observed: (1) An initial nucleation or stable growth phase at a rate of ~ 1.3 mm/s, (2) a sudden increase to a constant or slowly accelerating propagation speed of ~ 18 mm/s, and (3) unstable, accelerating propagation. In the ~ 100 ms before rupture, the high level of AE activity (as seen on the continuous record) prevented the location of discrete AE events. A lower bound estimate of the average propagation velocity (using the time-to-rupture and the existing fracture length) suggests values of a few m/s. However from a low gain acoustic record, we infer that in the final few ms, the fracture propagation speed increased to 175 m/s. These results demonstrate similarities between fracture nucleation in intact rock and the nucleation of dynamic instabilities in stick slip experiments. It is suggested that the ability to constrain the size of an evolving fracture provides a crucial tool in further understanding the controls on fracture nucleation.

Key words: Shear-fracture, nucleation, earthquake, acoustic emission, rupture, granite.

1. Introduction

Shear failure of intact, brittle rock has long been experimentally investigated in order to better understand the mechanical properties of heterogeneous materials. In the case of earthquake physics, the presence of barriers or asperities is required on a pre-existing fault plane to store elastic strain energy which upon failure, generate strong motion seismic waves. The strength of these strong regions may approach that

---

[1]Department of Earth and Ocean Sciences, University Liverpool, Liverpool, UK
[2]Lassonde Institute, University Toronto, 170 College St, Toronto, ON, Canada.
E-mail: b.thompson@liv.ac.uk
[3]United States Geological Survey, Menlo Park, California, USA

of the intact material (OHNAKA, 2003). RECHES (1999) proposed that slip nucleation in earthquakes is analogous to rupture nucleation within an intact sample under triaxial conditions, based on: i) Field observations of fracture geometry complexity in fault zones, ii) partial or complete healing of crushed fault gouge at great depth, and iii) high values of energy release rate during earthquakes. OHNAKA (2003) considered earthquake rupture to be a mixed process of frictional slip failure and shear fracture of intact rock.

Previous AE studies have proven extremely useful in understanding brittle failure of rock. SCHOLZ (1968a) observed a temporal correlation between the onset of AE and dilation in a sample under triaxial loading, confirming the interpretation of BRACE *et al.* (1966) that dilation was caused by pervasive microcracking, primarily oriented parallel to the maximum compressive principal stress, σ_1. Insights have been provided into the nucleation phase of fracture, using AE source locations to map the temporal and spatial evolution of fracture. LOCKNER *et al.* (1991, 1992) slowed fracture to a quasi-static (stable) state using an AE feedback loading method, described by TEREDA *et al.* (1984). In the LOCKNER *et al.* study, AE source locations in granite were initially distributed uniformly throughout the sample, however close to peak stress, a clustering of AE occurred, forming a nucleus, from which fracture propagated as a process zone of intense activity. RECHES and LOCKNER (1994) and MOORE and LOCKNER (1995) used these experimental results to show that faults nucleate and propagate by the interaction of tensile microcracks. LEI *et al.* (2000) used a fast acquisition system to capture AE during triaxial tests on three hornblende schist samples under constant stress (creep) loading. In all tests, a fault nucleated close to the top of the sample and propagated through to cause rupture. The process zone is characterized as a region of intense tensile cracking, with the damaged fault zone behind this region characterized by shear events. LEI *et al.* (2003) also considered fracture initiation in samples with pre-existing planes of weakness. ZANG *et al.* (2000) used asymmetric uniaxial loading to propagate shear fracture in Aue granite. They examined fracture process zone characteristics using AE locations, and used AE first motion polarity to demonstrate that the predominant AE mechanism was shear-type cracking.

During periods of high AE activity, data sets collected using triggered AE systems can be incomplete due to system saturation. A continuous ultrasonic waveform acquisition system; the Giga RAM Recorder, has been developed to remove this limitation with the capacity to record a 268 s segment of waveform data on 16 channels (sampled at 5 MHz), typically about a significant occurrence, i.e., fracture of a sample. Discrete AE events are then extracted from this record, to provide the complete AE catalogue. In addition, the continuous record itself provides an extremely useful resource in terms of the timing of significant events in the experimental process, as demonstrated during stick slip tests by THOMPSON *et al.* (2005).

Although previous tests have demonstrated the characteristics of fracture nucleation in intact rock, the transition from nucleation to unstable fracture has not been clearly shown. Here, new observations of fracture are presented, which include a comparison of the nucleation of quasi-static and unstable fracture. Three experiments were performed in which fracture was induced in samples of Westerly granite under triaxial compression, using the same loading apparatus as LOCKNER et al. (1991). Firstly, the original LOCKNER et al. (1991) experiment was repeated, in which AE feedback loading was used to propagate a quasi-static fracture. Secondly, in order to demonstrate the evolution of unstable fracture, and the effect of loading rate on this process, two further tests were conducted under a relatively slow and a fast constant strain rate to demonstrate the evolution of shear fracture through the stages of nucleation, quasi-static (stable), and unstable propagation.

2. Experimental Procedure

2.1 Samples and Loading Conditions

Three triaxial compression tests are described featuring air-dried Westerly granite (cylindrical) samples with diameter 76.2 mm and length 190.5 mm. Sample ends were ground parallel to ± 0.05 mm. The grain size range of the samples is 0.05–2.2 mm, and a detailed description of the texture and microcrack morphology of Westerly granite can be found in MOORE and LOCKNER (1995). The samples were enclosed in a polyurethane jacket of thickness 3.2 mm, to prevent the entry of confining oil. Seventeen piezoelectric transducers were attached directly to the rock face with their surfaces ground to fit the curvature of the sample.

Testing was conducted at 50 MPa confining stress, with axial load being measured with an internal load cell to a precision of ± 0.02 MPa and axial shortening of the sample column measured outside the pressure vessel with a DCDT displacement transducer. In the first test, loading was controlled using an AE feedback loop to observe quasi-static fracture. In the second and third tests load was applied using constant piston displacement rates corresponding to constant axial strain rates of 2.5×10^{-6}/s and 5×10^{-5} /s. This was done to study the effect of loading rate on the onset of unstable fracture. These three experiments are hereafter referred to as the 'AEF' (AE Feedback), 'SL' (Slow Loaded) and 'FL' (Fast Loaded) tests.

The AE Feedback load control method was described by TERADA et al. (1984) and employed to provide AE source locations during quasi-static fracture propagation by LOCKNER et al. (1991). For the AEF test described in this paper, the procedure was software controlled. The AE rate was used to condition the computer-supplied reference voltage controlling the axial strain rate. The load control system was configured to maintain a constant rate of AE by reducing or reversing the axial load in response to increased AE activity, therefore enabling the sample to be failed under quasi-static conditions. In the early stages of fracture propagation, rapid decreases in deviatoric stress were required to maintain the quasi-static state of the fracture. To facilitate this, axial load was controlled by a fast-acting hydraulic valve. The system response was approximately 0.2 Hz. As will be shown, altering the sensitivity of the AE feedback regime enabled the speed of fracture growth to be controlled at rates between 13–50 mm/hour. The quasi-static fracture state was maintained for approximately 14 hours.

2.2 *Monitoring Acoustic Emission and Ultrasonic Velocities*

The seventeen piezoelectric transducers used in the experiments had a radial mode resonant frequency of 0.4 MHz. Thirteen of these transducers were used as AE receivers, the remaining four were used as ultrasonic sources for velocity measurements. Additional details of the transducer response characteristics can be found in STANCHITS *et al.* (2003). The signal from the receivers passed through 40 dB external pre-amplifiers (PAC model 1220A). In addition, the signals from two receivers were split and also amplified using 20 dB amplifiers (within the Giga RAM Recorder) to provide a lower gain record of the experiment. The waveforms recorded during the SL test were observed to clip the $\pm$ 2.5 V amplitude scale during failure on the 20 dB channels and so one receiver was used to record an un-amplified (0 dB) signal for the FL test.

The Giga RAM Recorder, a novel AE acquisition system, enables recording of continuous ultrasonic waveforms for a significant period of time. This system records data on 16 channels, with two channels per acquisition card. Each acquisition card contains 5 GB of Random Access Memory (RAM), in total the RAM buffer is 40 GB. Data is digitized at 14-bit resolution and the input voltage range is $\pm$ 2.5 V. Data can be sampled between 2 and 40 MHz. Each board contains two 16 Kb First In First Out (FIFO) memories for discrete, triggered data, and two 20 dB amplifiers. The sampling frequency was 5 MHz for these experiments, providing a 268 s segment of continuous waveform data on 16 channels. The 40 GB RAM acts as a continuous circular buffer that can be 'locked' following the recording of a significant event, i.e., fracture of the sample, and subsequently transferred to disk. The system also records triggered data throughout testing, with identical parameters to the continuous data. A triggered event was recorded if five or more waveforms had amplitudes exceeding 100 mV in a 50 μs time window. Triggered events have a length of 204.8 μs, and a maximum of 16 events per second was captured. In addition, the number of AE per second was recorded throughout the experiment, this is termed the AE 'Hit' count to distinguish between this and the numbers of triggered AE actually recorded. Figure 1 shows a schematic diagram of the triaxial load configuration and acquisition modes of the Giga RAM Recorder.

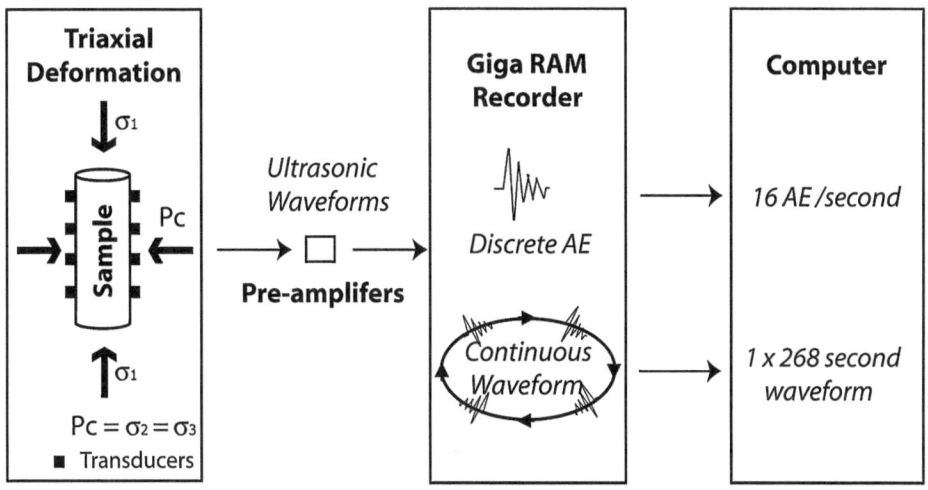

Figure 1
Schematic diagram of the experimental setup, including the triaxial load configuration and acquisition capabilities of the Giga RAM Recorder.

Following each experiment, the continuous ultrasonic record is replayed and the complete catalogue of discrete AE events is extracted using the trigger logic described above. Advances in technology have led to rapid AE waveform acquisition systems (LEI *et al.,* 2000). However, we believe this to be the first system to record continuous waveforms for significant time periods. The advantages of continuous waveform data include: (1) no 'mask' time, (i.e., the periods of recording downtime occurring when AE events are written to permanent storage), and (2) removal of the sampling bias imposed by trigger levels.

For ultrasonic velocity measurements, the four designated transducers were excited using a Panametrics (model 5072PR) pulse generator. A switch box rapidly connects the pulse generator to each source transducer. Stacked velocity surveys took 1–2 s to complete, with coverage along 52 raypaths.

2.3 AE Processing

P-wave first arrival times are obtained using an automated picking routine (ASC, 2003) and used to calculate AE source locations using a Simplex method (NELDER and MEAD, 1965). The residual difference between the measured and theoretical travel time from the source location is calculated for each channel, and if the residual exceeds 0.3 μs the channel is dropped from the location algorithm. Source locations calculated using fewer than 7 arrivals are not considered in this study. A Transversely Isotropic (TI) velocity structure was used in the location routine, defined by velocity measurements made throughout the experiments. An estimate of the average source location error on the samples' surface is provided by location of the source transducers used in velocity surveys to within ± 4 mm of their true location. PETTITT (1998) theoretically considered the case of a cubic specimen (50 mm length) with velocity 4000 m/s and an array of 12 receivers, sampled at 10 MHz. AE were hypothetically located at the edge and in the center of the cube, and the location errors were analyzed as a function of position within the sensor array. AE at the center of the sample had between 25 and 50% smaller location errors than AE at the edge of the sample. It is probable therefore that our ± 4 mm measured error on the sample edge overestimates the location error within the sample by a similar amount.

2.4 b-values

The power-law frequency magnitude relationship is calculated using AE magnitudes M, corresponding to the Gutenberg-Richter b-value relationship for earthquakes (1):

$$\text{Log } N(M) = a - bM, \tag{1}$$

where N is the number of events greater than M, and a and b are constants. b is estimated using the least-squares method. The AE magnitude M is calculated for each located event as (2) and (3):

$$M = \log\left(\frac{\sum_{m=1}^{n}\left(W_{\text{RMS}.d_m}\right)}{n}\right), \tag{2}$$

$$W_{\text{RMS}} = \sqrt{\frac{\sum_{i=1}^{x} W_i^2}{x}}, \tag{3}$$

where n is the number of sensors and d_m is the distance between receiver m and the AE location. W_{RMS} is the RMS waveform amplitude calculated on each receiver, using the waveform amplitude W_i, averaged over x data points. In the Earth, typically b ranges between 0.5 and 1.5 (VON SEGGERN, 1980).

3. Results

3.1 Mechanical Behavior

The mechanical data for the three tests are plotted in Figure 2. Figures 2a–c show the deviatoric stress (σ_1-P_c) versus time with cumulative AE count plotted on the secondary axis, for the AEF, the SL and FL experiments, respectively. Figure 2d shows the deviatoric stress versus axial strain measurements for the three experiments.

The AEF test (Fig. 2a) can be summarised as follows:

(Approximate times are given in minutes, exact timing in seconds)

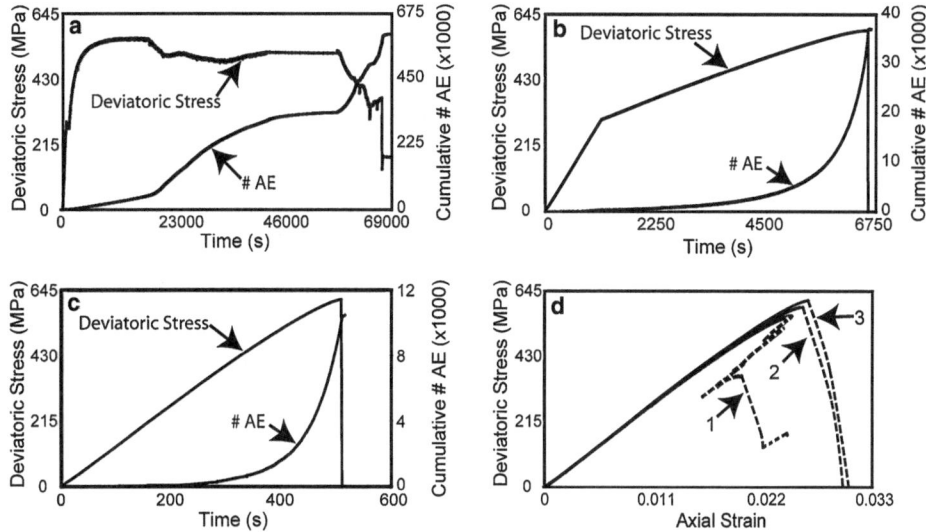

Figure 2
(a–c) Deviatoric stress and AE Hit count versus time for tests under the following load control methods: (a) AE Feedback (AEF), (b) slow constant strain rate (2.5×10^{-6} /s) (SL) and (c) fast constant strain rate (5×10^{-5}/s strain rate) (FL). (d) Deviatoric stress versus axial strain curves for the AEF (marked 1), SL (2) and FL (3) tests, with the unloading portions of the curves dashed.

1) Initial loading under a constant strain rate of 1×10^{-5}/s until approximately 390 MPa. From this point, the loading rate is moderated by the AE. The kink in stress curve at 800 s occurred when a velocity survey interfered with the AE feedback control. Such loading anomalies were subsequently avoided by momentarily turning off the AE feedback control prior to velocity surveys.

2) At 182 minutes (10889 s) a deviatoric stress 'plateau' was reached (560 MPa).

3) Peak stress (564 MPa) was reached after 260 minutes (15579 s). There was then a gradual stress drop to 559 MPa (18575 s).

4) The stress then drops at an approximate rate of 0.2 MPa per second to a minimum of 504 MPa after 358 minutes (21500 s).

5) Between 731 minutes (43869 s) and 979 minutes (58765 s) a constant axial strain of 0.0236 was maintained as the equipment was unattended.

6) The AE feedback control was restarted at 979 minutes (58765 s) and periodically adjusted to increase the speed of fracture propagation.

7) Finally at 1132 minutes (67936 s) the AE Feedback regime was adjusted to accelerate the fracture rate and the fracture became unstable and propagated completely through the sample, resulting in a 195 MPa stress drop.

The sample in the SL test (Fig. 2b) was loaded at an initial axial strain rate of 1×10^{-5}/s to 300 MPa to approximately 50 percent of failure stress and then reduced to 2.5×10^{-6}/s. The AE hit count increased exponentially towards failure, the peak stress of 592 MPa was reached after 6549 s. There was a 38 s period of strain weakening, after which the stress dropped abruptly. Hardware problems caused a periodic loss of AE location data up to failure, but fortunately this problem did not affect the final 21 s in which strain weakening and dynamic fracture occurred. The AE hit count was unaffected by the hardware problem and 35,000 AE hits were registered.

The FL test was loaded at a strain rate of 5×10^{-5}/s (20 times faster than the SL experiment), reaching the peak deviatoric stress of 615 MPa after 510 s. In this case, fracture ensued immediately without a period of strain weakening. Over 11,000 AE hits were recorded, compared with 35,000 during the SL test, and close to 500,000 during the AEF test. Total stress drops were experienced for both the SL and FL samples.

3.2 Velocity Data

Fourteen velocity surveys were conducted for the AEF test, constraining the evolution of damage within the sample. Initially, at the onset of deviatoric loading, maximum and minimum velocities were 6070 ± 83 and 5602 ± 83 m/s, respectively (7.7 % anisotropy). Under deviatoric loading, a TI structure evolves, as shown in Figure 3a, where velocity is plotted as a function of raypath angle at intervals during the test. For clarity, only six of the fourteen surveys are plotted. The TI structure can be modelled by:

$$V(\theta) = ((V_{max} + V_{min})/2) - ((V_{max} - V_{min})/2)\cos(180 - 2\theta)), \qquad (4)$$

where V is modelled P-wave velocity as a function of raypath angle, θ, and V_{max} and V_{min} are the maximum and minimum P-wave velocities. Velocity models of the form (4) are fitted to the measured velocity data and used to define the TI structure input into the AE source location routine. Figure 3b shows a similar plot for the SL test with four velocity models estimated between deviatoric stresses of 425 and 545 MPa. Fewer velocity surveys were conducted due to the relatively shorter period of the test. Indeed, the duration of the FL test meant surveys were only conducted up to 69% peak stress, these data are not shown. In Figure 3c, the percentage of anisotropy derived from the TI velocity models is plotted as a function of deviatoric stress for the AEF and SL tests. The two curves display close similarity. Between 350 MPa and 500 MPa there is a linear increase in anisotropy from 8% to 19%, from which point the increase appears nonlinear. By 545 MPa, the samples are 26% anisotropic, above which measurements only exist for the AEF test. At peak stress, the sample exhibits 35% anisotropy. The evolution of velocity for this test is simplified in Figure 4i, where the averaged velocity for four horizontal ray-paths is shown as a function of deviatoric stress and time. The TI data from the AEF test are used by SCHUBNEL *et al.* (this volume) to demonstrate the evolution of crack density in Westerly granite.

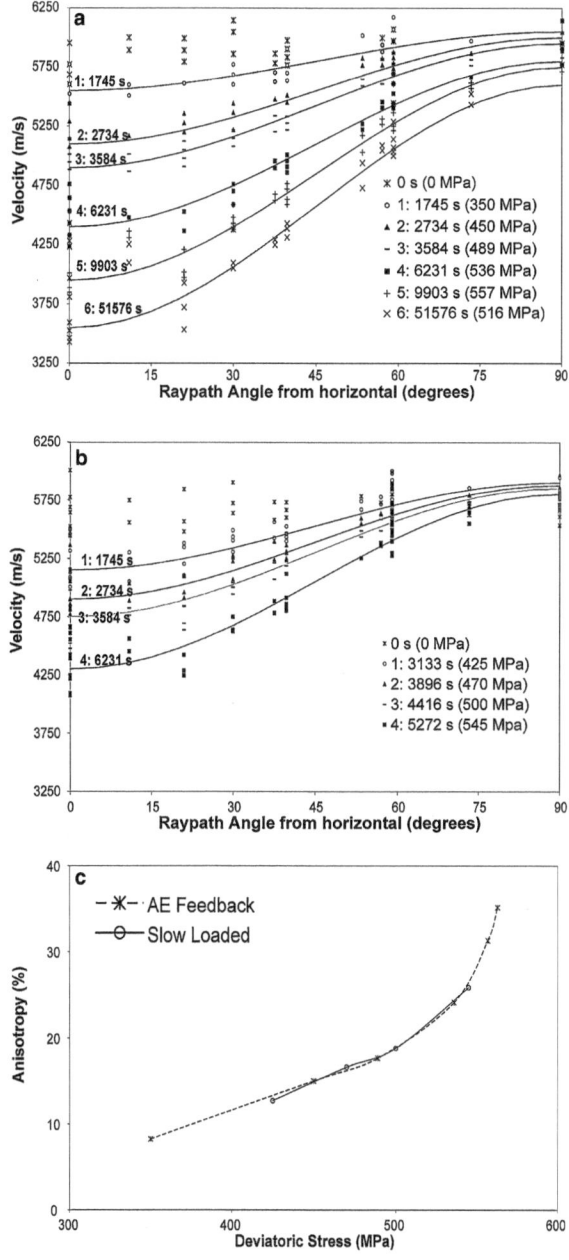

Figure 3

The evolution of the Transversely Isotropic velocity structures for (a) the AEF and (b) the SL test. Data points represent measured velocities as a function of raypath angle, measured from the horizontal, for a series of velocity surveys conducted at increasing deviatoric stresses (as marked, time of survey is also indicated). Velocity models are fitted to the data points (models are not fitted to the initial surveys), as detailed in the text. (c) The percentage anisotropy derived from the velocity models is plotted against deviatoric stress for these two tests, indicating a similar degree of damage was induced in the samples as a function of deviatoric stress.

(i)

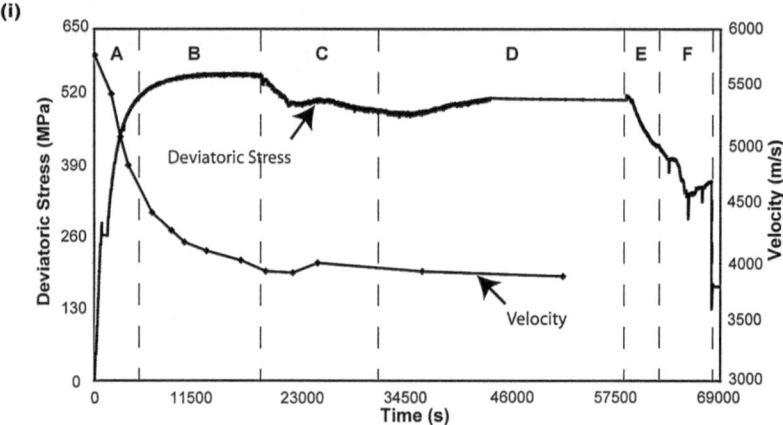

(ii) View along strike of eventual fracture:

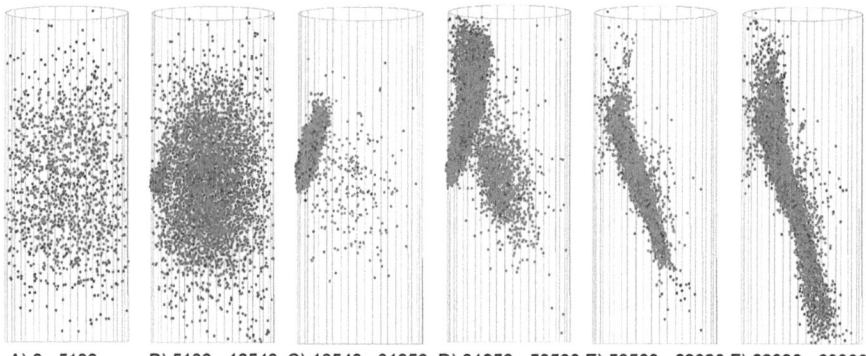

A) 0 - 5100   B) 5100 - 18540 C) 18540 - 31950 D) 31950 - 58500 E) 58500 - 62820 F) 62820 - 69300
Time (s)

(iii) View in plane of eventual fracture:

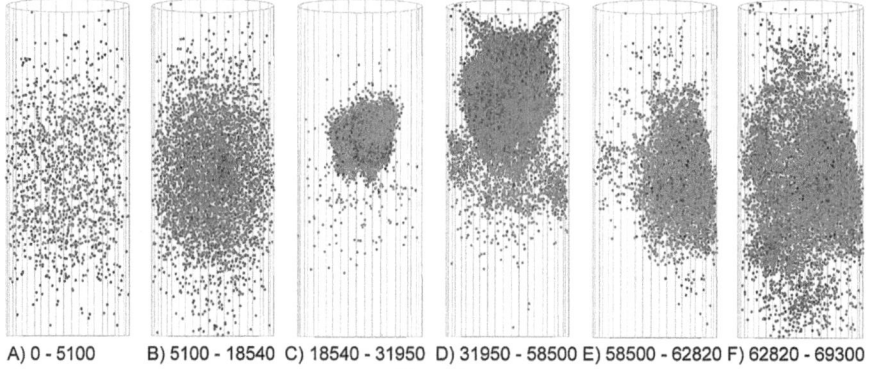

A) 0 - 5100   B) 5100 - 18540 C) 18540 - 31950 D) 31950 - 58500 E) 58500 - 62820 F) 62820 - 69300
Time (s)

Figure 4

(i) Deviatoric stress and average velocity (measured along four horizontal raypaths) during the AEF test, with time. AE locations are shown looking along the strike of the first fracture (ii), and into the plane of the eventual fracture. Periods A–F are marked on (i).

3.3 AE Locations

Over 70,000 triggered AE were located from the AEF test, and are presented in Figure 4. The locations are divided into six periods (A–F), as marked on Figure 4i. AE locations are shown along the approximate strike of the first fracture plane (Fig. 4ii), and secondly, perpendicular to the strike of the first fracture plane (Fig. 4iii). In Period A, AE locations are distributed throughout the sample. In Period B events are still distributed throughout the sample, however there is also a clustering at the edge of the sample. Period C shows a fracture which propagates from the nucleation site. Few events are located elsewhere in the sample. In Period D, the fracture has reached the top of the sample, intersecting the steel platen. The initial angle of the fracture is 70°, this steepens as the fracture propagates towards the loading platen, and reacts to the more complicated stress field produced by this nearly rigid boundary. A second fracture is then observed to nucleate and propagate (Periods D to F). The second fracture is shown in greater detail in Figure 5. The first fracture grows at an average rate of 12 mm/hour (3 μm/s) while the rate of growth for the second fracture was increased from 12 to 50 mm/hour (3 to 14 μm/s), by reducing the sensitivity of the AE feedback control.

For the SL and FL tests the AE events presented are extracted from the continuous record. The triggered events occurring outside the continuous record are not considered in this paper, although it is noted they show a similar distribution to events in the early stages of the AEF. A continuous waveform is shown in Figure 6i, including a 22 s period which includes rupture for the SL experiment, during which time the sample exhibits strain-weakening behavior. Periods A to E on the continuous waveform define the times for which AE locations are displayed. Region YY' is expanded to show a 0.8 s period about failure. A 40 ms region of this waveform is again expanded (ZZ') showing details of failure. 2100 events were located during the final 21 s of the SL test and are displayed in five time periods. Views are shown along strike of the eventual fracture plane (Fig. 6ii), and into the plane of the eventual fracture (Fig. 6iii). In Periods A and B events cluster on the sample edge. In Period C the cluster zone has grown slightly, and in Period D significant growth is observed. In Period E, AE are used to estimate the length of the fracture as 52 mm. These are the last AE that can be located, and the time of the last located AE is marked on the continuous record (Fig. 6i, Z-Z'). The continuous record shows a gradual rise in amplitude from this point, and 17 ms after the last AE located, the 2.5 V amplitude scale continuously clips, for a period of 6 ms. During this period, rupture is assumed to occur. Following rupture, there is approximately 200 ms of relative quiescence prior to large amplitude aftershocks.

A summary of continuous waveform for the final 178 s of the FL test is displayed in Figure 7i, and the final four seconds are expanded in waveform Z-Z'. Similar to the SL test, a rapid increase in activity is observed a few seconds before failure. 2700 AE are located from the continuous record and displayed for Periods A–E, as

(i) View along strike of fracture:

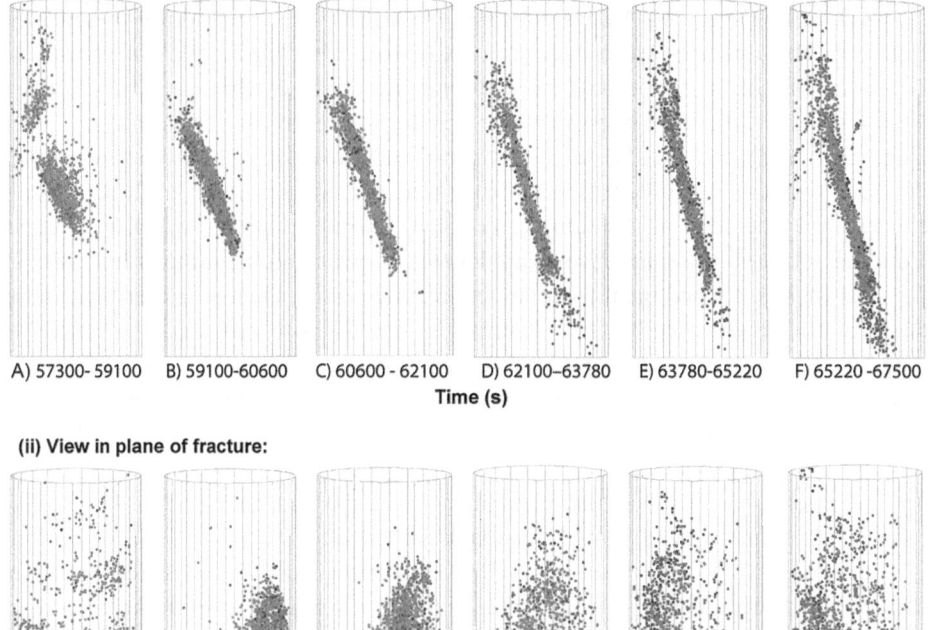

A) 57300- 59100   B) 59100-60600   C) 60600 - 62100   D) 62100–63780   E) 63780-65220   F) 65220 -67500

Time (s)

(ii) View in plane of fracture:

A) 57300- 59100   B) 59100-60600   C) 60600 - 62100   D) 62100–63780   E) 63780-65220   F) 65220 -67500

Time (s)

Figure 5

Detail of the second fracture propagating in the AEF test. The fracture nucleates and propagates as a process zone of intense activity.

marked in Figure 7i. In Periods A and B events are distributed throughout the sample. In Period C, events are also distributed throughout the sample, with a cluster of events developed in the center of the sample. In Period D, this cluster develops further. Period E shows the final AE before rupture. The AE is concentrated in two areas; firstly in the center of the sample, and secondly, a new cluster has developed on the edge of the sample. The fault angle and a physical examination of the sample indicate this second cluster is the source of the macro-fracture. Figure 8a is a continuous waveform showing a 262 ms period about failure, expanded from Figure 7i. The last AE locates at 510.164 s. For this experiment only, an unamplified

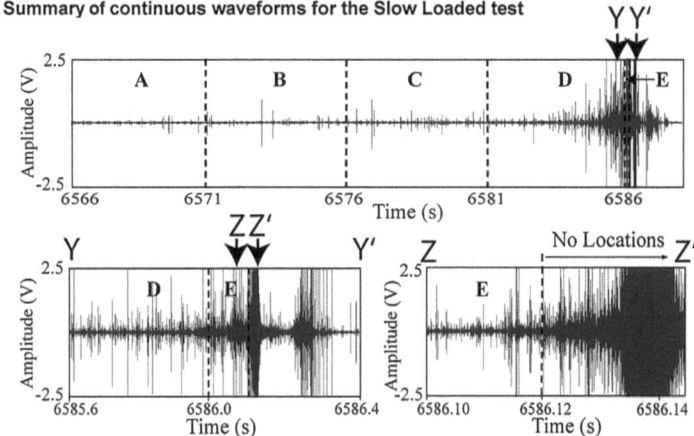

(i) Summary of continuous waveforms for the Slow Loaded test

(ii) View along strike of eventual fracture:

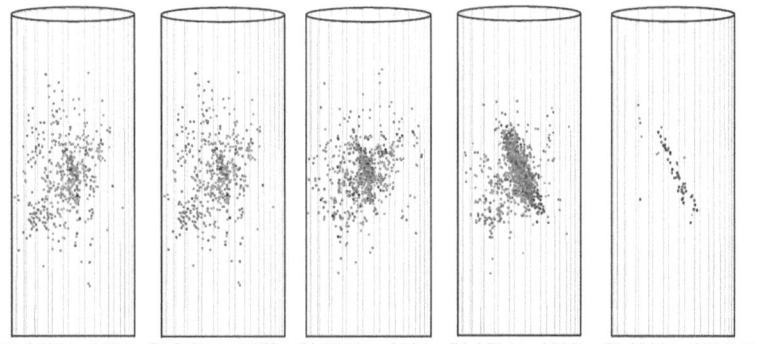

A) 6566 to 6571 s B) 6571 to 6576 s C) 6576 to 6581 s D) 6581 to 6586 s E) 6586 to 6586.12 s

(iii) View in plane of eventual fracture:

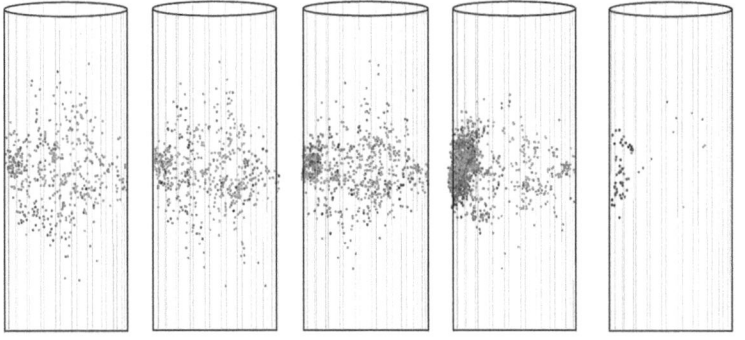

A) 6566 to 6571 s B) 6571 to 6576 s C) 6576 to 6581 s D) 6581 to 6586 s E) 6586 to 6586.12 s

Figure 6

Slow Loaded Test: (i) the continuous waveform is summarized for a 20 dB channel for a 22 s period. The period marked Y to Y' is expanded, showing 0.8 s of detail about the failure. The period marked Z to Z' is expanded, showing in greater detail a 40 ms period of failure. (ii–iii), AE locations for the 21 s period prior to failure are shown with location periods A–E marked on the continuous record. Following E, there is a short time prior to the sample rupture during which no locations are possible.

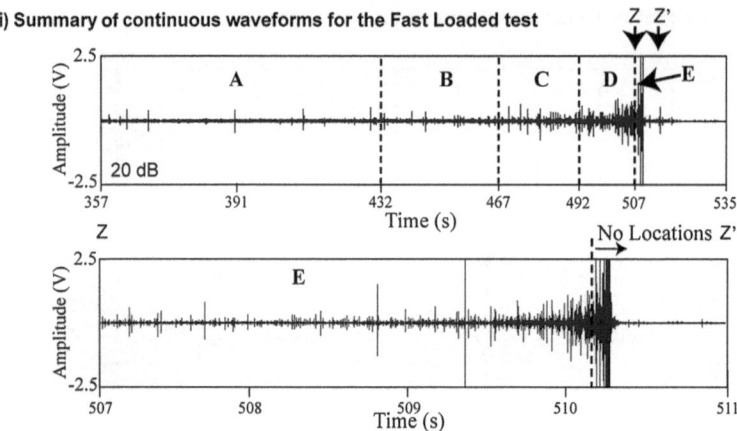

(i) Summary of continuous waveforms for the Fast Loaded test

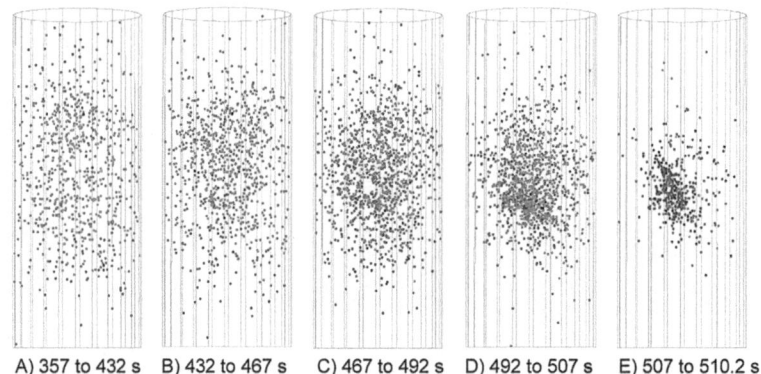

(ii) View along strike of eventual fracture:

A) 357 to 432 s   B) 432 to 467 s   C) 467 to 492 s   D) 492 to 507 s   E) 507 to 510.2 s

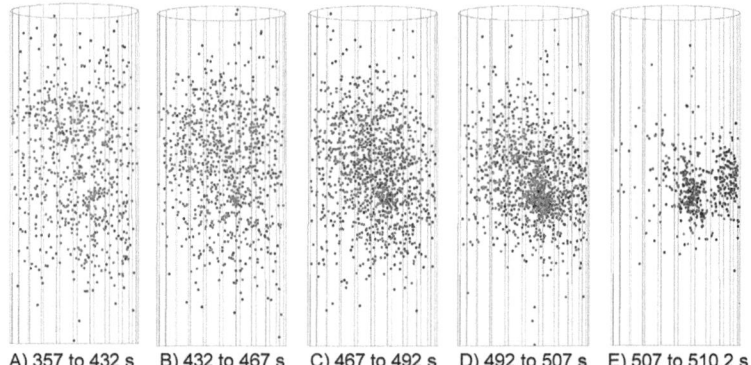

(iii) View in plane of eventual fracture:

A) 357 to 432 s   B) 432 to 467 s   C) 467 to 492 s   D) 492 to 507 s   E) 507 to 510.2 s

Figure 7

Fast Loaded Test: (i) the continuous waveform is summarized for a 20 dB channel for a 178 s period. The period marked Z to Z' is expanded, showing 4 s of detail about the failure. This region is further expanded in Figure 8. (ii-iii) AE locations for the 153.2 s period prior to failure are shown with location periods A-E marked on the continuous record. Following E, there is a short time prior to the sample rupture during which no locations are possible.

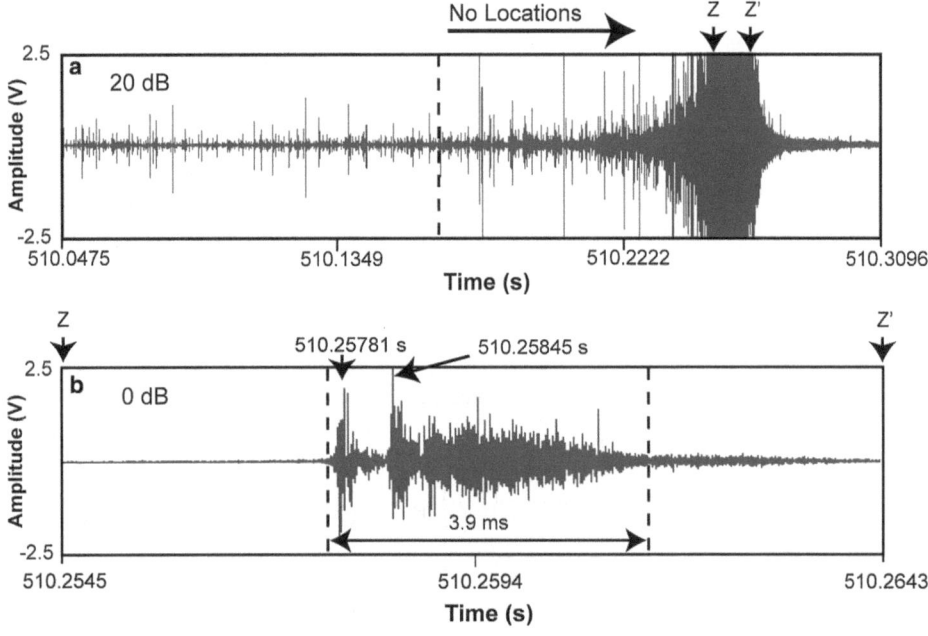

Figure 8
Continuous Waveform for Fast Loaded Test, expanded from Figure 7i to show detail of fracture. a) The amplitude scale of the 20 dB channel saturates, however on the 0 dB channel (b) the ultrasonic response of failure is unclipped. In (b), two pulses are observed over a 3.9 ms period, the point of sample rupture is interpreted to coincide with the maximum amplitude at 510.2585 s. The timing of waveform Z-Z' (b) is indicated on (a).

(0 dB) transducer recorded failure and Figure 8b shows the response for this channel. There are two pulses or bursts of activity: the first of which lasts 0.3 ms. The onset of the second is approximately 0.6 ms after the onset of the first. The point of highest energy release is at 510.2585 s and is assumed to represent the time of failure. From this we conclude that the final AE locates approximately 100 ms prior to rupture.

3.4 b-values

The b-values calculated for the three experiments are presented in Figure 9, along with deviatoric stress and time. b-values for the SL and FL tests were calculated using 400 event samples. Figure 9i shows results for the AEF test, calculated using 400 event samples until fracture propagation, and using 1000 event samples after this point (due to the large amount of data). Between 14,285 and 21,312 s, b-values are calculated using both 400 and 1000 event samples. The trends are nearly indistinguishable, which demonstrates that in this case, varying the sample size does not affect the results presented. b-values are initially about 2.5 ± 0.3. There is a

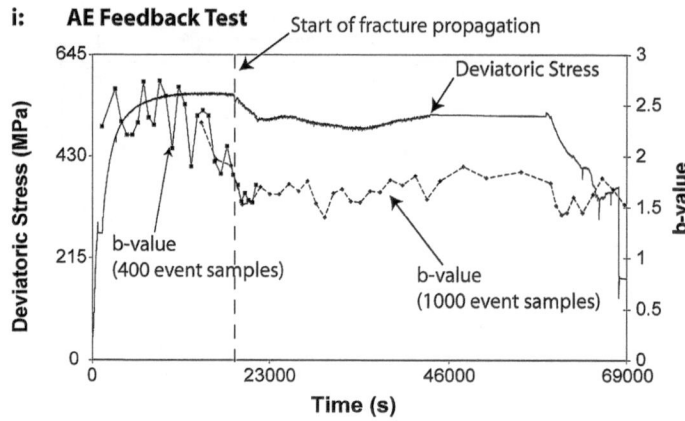

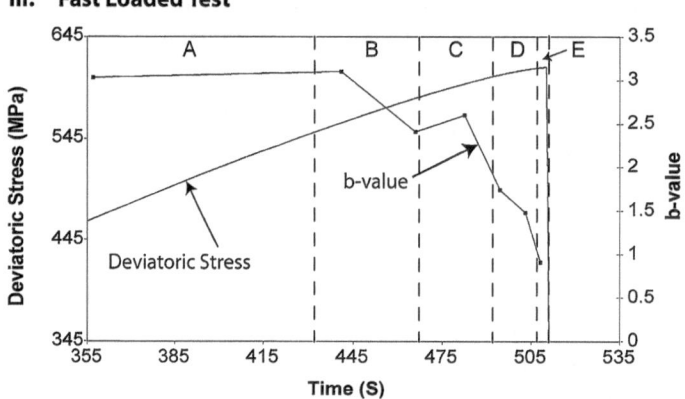

Figure 9

b-values marked with deviatoric stress and time for the three tests. The moment of fracture propagation is marked on the AEF test (i), and the 400 and 1000 event sample sizes show close agreement around this period. It is practical to use large event bin sizes for the duration of the AEF test, however, a comparison is provided for the SL (ii) and FL Tests (iii) by calculating b-values using smaller 400 event bins up to nucleation of fracture. A to E on (ii) correspond to the AE location periods for the SL Test in Figure 6, and A to E on (iii) correspond to AE location periods for the FL Test in Figure 7.

gradual decrease in the average b-value with increasing stress. The start of fracture propagation is marked by a dashed line in Figure 9i, from which point the large fluctuations observed in the b-value trend are not observed. The b-value reaches a low of 1.53 following the initial stress drop (21500 s). From this point until the end of the test the average b-value is 1.69.

The b-values from the SL test are shown in Figure 9ii, along with the deviatoric stress record for the 60 s period about failure. b-values average 1.23, decreasing suddenly to 0.53 about failure. b-values for the FL test are shown in Figure 9iii. Initially b-values are above 3. From 482 s there is a near linear decrease to a final b-value of 0.91.

4. Nucleation of Unstable Fracture

AE locations from the SL and FL tests demonstrate fracture nucleation, and the initial propagation of a fracture. The continuous record is used to estimate the time of sample rupture, which is assumed to be the point of highest amplitude. For the FL test, the 0 dB channel recorded the maximum amplitude at 510.2585 s, which is 1.2 ms after the $\pm$ 2.5 V amplitude scale continuously clipped on the 20 dB amplified channels. We assume that the point of rupture for the SL test (which did not feature a 0 dB channel) occurred approximately 1.2 ms after saturation of the 20 dB channels, i.e., at 6586.138 s (the potential errors associated with this assumption are considered later).

Having estimated the time of sample rupture, we will demonstrate the spatial and temporal evolution of nucleation and initial fracture propagation with respect to this event for the constant strain rate tests. To provide accurate measurements of fault growth, AE location coordinates (X,Y,Z) are rotated so the new Y axis is oriented along the strike of the fracture. This is shown in Figure 10. AE locations only in the immediate region defined by the respective fracture zones are considered. The length of the fracture (defined as the long axis of the main AE cluster) is directly measured. For the SL test, 600 events are sequentially displayed in samples of 100 (Fig. 10a) while in the FL test, four groups of 25 events are displayed (Fig. 10b). In order to provide a time resolution for Figure 10, each subsequent sample is stepped along the horizontal (X) axis, by 15 mm per sample for the SL test, and 25 mm per sample for the FL test. The nucleation or fracture zone is shaded and was visually estimated as the zone of dense AE activity for each sample of events. For each nucleation or fracture zone the long axis is measured using the coordinates of the two outlying AE locations. Figure 11a shows the measured fracture length with time (the time of the final AE located in that sample of events). Figure 11b is plotted using a logarithmic time scale in order to emphasize the length-time relationship in the second prior to dynamic fracture. The interpreted stages of fracture are illustrated for the SL test (as defined in the discussion).

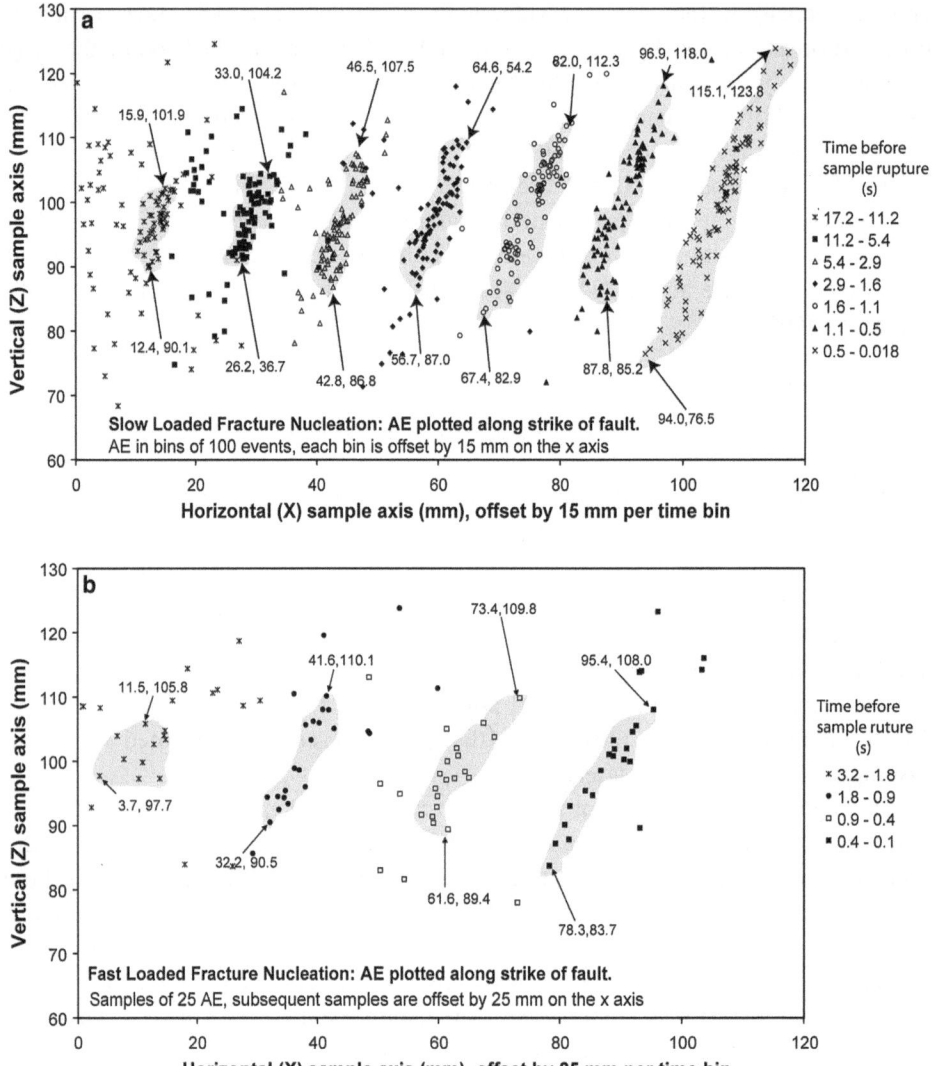

Figure 10
Comparison of Nucleation/Fault Length for the SL (a) and FL (b) tests. AE coordinates (X,Y,Z) are rotated so the Y axis is aligned with the strike of the eventual fault plane, and plotted for the rotated X and Z axes. Time resolution is provided by stepping each event bin along the X axis by 15 or 25 mm (for the SL and FL Tests respectively). The time prior to sample rupture for each group of AE is indicated.

For the SL test, nucleation can be accurately defined from AE locations between 17.2 and 11.2 s prior to rupture of the sample. It is possible that the nucleation zone existed before 17.2 s, however it is difficult to define from the AE locations. At 11.2 s the initial length is 11 mm, which slowly increases to 24 mm, 1.62 s prior to rupture

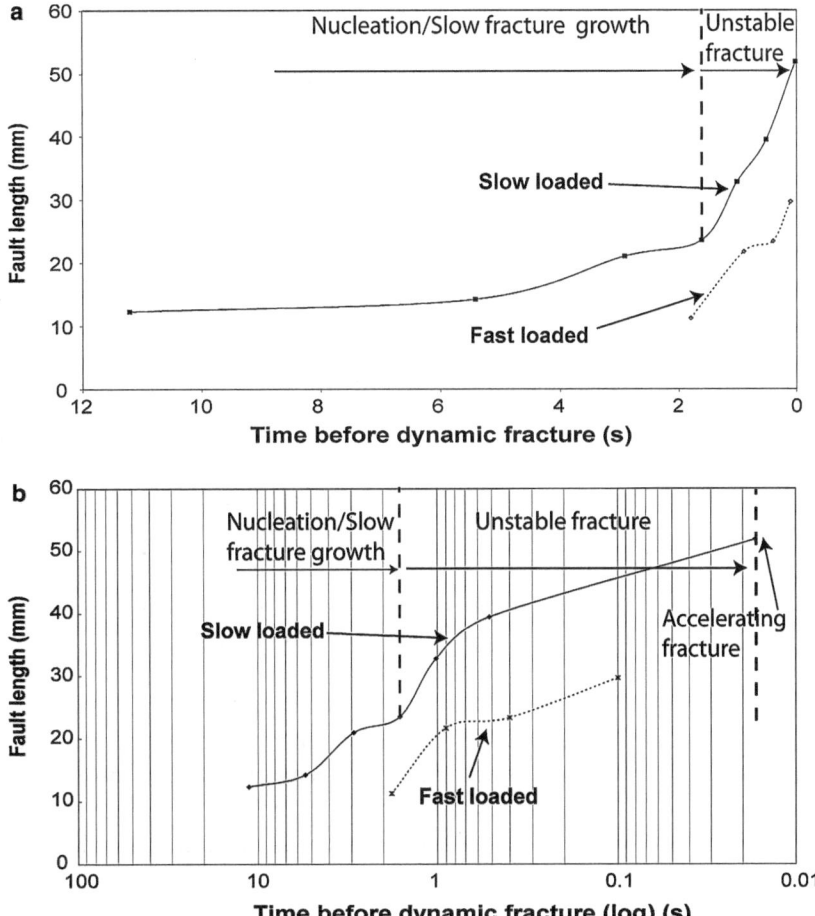

Figure 11
Nucleation or fracture lengths inferred from Figure 10 are plotted (a) against time prior to sample rupture for the FL and SL tests. (b) A logarithmic time scale is used to emphasize the differences between the two tests immediately prior to the dynamic rupture, the timing of which is inferred from the continuous records. The interpreted stages of nucleation and fracture growth are marked for the SL test.

(a rate of 1.3 mm/s). The fracture is then measured to grow at 17.7 mm/s until 18 ms prior to rupture, when the fracture length is 52 mm.

For the FL test, nucleation is observed between 3.2 and 1.8 s prior to the sample rupture, with an approximate diameter of 11 mm. The fracture evolves into a planar feature in the period 1.8 and 0.9 s before rupture. There is a small increase in fault length between 0.9 and 0.4 s before rupture. At 100 ms before rupture, the fracture length increases to a length of 30 mm at an average velocity of 21 mm/s. No events are located between 100 ms and the sample rupture. The inability to locate further AE in the SL and FL tests is due to an increase in activity in the sample, as when many events occur in a short time period, AE first arrivals are masked by the coda of earlier events.

5. Discussion

In the previous section AE locations are used to define a propagating fracture. At a critical point, no AE are located due to the high level of activity in the sample. This occurs 18 and 100 millisecond periods prior to fracture, for the SL and FL tests respectively. An important question is what happens to the fracture growth in this final period in which no AE locations are possible? The simplest explanation is that the increased AE activity preventing source location is purely a function of the larger fault size, and that the length of the fault continues to grow at 17.7 mm/s (in the SL test). This is discounted by a simple calculation of average velocities required for the fault to reach the sample dimensions (which is 136 mm, taking the final 56° fracture angle measured from Figure 10a). Given that the fault must grow a further 84 mm over a period of 18 ms, the average velocity must accelerate to 4.7 m/s. This value is based on the assumption that similar to the FL test, rupture occurs 1.2 ms after the 20 dB waveform continuously clips, with the continuous clipping persisting for 6 ms. If rupture occurred later, i.e., midway through the clipped period, the 21 ms propagation period would suggest a 4.0 m/s average velocity. To illustrate the extreme case, if rupture occurred at the end of the clipped period, then the 24 ms propagation period would suggest an average velocity of 3.5 m/s. However, it would seem counterintuitive for rupture to coincide with unclipping of the waveform and the fracture rate is thought to be in the 4.0 m/s to 4.7 m/s range. Furthermore, it is emphasized that these estimates represent the average fracture velocity over the final ~20 ms rather than the ultimate propagation velocity.

In the case of the FL test, the fracture is required to propagate a further 106 mm in 100 ms, requiring acceleration to an average velocity of 1.0 m/s. However in this case the unclipped waveform offers further evidence. As shown on Figure 8b, there is an initial pulse followed by a second, higher magnitude pulse with longer duration, occurring 0.6 ms after the onset of the first. It would be plausible to assume the first represents the onset of dynamic fracture, and the second the point at which the fracture has completed its growth phase and slip occurs on the entire fault. Indeed, fracture mechanics analysis (RICE, 1980) and microstructural analysis of a quasi-statically propagated fault (MOORE and LOCKNER, 1995) suggest that for a crack propagating into unfractured rock, there is very little net slip on the fault behind the fracture tip, with the unbroken rock ahead of the fracture resisting the applied load. However, once the fracture has propagated completely, only the frictional resistance remains and slip on the entire fault plane occurs as in a spring-slider system. Such slip will involve considerable AE activity, as observed by the larger magnitude and duration of the second pulse. Assuming that the fault continues to accelerate at 21 mm/s between the last locatable AE and the unstable propagation (a period of 0.1 s), the fault will have a length 31 mm. A further growth of 105 mm is required, which must then occur in the 0.6 ms period between the onset of the first and second pulse. This suggests the average unstable fracture speed is 175 m/s. High resolution strain measurements will be employed to test the assumptions involved in this calculation.

The fracture process observed for the SL test is interpreted as a three stage process. Firstly, a nucleation or slow growth stage is observed to extend at an average velocity of 1.3 mm/s. Secondly, the nucleating fracture reaches a critical size or propagation velocity and a sudden increase in propagation velocity to 17.7 mm/s is observed. It is beyond the resolution of the techniques used here to determine whether the velocity is constant or accelerating in this stage. As acceleration in stress drop is observed (Fig. 9ii) we infer that the fracture could not be halted at this stage even with a release in the applied load, and so we interpret this stage as unstable fracture. The third stage occurs after the final AE locates, when it is demonstrated that acceleration in propagation speed to the order of m/s must occur. The interpreted stages of nucleation and fracture growth are marked on Figure 11. In the FL test, failure occurs at peak stress, i.e., strain weakening is not observed. We consider that the process observed over 17 s for the Slow Loaded test is accelerated into a period of 3 s for the Fast Loaded test. Furthermore, by taking into account the additional constraints of the unclipped waveform about fracture for the FL test, it is suggested that the final unstable velocity is approximately 175 m/s.

These observations can be compared to an earthquake nucleation model proposed by OHNAKA (2000), based upon results from stick slip experiments conducted on preexisting fault planes under biaxial loading, (OHNAKA and SHEN, 1999). In the first phase of this model, rupture nucleation initially proceeds stably and quasi-statically to a critical length, the second phase is a spontaneous acceleration of the nucleation to a second critical length, from which point rupture will extend at a constant velocity close to the shear-wave velocity (phase 3). Indeed OHNAKA (2003) links fracture initiation, frictional slip, and earthquake nucleation in a constitutive law. To apply our observations to this model; the first and sudden acceleration in fracture speed when the fracture length reached 24 mm would correspond to the critical length (or propagation velocity) above which the fracture becomes unstable. Similarly, the 52 mm fracture length at the point from which AE were not located represents a lower bound of the critical dimension (or propagation rate) above which the final propagation velocity ($\sim$ 175 m/s) is reached. That the suggested ultimate propagation velocity is lower than that of the shear wave-velocity, may be due to the limited sample size in these tests.

LEI et al. (2000) demonstrated the processes of fracture under constant stress (creep) loading using a fast AE acquisition system. They observed two stages of nucleation with nucleation initially growing to a size of 2 cm at a speed of 3 cm/s. There was then a pause in growth for 12 s, while activity in this area continued. The nucleating fault then grew to the edge of the sample at 10 cm/s. A 2 s delay ensued before the sample rupture occurred. There are differences between LEI *et al.*'s observations and those presented here; specifically the complete propagation of the nucleating fracture, its rapid propagation rate and the two-second pause between propagation and rupture were not observed in our experiments. There are possible reasons for these discrepancies. Firstly, the loading rate may influence the critical fault dimensions (or propagation rate) above which fracture propagates at velocities of order m/s, causing a larger critical size above which an accelerating fault is observed. Secondly the sample size may be an issue as those used in this study are larger, with a length of 190.5 millimeters compared to 100 millimeters used by LEI *et al.*, with an aspect ratio of 2.5:1 compared to 2:1 in their experiments. Use of a larger aspect ratio increases the region in the center of the sample that has low stress gradients (MOGI, 1962; PATERSON and WONG, 2005) and can help reduce problems related to end effects. Unlike in LEI *et al.*'s experiments, we find fracture nucleates consistently away from the loading platens. It could be that the fracture processes are complicated by stress concentrations when nucleation occurs close to the loading platen. Separating the relative importance of these geometrical differences from differences in loading conditions (i.e., constant strain rate versus constant stress) will require additional experiments.

The nucleation prior to quasi-static propagation in the AE feedback test is measured to have an approximate long axis of 18 mm. This is close to the 24 mm fault length prior to the 17 mm/s propagation measured for the SL test, and is consistent with that previously measured in AE feedback tests (LOCKNER *et al.*, 1991) and creep tests (LOCKNER and BYERLEE, 1980) on Westerly granite, and by LEI *et al.*, (2000). Therefore, we conclude that the initial nucleation size of fracture initiation under these experimental conditions is not greatly influenced by the loading rate, and it is in the initial propagation stage that the fracture characteristics diverge according to loading rate.

SCHOLZ (1968b) demonstrated that *b*-values were negatively correlated with the level of applied stress. MAIN and MEREDITH (1989) related intermediate and short-term *b*-value anomalies to stress intensity during fracture. LEI *et al.* (2000) concluded that variations in *b*-value occurred as result of hierarchical fault growth. The *b*-values presented here are broadly similar to LEI *et al.*'s experimental results, although *b*-values during quasi-static fracture propagation have not previously been measured over an extended period (i.e., AEF). For this test, the *b*-values fall to about 1.5 to coincide with stable fracture propagation. For the FL test, the *b*-values fall to about 1, however in this case large events are masking smaller events, lowering the measured *b*-value. For the SL test, locations are only shown for the post peak region of the test, and therefore nucleation has already developed, explaining the initial *b*-value of between 1 and 1.5. However as the fracture develops, the *b*-value falls to 0.5, which is a value associated with earthquake foreshock nucleation (VON SEGGERN, 1980).

6. Conclusion

This paper presents results of three triaxial compression tests in which the process of fracture nucleation is compared under AE Feedback, and 'Slow' and 'Fast' constant strain rate loading. A continuous ultrasonic waveform recorder was used to capture the entire AE catalogue about failure, and this continuous waveform enables the exact time of rupture to be estimated. By repeating the AE feedback controlled test of LOCKNER et al. (1991), quasi-static fracture is propagated through a sample over 14 hours, at rates of between 3 and 14 μm/s. As shown by LOCKNER et al., fracture nucleates at a site of intense cracking, at the side of the sample, and propagates through the sample as a process zone of intense activity.

By loading under a constant strain rate, the evolution of unstable fracture nucleation has been demonstrated. Firstly, a slowly expanding nucleation site is observed, followed by a sudden increase in fracture speed (to mm/s or cm/s) up to a size (or propagation rate) above which AE are not located. A lower bound estimate of the propagation velocity from this point (using the time-to-rupture, the existing fracture length, and assuming a constant velocity) suggests values of a few m/s. However, from the low gain acoustic record, we infer that the final speed of rupture propagation had increased to hundreds of meters per second. For both slow and fast constant strain loaded experiments, we observe a monotonically increasing fracture growth rate similar to behavior predicted by avalanche models for fault growth.

It is accepted that earthquake physics can be modelled largely as a frictional process, but in situations where nucleation occurs in fault step-overs and on nonplanar surfaces at high normal stresses, earthquake occurrence may involve a component of rock fracture. Indeed, under special circumstances, such as in some mining-induced earthquakes, nucleation and initial propagation may occur entirely in intact rock. Here we have shown there are similarities between the nucleation of dynamic instabilities during intact rock failure and frictional instabilities on pre-existing fault planes in the laboratory. The critical dimensions of the nucleating fault at the transition to unstable rupture appear to be easily observed under the experimental conditions described here, using continuous AE recording. It is suggested that a better understanding of the controls on fracture nucleation, and the transition to unstable rupture in the laboratory could improve the understanding of earthquake nucleation.

Acknowledgments

We thank A. Schubnel, D. Collins, R. Bowes (Engineering Seismology Group, Inc.) W. Pettitt (Applied Seismology Consultants, Ltd.), A. McGarr and N. Beeler for helpful discussions, and A. Zang and X-L. Lei for useful reviews. This work was made possible by a National Environment Research Council equipment grant, in partnership with ESG, and was partially funded by the University of Liverpool and the Natural Sciences and Engineering Research Council of Canada (Discovery Grant to Young).

REFERENCES

ASC (2003), *InSite Seismic Processor Users Manual*, Version 2.10, Applied Seismology Consultants, Ltd., Shrewsbury, UK.

BRACE, W.F., PAULDING, B.W., and Scholz, C. (1966), *Dilatancy in the fracture of crystalline rocks*, J. Geophys. Res. *71*, 3939–3953.

LEI, X.-L., KUSUNOSE, K., RAO, M.V.M.S., NISHIZAWA, O., and SATOH, T. (2000), *Quasi-static fault growth and cracking in homogenous brittle rock under triaxial compression using acoustic emission monitoring*, J. Geophys. Res. *105*, 6127–6139.

LEI, X.-L., KUSUNOSE, K., SATOH, T., and NISHIZAWA, O. (2003), *The hierachical rupture process of a fault: An experimental study*, Phys. Earth Planet. Inter. *137*, 213–228.

LOCKNER, D. A., BYERLEE, J. D., KUKSENKO, V., PONOMAREV, A. and SIDORIN, A. (1991), *Quasi-static, fault growth and shear fracture energy in granite*, Nature, 350, 39–42.

LOCKNER, D. A., BYERLEE, J. D., KUKSENKO, V., PONOMAREV, A., and SIDORIN, A., *Fault mechanics and transport properties of rocks* (eds. Evans, B., and Wong, T-F) (Academic, London 1992).

LOCKNER, D. A. and BYERLEE, J. D., *Development of fracture planes during creep in granite*, in *Proceedings, Second Conference on Acoustic Emission/Microseismic Activity in Geological Structures and Materials* (eds. H. R. Hardy and W.F. Leighton) pp. 11–25 (Trans-Tech Publications, Clausthal- Zellerfeld, Germany 1980).

MAIN, I. G. and MEREDITH, P.G. (1989), *Classification of earthquake precursors from a fracture mechanics model*, Tectonophysics *167*, 273–283

MOORE, D. E. and LOCKNER, D.A. (1995), *The role of microcracking in shear-fracture propagation in granite*, J. Struct. Geol. *17*, 95–114.

MOGI, K. (1962), *The influence of the dimensions of specimens on the fracture strength of rocks*, B. Earthquake Res. I, Tokyo, 40, 175–185.

NELDER, J. and MEAD, R. (1965), *A simplex method for function minimization*, Computer J. *7*, 308–312.

OHNAKA, M. (2000), *A physical scaling relation between the size of an earthquake and its nucleation zone size*, Pure Appl. Geophys. *157*(11–12), 2259–2282.

OHNAKA, M., (2003), *A constitutive scaling law and a unified comprehension for frictional slip failure, shear fracture of intact rock, and earthquake rupture*, J. Geophys. Res. *108* Art. No. 2080.

OHNAKA, M. and SHEN, L. F. (1999), *Scaling of the shear rupture process from nucleation to dynamic propagation: Implications of geometric irregularity of the rupturing surfaces*, J. Geophys. Res. *104*, 817–844.

PETTITT, W. S. (1998), *Acoustic emission source studies of microcracking in rock*, PhD, Keele University, UK.

PATERSON, M.S. and WONG, T.-f., *Experimental Rock Deformation - The Brittle Field*, (Springer, New York 2005), 347 pp.

RECHES, Z. and Lockner, D.A. (1994), *Nucleation and growth of faults in brittle rocks*, J. Geophys. Res. *99*, 18159–18173.

RECHES, Z., (1999), *Mechanisms of slip nucleation during earthquakes*, Earth Planet. Sci. Lett. *170*, 475–486.

RICE, J. R., *The mechanics of earthquake rupture*, In *Physics of the Earth's Interior* (Proc. Int'l School of Physics "E. Fermi", course 78,) (Amsterdam, Italian Physical Society/North Holland Publ. Co. 1980).

SCHOLZ, C. H. (1968a), *Microfracturing and the inelastic deformation of rock in compression*, J. Geophys. Res. *73*, 1417–1432.

SCHOLZ, C. H. (1968b), *The frequency magnitude relation of micro-fracturing in rock and its relation to earthquakes*, Bull. Seismol. Soc. Am. *58*, 399–415.

SCHUBNEL,. A. BENSON, P., THOMPSON, B. D., HAZZARD, J., and YOUNG, R. P. (2006), *Quantifying damage, saturation and anisotropy in cracked rocks by inverting elastic wave velocities*, Pure Appl. Geophys. this volume.

STANCHITS, S. A., LOCKNER, D. A., and PONOMAREV, A. V. (2003), *Anisotropic changes in P-wave velocity and attenuation during deformation and fluid infiltration of granite*, Bull. Seismol. Soc. Am. *93*, 1803–1822.

TERADA, M., YANIGADANI, T., and Ehara, S., *AE Rate controlled compression test of rocks*. In *Proc. 3ʳᵈ Conf on Acoustic Emission/Microseismic Activity in Geol Structures and Materials* (eds. Hardy, H.R. and Leighton F.W.) pp. 159–171 (Trans-Technical, Clausthal- Zellerfeld 1984).

THOMPSON, B. D, Young, R. P., Lockner, and D. A. (2005), *Observations of premonitory acoustic emission and slip nucleation during a stick slip experiment in smooth faulted Westerly granite*, Geophys. Res. Letts. *32*, L10304, doi:10.1029/2005GL022750.

VON SEGGERN, D. (1980), *A random stress model for seismicity statistics and earthquake prediction*, Geophys. Res. Letts. *7*, 637–640.

ZANG, A., WAGNER, F. C., STANCHITS, S., JANSSEN, C., and DRESEN, G. (2000), *Fracture process zone in granite*, J. Geophys. Res. *105*, 23651–23661.

(Received May 24, 2005, revised October 31, 2005, accepted November 7, 2005)
Published Online First: May 12, 2006

 To access this journal online:
http://www.birkhauser.ch

Pure appl. geophys. 163 (2006) 1021–1029
0033–4553/06/061021–9
DOI 10.1007/s00024-006-0063-9

© Birkhäuser Verlag, Basel, 2006

| Pure and Applied Geophysics

Stress Sensitivity of Seismic and Electric Rock Properties of the Upper Continental Crust at the KTB

AXEL KASELOW,[1,3] KATHARINA BECKER,[2] and SERGE A. SHAPIRO[2]

Abstract—We test the hypothesis that the general trend of *P*-wave and *S*-wave sonic log velocities and resistivity with depth in the pilot hole of the KTB site Germany, can be explained by the progressive closure of the compliant porosity with increasingly effective pressure. We introduce a quantity θ_c characterizing the stress sensitivity of the mentioned properties. An analysis of the downhole measurements showed that estimates of the quantitiy θ_c for seismic velocities and electrical formation factor of the *in situ* formation coincide. Moreover, this quantity is 3.5 to 4.5 times larger than the averaged stress sensitivity obtained from core samples. We conclude that the hypothesis mentioned above is consistent with both data sets. Moreover, since θ_c corresponds approximately to the inverse of the effective crack aspect ratio, larger *in situ* estimates of θ_c might reflect the influence of fractures and faults on the stress sensitivity of the crystalline formation in contrast to the stress sensitivity of the nearly intact core samples. Finally, because the stress sensitivity is directly related to the elastic nonlinearity we conclude that the elastic nonlinearity (i.e., deviation from linear stress-strain relationship i.e., Hooke's law) of the KTB rocks is significantly larger *in situ* than in the laboratory.

Key words: KTB, seismic velocities, resistivity, stress.

Introduction

In this work we present an analysis of the dependence of *P*- and *S*-wave velocities and electrical resistivity of metamorphic crustal rocks on differential pressure (depth) at the German Continental Deep Drilling Site (KTB) down to 4000 m. We involve downhole sonic and deep lateral resistivity log data from the pilot hole and laboratory derived stress-dependent *P*- and *S*-wave velocities into our analysis. The used rock samples were also recovered from the KTB pilot hole.

The pilot (4000 m) and the main borehole (9101 m) and the surrounding area of the KTB site have been the subject of geoscientific research for more than a decade

---

[1]Fachrichtung Geophysik, Freie Universität Berlin, Berlin, Germany
[2]Fachrichtung Geophysik, Freie Universität Berlin, Malteserstrasse 74-100,Build. D, 12249, Berlin, Germany E-mail: becker@geophysik.fu-berlin.de
[3]Seismic Micro-Technology Alps, Rosegger Strasse 15, 8700, Leoben, Austria

(EMMERMANN and LAUTERJUNG, 1997). The project was dedicated to understand the geological, hydrogeological and geophysical settings of the upper continental crust. The suite of geophysical monitoring covers downhole measurements, core and cutting analyses, potential field analyses, VSP, and numerous 2-D and 3-D seismic surveys (see special section on KTB in J. Geophys. Res., 102 (B8)). The hydraulic system of the drilled crustal section which shows many prominent hydraulically conducting fracture systems and hydrostatic pore pressure down to the final depth of the pilot hole was tested by pumping and injection tests. A comprehensive summary of the KTB research and further references are given in HAAK and JONES (1997). In summary, the results from the KTB research project provide a huge and unique data base to obtain a deeper insight into the geoscientific properties of the upper continental crust.

The goal of this work is to use sonic downhole velocities, resistivity log and ultrasonic laboratory data from rocks of the KTB pilot hole down to 4000 m in order to (a) test our hypothesis that the general depth trend of the mentioned properties results from compliant pore space closure with increasingly effective pressure (i.e., depth) and (b) to give a quantitative measure for the stress sensitivity of the KTB rocks. Therefore we use the Stress-Sensitivity-Approach for isotropic rocks as introduced by SHAPIRO (2003) and the extension of the approach to the stress sensitivity of electrical resistivity (KASELOW and SHAPIRO, 2004). The stress-sensitivity approach enables a rock physical interpretation of seismic velocity observations as a function of confining stress and pore pressure. The basic assumption of the approach is that the stress dependence of seismic velocities results from stress-induced changes of the dry rock compliances, which are closely related to changes in the pore space geometry. The most important characteristic for the stress dependences of various rock properties is the tensor of stress sensitivity as introduced by SHAPIRO and KASELOW (2005).

The Data Sets

The laboratory measurements of *P*- and *S*-wave velocities were conducted on dry cubic samples in a true tri-axial pressure apparatus up to 600 MPa confining pressure. Three *P*- and six corresponding *S*-wave velocities were simultaneously measured in three orthogonal directions. The measurement coordinate system was oriented with respect to the macroscopically visible foliation plane and lineation elements. For details about the measurements see KERN and SCHMIDT (1990), KERN *et al.* (1991, 1994). These laboratory observations showed a clear correlation between the lithology and seismic velocities (e.g., KERN *et al.*, 1991).

According to EMMERMANN and RIESCHMÜLLER (1990) the rocks of the KTB pilot hole can lithologically be subdivided into nine units. In agreement with the laboratory data, the sonic *P*- and *S*-wave profiles shown in Figure. 1 indicate a

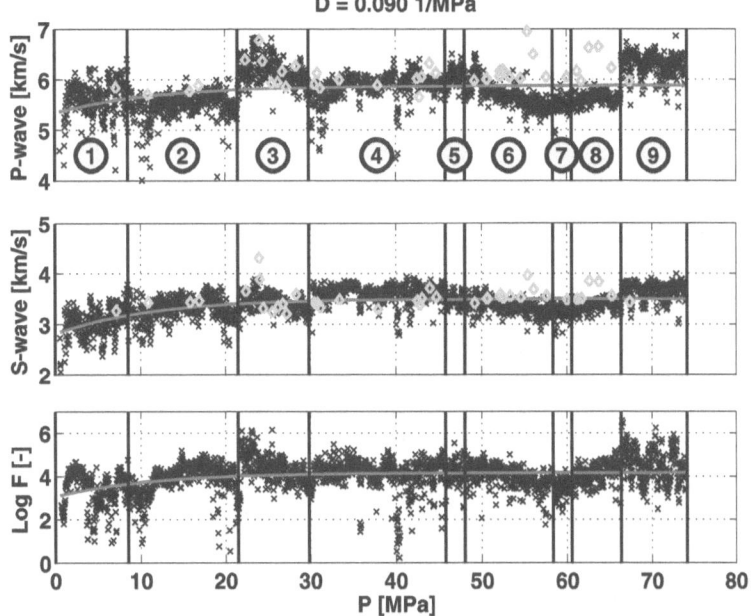

Figure 1
Logging (black crosses), laboratory derived *in situ* velocities (grey diamonds), and best fit (grey line) after the second iteration with the same parameter D = 0.090 1Mpa for all three physical properties *P*-wave velocities (top), *S*-wave velocities (middle) and logarithmic formation factor. Numbers in the upper diagram denote the 9 different units subdivided after EMMERMANN and RIESCHMÜLLER (1990), which are mainly amphibolites/ gneisses for units 1, 5 and 7, biotite gneisses for units 2, 4, 6, 8 and amphibolites/ metagabbros for units 3 and 9.

correlation between the lithological units and wave velocities. The correlation seems to be stronger for *P*-wave than for *S*-wave velocities, especially in the depth range above 1610 m (unit 1–3) and below 3575 m (unit 9).

The resistivity logs of the KTB location were investigated by numerous researchers. A comprehensive summary is given by ELEKTB-GROUP (1997). They found that the mean resistivity (in logarithmic scale) increases almost linearly with depth and reflects practically no correlation with lithology. Moreover, some very pronounced low resistivity anomalies could be identified which are also clearly not linked to lithology.

The observation that resistivity and lithology are not correlated was confirmed by laboratory measurements (RAUEN, 1991). An observed decrease of resistivity with increasing recovery depth is assumed to be caused by pressure release effects during and after recovery. The influence of pressure on the resistivity was investigated by HUENGES *et al.* (1990), NOVER and WILL (1991), DUBA *et al.* (1994), NOVER *et al.* (1995). Although most samples showed an expected increasing resistivity with increasing applied pressure some samples showed the opposite behavior. Their decreasing resistivity with increasing pressure was interpreted by DUBA *et al.* (1994)

as the result of reconnection of metallic bounds that might have been disrupted by pressure release during core recovery. Long-term observations (>300–600 h) on anomalous samples showed that some of them returned to normal behavior. This was interpreted by RAUEN *et al.* (1994) as caused by pressure-induced internal redistribution of fluids. Thus the anomalous pressure effect seems to indicate the reconnection of metallic bounds and the dominance of electrolytic conductivity.

Method

In a first step we assume that the KTB rock pile can be represented as a homogeneous isotropic porous medium. Thus we neglect the influence of different lithologies on the logs. This is a rough approximation since the KTB rocks are known to be anisotropic and at least *P*-wave logging shows a correlation between velocity and lithology. However, this correlation seems to be remarkably weaker for *S*-wave velocities, and resistivity is known to be practically independent of lithology. Since we try to jointly interpret the velocity and the resistivity profile, and since we are interested in the general depth-dependent trend rather than in a detailed analysis, we assume that this simplification is still reasonable.

In isotropic saturated and dry rocks the dependence of *P*- and *S*-wave velocities on effective pressure P_{eff} is given by equations of the form (SHAPIRO, 2003):

$$V(P_{\text{eff}}) = A_V + K_V P_{\text{eff}} - B_V \exp(-DP_{(\text{eff})}) \tag{1}$$

Here *V* represents *P*- or *S*-wave velocities and a subscript denotes that the physical meaning of the adjustable parameters (A, K, and B) actually depends on whether *P*- or *S*-wave velocities are considered. In isotropic rocks parameter D should be the same for both wave modes. This parameter is a measure for the sensitivity of the rock properties to load induced crack closure and is given by

$$D = \frac{\theta_c}{K_{\text{drys}}} \tag{2}$$

Parameter θ_c was introduced by SHAPIRO (2003) as the piezosensitivity. K_{drys} is a reference matrix bulk modulus of a rock with no compliant and undeformed stiff porosity.

For isotropic porous rocks where mainly electrolytic charge transport occurs, KASELOW and SHAPIRO (2004) have shown that the dependence of the logarithmic electrical formation factor *F* upon effective pressure is also given by an equation of the form of eq. (1) namely:

$$\log F(P_{\text{eff}}) = A_F + K_F P_{\text{eff}} - B_F \exp(DP_{\text{eff}}) \tag{3}$$

Again, parameter D is given by equation (2). Note that this derivation is valid for rocks where only electrolytic charge transport occurs.

From the analysis of stress-dependent velocity observations on numerous core samples from the KTB pilot hole we found that parameter K in equation (1) is usually <0.001 ms^{-1}Pa^{-1} and the second term can thus be neglected. Parameter K reflects the influence of stiff porosity closure on the considered property. Vanishing small estimates of parameter K are in agreement with former interpretations that the KTB rocks show a very slight or even no stiff porosity at all (POPP, 1994). Note in this case K_{drys} is equal to the bulk modulus of the grain material. Consequently, we simplify equations (1) and (2) to:

$$\Gamma(P_{\text{eff}}) = A_\Gamma - B_\Gamma \exp(-DP_{\text{eff}}) \tag{4}$$

where Γ stands for V_p, V_s, or F (in logarithmic scale). This equation is used for further analysis.

Another simplification is introduced by using isotropic depth-to-pressure transforms for confining and pore pressure, neglecting the anisotropy of confining stress at the KTB site. Confining pressure P_c and pore pressure P_{fl} at a depth d are calculated using a constant rock density ρ_r and pore fluid density ρ_{fl}, namely 2767 kg/m^3 and 1010 kg/m^3, respectively. The resulting depth-to-differential pressure transform reads

$$P_{\text{diff}} = P_c - P_{fl} = gd(\rho_r - \rho_{fl}) \tag{5}$$

It can be shown that the effective pressure P_{eff} for pore space deformation is just the differential pressure if the grain matrix material is homogeneous and elastic (e.g., ZIMMERMAN et al., 1986; SHAPIRO, 2003; GUREVICH, 2004), and/or the bulk porosity is low (SHAPIRO and KASELOW, 2003; KASELOW and SHAPIRO, 2004). As the in situ porosity of the metamorphic KTB rocks is below 1% (e.g., KERN et al., 1991) at least the last condition seems to be satisfied. Thus we assume that the effective pressure can be reasonably approximated by the differential pressure.

A simultaneous application of eq. (4) to the P-wave log, the S-wave log and to the formation factor profile requires a two-step fit procedure which is described in detail by KASELOW (2004). In the first step the data sets are separately fitted using the nonlinear least-squares Marquardt-Levenberg algorithm with A, B, and D as adjustable parameters. Due to measurement errors, anisotropy and numerical artefacts of the iterative fit algorithm, one cannot expect that values of parameter D obtained from P wave (DP), S wave (DS), and formation factor fit (DFF), respectively, to exactly coincide as theoretically predicted. Thus, the mean D of the three estimates is calculated and the fit procedure is repeated with D kept fixed.

Analysis and Results

We obtained for the first data fit the parameters given in Table 1. The D parameters agree quite well, especially in the case of S-wave log and formation factor.

Table 1

Best-fit parameters of P wave, S wave, and formation factor with asymptotic standard errors for first and second fitting iterations. Parameters A and B in case of velocities are in km/s, and in the case of formation factor F are dimensionless. The rms of the residuals of P, S waves, and formation factor for the first fit are 0.380, 0.210, and 0.695, respectively and for the second fit are 0.381, 0.210, and 0.695, respectively

	Data	A	B	D [1/MPa]
Results for first fitting	VP	5.906 ± 0.017	0.500 ± 0.035	0.060 ± 0.010
	VS	3.494 ± 0.006	0.758 ± 0.029	0.111 ± 0.007
	F	4.171 ± 0.020	1.147 ± 0.084	0.099 ± 0.012
Results for second fitting	VP	5.877 ± 0.009	0.547 ± 0.036	0.090
	VS	3.507 ± 0.005	0.695 ± 0.020	
	F	4.180 ± 0.016	1.106 ± 0.063	

The repeated fit with D = mean (DP, DS, DFF) = $(0.090 \pm 0.009 \, 1\text{MPa})$ revealed the final set of fit parameters given in Table 1. Obviously, the parameters A and B change only slightly due to an averaging of parameter D. We conclude that the agreement between the D-values supports our hypothesis of the role of crack closure on the depth trend.

Figure 1 illustrates a comparison between the profiles and the best fit results. In addition the green diamonds represent P- and S-wave velocities calculated for the *in situ* pressure-from pressure-dependent dry rock laboratory measurements. The saturated velocities were calculated from dry rock velocities using Gassmann's equations (GASSMANN, 1951).

Although most of the laboratory velocities, correspond to the sonic log velocities, they are generally at the upper limit of the log velocities and remarkably higher between 55 and 65 MPa (2600–3500 m). The discrepancy between laboratory derived velocities and logging results in this specific depth range might be related to the heterogeneity of the KTB rocks. An additional reason could be an anisotropy caused by such a heterogeneity as cracks and foliation. It is also known from further studies that the foliation dips from near vertical to horizontal orientation in this depth interval (KERN *et al.*, 1991). Thus a decrease of log-derived velocity in comparison to laboratory measurements can be caused by the wave propagation along the slow direction in effectively anisotropic rocks.

Generally, the enhanced seismic laboratory velocities might be caused by scattering and/or scaling effects. One of the cores investigated in the laboratory reveals macroscopically visible cracks. Thus we assume that the rock samples reflect the rock properties while the log velocities approach the *insitu* formation velocities, i.e., that the crack size distribution of the intact rock and the formation differ due to larger fractures and faults within the formation. SHAPIRO (2003) has shown that the piezosensitivity θ_c is approximately the inverse of the effective crack aspect ratio. If this is valid, a smaller crack aspect ratio should result in a higher θ_c.

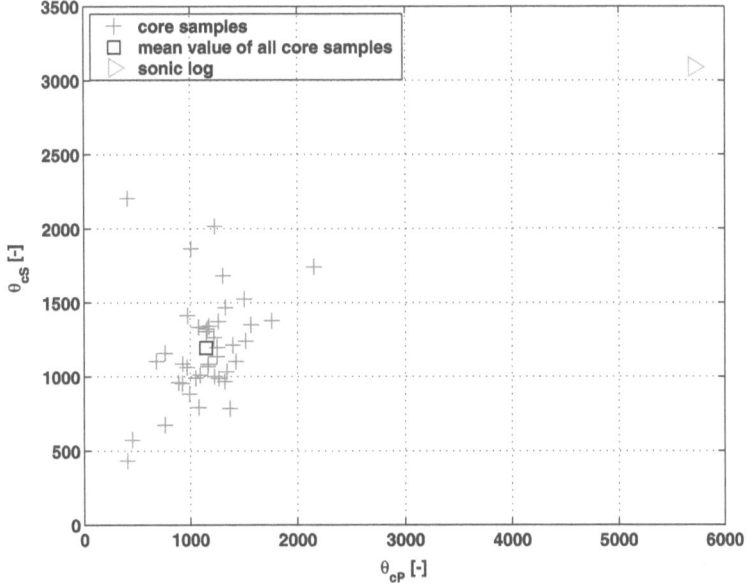

Figure 2

Cross-plot of stress sensitivity parameter θ_c obtained from the first fitting iteration for P-wave (θ_{cP}) and S-wave (Θ_{cS}) velocities of 42 core samples (crosses) and sonic logging (triangle). Green square indicates mean θ_c from laboratory data.

A plot of the θ_c parameters obtained from the first fit results (θ_{cP} from DP and θ_{cS} from DS) of 42 samples of the KTB pilot hole and the corresponding values from the sonic logs is shown in Figure 2. Obviously, θ_c obtained from the sonic logs is approx. 3.5 to 4.5 times larger than the mean θ_c obtained from the core samples. This observation might be understood as a hint for the influence of fractures and faults on the *in situ* sonic velocities. It possibly indicates that the averaged crack aspect ratio of the formation around the bore hole is 3.5 times smaller than the one of the laboratory samples. Again, we observe that the *in situ* stress sensitivity of the formation is higher than that of the core samples. However, SHAPIRO (2003) has shown that the stress sensitivity is directly related to nonlinear elastic moduli through the coefficient β_k (see ZAREMBO and KRASILNIKOV, 1966, pages 299–309), namely

$$K_{\text{dry}}(P_{\text{eff}}) = K_{\text{dry}}(0)[1 + \beta_k P_{\text{eff}}] \tag{6}$$

where the coefficient β_k is

$$\beta_k = \frac{\theta_c^2 \phi_{co}}{K_{\text{dry}}(0)} \tag{7}$$

and ϕ_{c0} is the crack porosity at $P_{\text{eff}} = 0$. This means also that the elastic nonlinearity of the KTB rocks is higher for the formation than for the core samples.

Conclusions

We have analyzed sonic P wave, S wave and deep lateral resistivity downhole loggings from the pilot hole of the German Continental Deep Drilling Site (KTB) using the isotropic piezosensitivity approach with respect to the general dependence of the logging results on increasing effective pressure with depth. We have neglected anisotropy and lithology. The latter influences P-wave velocities more than S-wave velocities and only weakly influences resistivity. The used data set was completed with stress-dependent velocity data obtained from core samples of the bore hole. We were able to fit pressure-dependent sonic velocities and formation factor with equations of the form $\Gamma(P_{eff}) = A_\Gamma - B_\Gamma \exp(-DP_{eff})$. Best fits of the data revealed a parameter $D = (0.090 \pm 0.009)$ 1/MPa identical for all mentioned logging data.

This is in agreement with the theoretically derived universality of D for isotropic rocks. This universality of D supports our assumption that the general depth-dependence of P- and S-wave velocities and resistivity can be attributed to the closure of the compliant porosity. The dependence can be quantified in terms of the stress-sensitivity θ_c. A comparison between the formation and the rock stress-sensitivity obtained from downhole measurements and laboratory observations, respectively, revealed 3.5 to 4.5 times higher stress-sensitivity of the formation rocks than of the apparently intact core samples, although the latter clearly show enhanced crack porosity due to pressure relaxation during and after recovery. We ascribe this to the smaller effective crack aspect ratio of the formation in comparison to that of the core samples. This difference might be due to larger fractures and faults which are absent in the macroscopically almost intact cores.

References

DUBA, A., HEIKAMP, S., MEURER, W., NOVER, G., and WILL, G. (1994), *Evidence from borehole samples for the role of accessory minerals in lower-crustal conductivity,* Nature. *367*, 59–61.

ELEKTB-GROUP (1997), *KTB and the electrical conductivity of the crust,* J. Geophys. Res. *102*(B8), 18,289–18,305.

EMMERMANN, R. and LAUTERJUNG, J. (1997), *The German Continental Deep Drilling Program KTB,* J. Geophys. Res. *102*(B8), 18,179–18,201.

EMMERMANN, R. and RIESCHMÜLLER, H. (1990), *Das kontinentale Tiefbohrprogramm der Bundesrepublik Deutschland (KTB) - Aktueller Stand der Planung der Hauptbohrung,* Die Geowissenschaften 9, 241–257.

GASSMANN, F. (1951), *Über die Elastizität poröser Medien,* Vierteljahresschrift der Naturforschenden Gesellschaft in Zürich 96, 1–23.

GUREVICH, B. (2004), *A simple derivation of the effective stress coefficient for seismic velocities in porous rocks,* Geophysics 69(2), 393–397.

HAAK, V. and JONES, A.G. (eds.) (1997), *The KTB deep drill hole,* J. Geophys. Res. . . *102*, Special section, AGU.

HUENGES, E., BUNTEBARTH, G., KERN, H., NOVER, G., and RAUEN, A. (1990), *Petrophysikalische Untersuchungen an Bohrkernen unter simulierten insitu Bedingungenmeine Brücke zwischen Feldlabor und Bohrlochmessungen,* KTB Rep. 90–4, Niedersächsisches Landesamt für Bodenforschung, Hannover, Germany.

KASELOW, A. (2004), *The stress-sensitivity-approach: Theory and application*, Ph.D. Thesis, Freie Universität Berlin, Germany, http://jorge.ub.fu-berlin.de/work/DissSearch.

KASELOW, A. and SHAPIRO, S. (2004), *Stress sensitivity of elastic moduli and electrical resistivity in porous rocks*, J. Geophys. Eng. *1*(1), 1–11.

KERN, H., and SCHMIDT R. (1990), *Physical properties of the KTB core samples at simulated in situ conditions*, Scientific Drilling *1*, 217–223.

KERN, H., SCHMIDT, R., and POPP, T. (1991), *The velocity and density structure of the 4000 m crustal segment of the KTB drilling site and their relation to lithological and microstructural characteristics of the rock: An experimental approach*, Scientific Drilling *2*, 130–145.

KERN, H., POPP, T., and SCHMIDT, R. (1994), *The effect of deviatoric stress on the rock properties: An experimental study simulating the in situ stress field at the KTB drilling site, Germany*, Surveys in Geophysics *15*, 467–479.

NOVER, G. and WILL, G. (1991), *Laboratory measurements on KTB core samples: Complex resistivity, zeta potential, permeability, and density as a tool for the detection of flow phenomena*, Scientific Drilling *2*, 90–100.

NOVER, G., HEIKAMP, S., KONTNY, A., and DUBA, A. (1995), *The effect of pressure on the electrical conductivity of KTB rocks*, Surveys in Geophys. 63–81.

POPP, T. (1994), *Der Einfluss von Gesteinsmatrix, Mikrorissgefüge und intergranularen Fluiden auf die elastischen Wellengeschwindigkeiten und die elektrische Leitfähigkeit krustenrelevanter Gesteine unter PT- Bedingungen*, Ph.D. Thesis, Kiel University Kiel, Germany

RAUEN, A. (1991), *Untersuchungen des komplexen elektrischen Widerstandes, insbesondere dessen Anisotropie und Frequenzabhängigkeit, von Proben des kontinentalen Tiefbohrprogramms der Bundesrepublik Deutschland (KTB)*, Ph.D. Thesis, Fak. für Geowiss., Univ. München, München, Germany.

RAUEN, A., DUYSTER, J., HEIKAMP, S., KONTNY, A., NOVER, G., and RÖCKEL, T. (1994), *Electrical conductivity of a KTB core from 7000 m: Effects of cracks and ore minerals*, Scientific Drilling *4*,197–206.

SHAPIRO, S.A. (2003), *Piezosensitivity of porous and fractured rocks*, Geophys. 68(2), 482–486.

SHAPIRO, S.A. and KASELOW, A. (2005), *Porosity and elastic anisotropy of rocks under tectonic stresses and pore pressure changes*, Geophysics, 70(5), N27–N38.

ZAREMBO, L.K. and KRASILNIKOV, V.A. (1966), *Introduction to nonlinear acoustics*, Nauka, Glavnaja Redaktsija Phys.-Math. Lit. (in Russsian).

ZIMMERMAN, R.W., SOMERTON, W., and KING, M. (1986), *Compressibility of porous rocks*, J. Geophys. Res. *91*, 12,765–12,777.

(Received April 20, 2005, revised September 16, 2005, accepted September 17, 2005)

To access this journal online:
http://www.birkhauser.ch

Pure appl. geophys. 163 (2006) 1031–1045
0033–4553/06/061031–15
DOI 10.1007/s00024-006-0058-6

© Birkhäuser Verlag, Basel, 2006

| Pure and Applied Geophysics

Can Damage Mechanics Explain Temporal Scaling Laws in Brittle Fracture and Seismicity?

DONALD L. TURCOTTE,[1] and ROBERT SHCHERBAKOV [2]

Abstract—Time delays associated with processes leading to a failure or stress relaxation in materials and earthquakes are studied in terms of continuum damage mechanics. Damage mechanics is a quasi-empirical approach that describes inelastic irreversible phenomena in the deformation of solids. When a rock sample is loaded, there is generally a time delay before the rock fails. This period is characterized by the occurrence and coalescence of microcracks which radiate acoustic signals of broad amplitudes. These acoustic emission events have been shown to exhibit power-law scaling as they increase in intensity prior to a rupture. In case of seismogenic processes in the Earth's brittle crust, all earthquakes are followed by an aftershock sequence. A universal feature of aftershocks is that their rate decays in time according to the modified Omori's law, a power-law decay. In this paper a model of continuum damage mechanics in which damage (microcracking) starts to develop when the applied stress exceeds a prescribed yield stress (a material parameter) is introduced to explain both laboratory experiments and systematic temporal variations in seismicity.

Key words: Fracture, seismicity, damage mechanics, aftershocks, power-law scaling.

1. Introduction

Time delays are generally observed in processes of rock fracture and earthquake occurrences. The time delay associated with the initiation and propagation of a single fracture can be attributed to stress corrosion and a critical stress intensity factor (DAS and SCHOLZ 1981; FREUND 1990). Usually, however, the fracture of a brittle material, such as rock, results from the coalescence and growth of microcracks (MOGI, 1962; HIRATA, 1987; HIRATA *et al.*, 1987; LOCKNER *et al.*, 1992; LOCKNER, 1993). In these processes it is possible to distinguish two types of time delays. One type is associated with processes leading to the failure of a material and is realized as a process of accumulation and coalescence of microcracks and microdefects or in the case of earthquakes the occurrence of foreshocks before a main shock. The second type of

[1]Department of Geology, University of California, Davis, CA, 95616, U.S.A.
E-mail: turcotte@geology.ucdavis.edu
[2]Center for Computational Science and Engineering, University of California, Davis, CA, 95616, U.S.A. E-mail: roshch@cse.ucdavis.edu

time delays is associated with the relaxation processes in materials or in the Earth's crust. A specimen subjected to a sufficient external load yields and, as a result, its mechanical properties degrade over time. Time delays are also associated with the occurrence of aftershocks that is a relaxation process.

In the case of fracture phenomena, time delays were observed prior to the failure of a material (SORNETTE and ANDERSEN, 1998; JOHANSEN and SORNETTE, 2000; GLUZMAN and SORNETTE, 2001). We will consider in some detail experiments on the fracture of fiberboard panels (GUARINO et al., 1998, 1999, 2002). When the panels were subjected to rapid loading, microcracks developed randomly and then coalesced until a throughgoing rupture developed. Experiments with very rapid loading, showed a systematic power-law decrease in the delay time to failure as a function of the increasing difference between the applied stress and a yield stress. These experiments also gave a power-law increase in the energy associated with acoustic emissions as a function of time prior to rupture. Also, the frequency-strength statistics of the acoustic emissions satisfied the power-law Gutenberg-Richter scaling applicable to earthquakes.

Another example of systematic delays in rock failure is the occurrence of earthquake aftershocks. Aftershocks are attributed to the increase in stress in some regions near an earthquake rupture. This increase is very rapid, on the scale of the earthquake rupture which is typically a few minutes. They also could result from the changes in pore fluid pressure associated with the migration of water in the damaged zone after a main shock. However, aftershocks occur days, months, and years later. This temporal decay of the rate of occurrence of aftershocks is extremely systematic and satisfies the modified Omori's law to a good approximation (UTSU et al., 1995; SCHOLZ, 2002).

The principal purpose of this paper is to demonstrate that the temporal scaling laws in material fracture and earthquakes described above can be explained by continuum damage mechanics. We will first provide a brief description of a damage-mechanics model that captures essential aspects of processes leading to the failure of a specimen and also processes of relaxation.

2. A Model of Continuum Damage Mechanics

Damage refers to the irreversible deformation of solids. Some examples include plasticity, brittle microcracking, and thermally activated creep [KRAJCINOVIC, 1996]. In order to quantify the deformation of solids associated with microcracking, several empirical continuum damage mechanics models were introduced and are widely used in civil and mechanical engineering [KACHANOV 1986; KRAJCINOVIC, 1989, 1996; HILD, 2002; KATTAN and VOYIADJIS 2002; SHCHERBAKOV and TURCOTTE, 2003). Damage mechanics has also been applied to the brittle deformation of the Earth's crust by a number of authors (LYAKHOVSKY et al., 1997, 2001; BEN-ZION and LYAKHOVSKY, 2002; TURCOTTE et al., 2003).

The application of continuum damage mechanics can be illustrated by considering a rod in a state of uniaxial stress $\sigma_{xx} \neq 0$, $\sigma_{yy} = \sigma_{zz} = 0$. For an elastic material Hooke's law is applicable and is written in the form

$$\sigma = E_0 \epsilon, \tag{2.1}$$

where ϵ is a strain and E_0 is the Young's modulus of the undamaged material.

In this paper we will consider a model of continuum damage mechanics as introduced by (SHCHERBAKOV*et al.*, 2005). If the stress is less than the yield stress $\sigma \leq \sigma_y$, (2.1) is assumed to be valid. If the stress is greater than the yield stress, $\sigma > \sigma_y$, a damage variable α is introduced according to

$$\sigma - \sigma_y = E_0(1 - \alpha)(\epsilon - \epsilon_y), \tag{2.2}$$

where $\sigma_y = E_0 \epsilon_y$. When $\alpha = 0$, (2.2) reduces to (2.1) and linear elasticity is applicable; as $\alpha \to 1 (\epsilon \to \infty)$ failure occurs. Increasing values of α in the range $0 \leq \alpha < 1$ quantify the weakening (decreasing E) associated with the increase in the number and size of microcracks in the material. Several authors (KRAJCINOVIC, 1996; TURCOTTE *et al.*, 2003) have shown a direct correspondence between the damage variable α in a continuum material and the number of surviving fibers N in a fiber bundle that originally had N_0 fibers, $\alpha = 1 - N/N_0$.

To complete the formulation of the damage problem it is necessary to specify the kinetic equation for the damage variable. In analogy to Lyakhovsky *et al.* (1997) we take

$$\frac{d\alpha(t)}{dt} = 0, \quad \text{if} \quad 0 \leq \sigma \leq \sigma_y \tag{2.3}$$

$$\frac{d\alpha(t)}{dt} = \frac{1}{t_d}\left[\frac{\sigma(t)}{\sigma_y} - 1\right]^\rho \left[\frac{\epsilon(t)}{\epsilon_y} - 1\right]^2, \quad \text{if} \quad \sigma > \sigma_y, \tag{2.4}$$

where t_d is a characteristic time scale for damage and ρ is a constant to be determined from experiments. The power-law dependence of $d\alpha/dt$ on stress (and strain) given above must be considered empirical in nature. However, a similar power dependence of dN/dt on stress is widely used in the analysis of fiber failures in fiber bundles (NEWMAN and PHOENIX, 2001). It is important to note the introduction of a yield limit in (2.3) and (2.4). If the stress is less than the yield stress, $\sigma < \sigma_y$, no damage occurs. The monotonic increase in the damage variable α given by (2.4) represents the weakening of the brittle solid by the nucleation and coalescence of microcracks. Microcracks are initiated only when $\sigma > \sigma_y$.

As a first example we considered a rod to which a constant uniaxial tensional stress $\sigma_0 > \sigma_y$ is applied instantaneously at $t = 0$ and held constant until the sample fails. The applicable kinetic equation for the rate of increase of damage with time is obtained from (2.2)–(2.4) with the result

$$\frac{d\alpha}{dt} = \frac{1}{t_d}\left(\frac{\sigma_0}{\sigma_y} - 1\right)^{\rho+2} \frac{1}{[1 - \alpha(t)]^2}.$$

(2.5)

Integrating with the initial condition $\alpha(0) = 0$ gives

$$\alpha(t) = 1 - \left[1 - 3\frac{t}{t_d}\left(\frac{\sigma_0}{\sigma_y} - 1\right)^{\rho+2}\right]^{1/3}.$$

(2.6)

Substituting (2.6) into (2.2) gives the strain in the sample as a function of time t

$$\frac{\epsilon(t)}{\epsilon_y} = 1 + \frac{(\sigma_0/\sigma_y - 1)}{\left[1 - 3\frac{t}{t_d}(\sigma_0/\sigma_y - 1)^{\rho+2}\right]^{1/3}}.$$

(2.7)

This behavior is illustrated in Fig. 1 for the case $\sigma_0/\sigma_y = 2.0$. Initially, at $t = 0$, we have $\epsilon(0)/\epsilon_y = 2.0$ at point A. The state of the rod moves along the constant stress path AB until it fails. Failure occurs at the time t_f when $\alpha \to 1 (\epsilon \to \infty)$. From (2.7) this failure time is given by

$$t_f = \frac{t_d}{3}\left(\frac{\sigma_0}{\sigma_y} - 1\right)^{-(\rho+2)}.$$

(2.8)

The time to failure tends to infinity as a power-law as the applied stress approaches the yield stress $\sigma_0 \to \sigma_y$. Substitution of (2.8) into (2.6) gives the time evolution of damage as

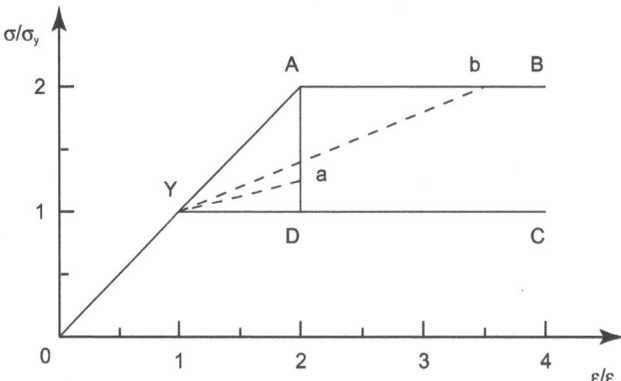

Figure 1

Dependence of the nondimensional stress σ/σ_y on nondimensional strain ϵ/ϵ_y (σ_y and ϵ_y are the yield stress and strain). Path OA is linear elasticity. If a stress $\sigma_0/\sigma_y = 2.0$ is applied instantaneously failure and damage occur along the path AB. If the stress is removed instantaneously the sample follows the path bYO. At point b the damage α is the slope of the straight line connecting point b with the yield point Y as illustrated. If a strain $\epsilon_0/\epsilon_y = 2.0$ is applied instantaneously stress relaxation and damage occur along the path AD. If the stress on the sample is removed instantaneously at the point a the sample follows the path aYO.

$$\alpha(t) = 1 - \left(1 - \frac{t}{t_f}\right)^{1/3} \tag{2.9}$$

and the corresponding time dependence of strain from (2.7)

$$\frac{\epsilon(t)}{\epsilon_y} = 1 + \frac{(\sigma_0/\sigma_y - 1)}{(1 - t/t_f)^{1/3}}. \tag{2.10}$$

The approach to failure is in the form of a power-law.

A particularly interesting set of experiments on brittle failure were carried out by GUARINO *et al.* (1999). These authors studied the failure of circular panels (222 mm diameter, 3–5 mm thickness) of chipboard panels. A differential pressure was applied rapidly across a panel and it was held constant until failure occurred. For these relatively thin panels, bending stresses were negligible and the panels failed under tension (a mode I fracture). Initially the microcracks appeared to be randomly distributed across the panel, as the number of microcracks increased they tended to localize and coalesce in the region where the final rupture occurred. The times to failure t_f of these chipboard panels were found to depend systematically on the applied differential pressure P. Taking a yield pressure (stress) $P_y = 0.038$ MPa their results are reinterpreted in Fig. 2 assuming $\sigma/\sigma_y = P/P_y$. Their results correlate very well with our failure condition (2.8) taking $\rho = 0.25$ and $t_d = 168$ s.

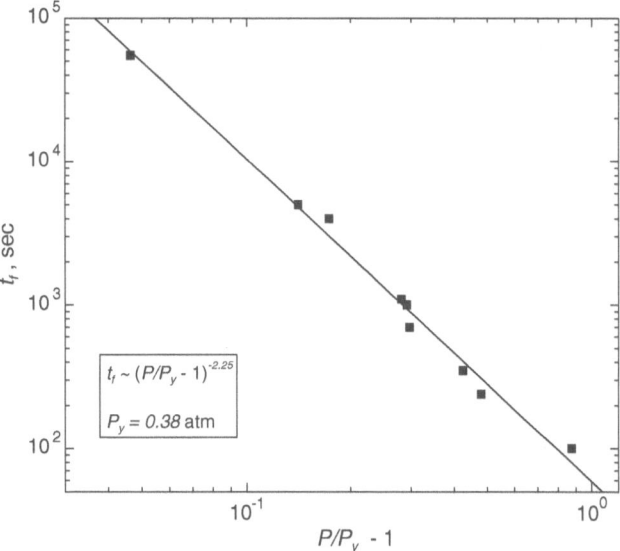

Figure 2

Times to failure of chipboard panels subjected to differential pressure (GUARINO *et al.*, 1999). Failure times t_f are given as a function of the ratio of the excess stress over the yield stress $\sigma/\sigma_y - 1$ (assumed to be equal to $P/P_y - 1$ with $P_y = 0.038$ MPa). The straight line correlation is with (2.8) taking $\rho = 0.25$ and $t_d = 168$ s.

GUARINO et al. (1999) also determined the cumulative energy in the acoustic emission (AE) events e_{AE} as a function of the time t. The total AE energy at the time of rupture is e_{tot}. They studied the dependence of e_{AE}/e_{tot} on $(1 - t/t_f)$. After an initial transient period good power-law scaling was observed that can be correlated with $e_{AE} \propto (1 - t/t_f)^{-0.27}$. This equivalent to having $de_{AE}/dt \propto (1 - t/t_f)^{-1.27}$.

We now relate the rate of increase of the damage variable $d\alpha/dt$ to the rate of AE events. The rod extends under the constant stress $\sigma_0 = 2\sigma_y$ from point A to point b as shown in Figure 1. The slope of the straight line connecting the point b and the yield point Y is $1 - \alpha$. We assume that when the stress at point b is removed instantaneously the sample will follow the path bYO. Thus the energy lost in acoustic emissions e_{AE} is equal to the work done on path Ab so that we have

$$e_{AE} = \frac{1}{2}(\sigma_0 - \sigma_y)(\epsilon_b - \epsilon_A) = \frac{\sigma_y \epsilon_y}{2}\left(\frac{\sigma_0}{\sigma_y} - 1\right)^2 \left[\frac{1}{(1 - t/t_f)^{1/3}} - 1\right], \qquad (2.11)$$

where we used (2.9). Taking the derivative of (2.11) with respect to t we obtain the rate of AE

$$\frac{de_{AE}}{dt} = \frac{\sigma_y \epsilon_y}{6t_f}\left(\frac{\sigma_0}{\sigma_y} - 1\right)^2\left(1 - \frac{t}{t_f}\right)^{-4/3}. \qquad (2.12)$$

This result compares with the observed time dependence of the rate of the acoustic energy release in experiments on chipboard panels described above.

It should be noted that the relaxation path from b to Y to O in Figure 1 is not unique. An alternative path would be parallel to the initial path O to A. The path we have chosen corresponds to the closure of type I extensional microcracks along the path b to Y and the subsequent elastic behavior from Y to O. The alternative parallel path would be appropriate for shear displacements that do not relax. The entire relaxation process is elastic. It should be noted however, that the time dependence given in (2.12) can be obtained taking either path.

As a second example we consider a rod under a tensional stress that increases linearly in time, that is

$$\sigma(t) = \sigma_y \frac{t}{t_y}. \qquad (2.13)$$

Damage begins to occur when $\sigma = \sigma_y$ at $t = t_y$. The applicable equation for the subsequent increase in damage with time is obtained from (2.2)–(2.4) and (2.13) with the result

$$\frac{d\alpha}{dt} = \frac{1}{t_d}\left(\frac{t}{t_y} - 1\right)^{p+2} \frac{1}{[1 - \alpha(t)]^2}. \qquad (2.14)$$

Integrating with the initial condition $\alpha(t_y) = 0$ gives

$$\alpha(t) = 1 - \left[1 - \frac{3t_y}{(\rho+3)t_d} \left(\frac{t}{t_y} - 1 \right)^{\rho+3} \right]^{1/3}.$$

$$(2.15)$$

Substituting (2.13) and (2.15) into (2.2) gives the time dependence of the strain

$$\frac{\epsilon(t)}{\epsilon_y} = 1 + \frac{(t/t_y - 1)}{\left[1 - \frac{3t_y}{(\rho+3)t_d} (t/t_y - 1)^{\rho+3} \right]^{1/3}}.$$

$$(2.16)$$

Failure occurs at the time t_f when $\alpha = 1 (\epsilon \to \infty)$. From (2.15) this failure time is given by

$$\frac{t_f}{t_y} = 1 + \left[\frac{(\rho+3)t_d}{3t_y} \right]^{\frac{1}{\rho+3}}$$

$$(2.17)$$

and the corresponding time dependent strain from (2.16) is given by

$$\frac{\epsilon(t)}{\epsilon_y} = 1 + \frac{(t/t_y - 1)}{\left[1 - \left(\frac{t-t_y}{t_f-t_y} \right)^{\rho+3} \right]^{1/3}}.$$

$$(2.18)$$

Again the approach to failure takes the form of a power law.

We will now apply the results derived above to the problem of seismic activation prior to earthquakes. Systematic increases in intermediate levels of seismicity prior to large earthquakes have been documented by several authors (SYKES and JAUMÉ, 1990; KNOPOFF et al., 1996; JAUMÉ and SYKES, 1999). It has also been observed that an increase in the seismic activity prior to large earthquakes takes the form of a power law. This was first proposed by BUFE and VARNES (1993). They considered the cumulative Benioff strain in a region defined as

$$\epsilon_B(t) = \sum_{i=1}^{N(t)} \sqrt{e_i},$$

$$(2.19)$$

where e_i is the seismic energy release in the ith precursory earthquake and $N(t)$ is the number of precursory earthquakes considered up to time t.

The precursory increase in seismicity is referred to as accelerated moment release (AMR). In terms of the cumulative Benioff strain it is quantified as

$$\epsilon_B(t) = \epsilon_B(t_f) - B \left(1 - \frac{t}{t_f} \right)^s,$$

$$(2.20)$$

where $\epsilon_B(t_f)$ is the final cumulative Benioff strain when the large earthquake occurs, t_f is time measured forward from the beginning of AMR, B is a constant, and s is the exponent. Examples of AMR have been given by BOWMAN et al. (1998), BUFE et al. (1994), VARNES and BUFE (1996), BREHM and BRAILE (1998, 1999a,b), ROBINSON (2000), ZÖLER et al. (2001), MAIN (1999), BOWMAN and KING (2001), YANG et al. (2001), KING and BOWMAN (2003), BOWMAN and SAMMIS (2004), and SAMMIS et al. (2004). RUNDLE et al. (2000) found that the distribution of values of the power-law exponent s for 12 earthquakes was $s = 0.26 \pm 0.15$.

We assume that the stress in a seismological zone increases linearly with time as given by (2.13). This increase in the tectonic stress is very slow compared with the development of damage so that it is appropriate to assume $t_y/t_d \gg 1$. We associate the AMR energy e_{AMR} with the energy added to the rod by the time t. Using (2.13) and (2.18) we can obtain the rate of energy release

$$\frac{de_{AMR}}{dt} = \frac{1}{2}\frac{d}{dt}[\sigma(t)\epsilon(t)] = \frac{\epsilon_y\sigma_y}{2t_y}\left\{1 + \frac{6t - 3t_y + [3t_y + (\rho - 3)t]\left(\frac{t-t_y}{t_f-t_y}\right)^{\rho+3}}{3t_y\left[1 - \left(\frac{t-t_y}{t_f-t_y}\right)^{\rho+3}\right]^{4/3}}\right\}. \quad (2.21)$$

It is also possible to analyze the behavior of the rate near the failure time t_f. The expansion of (2.21) around $(t - t_f)$ gives

$$\frac{de_{AMR}}{dt} = \frac{\epsilon_y\sigma_y}{2t_y}\left\{1 + \frac{t_f}{3t_y(\rho+3)^{1/3}}\frac{1}{\left(\frac{t-t_f}{t_y-t_f}\right)^{4/3}}\right.$$

$$\left. + \frac{t_f(10-\rho) - 6t_y}{9t_y(\rho+3)^{1/3}}\frac{1}{\left(\frac{t-t_f}{t_y-t_f}\right)^{1/3}} + O(t - t_f)^{2/3}\right\}. \quad (2.22)$$

The result $de_{AMR}/dt \propto (1 - t/t_f)^{-4/3}$ was previously derived from damage mechanics by BEN-ZION and LYAKHOVSKY (2002) who used the assumption of the constant applied stress.

To make the association with Benioff strain we use a derived expansion for the energy release rate (2.22)

$$\frac{d\epsilon_B}{dt} = \sqrt{\frac{de_{AMR}}{dt}}. \quad (2.23)$$

The cumulative Benioff strain can be obtained by integrating (2.23) with the result

$$\epsilon_B(t) = \epsilon_B(t_f) - \int_t^{t_f} \frac{d\epsilon_B}{dt} dt \tag{2.24}$$

$$= \epsilon_B(t_f) - \sqrt{\frac{3t_f(t_y - t_f)^{4/3}\sigma_y\epsilon_y}{2t_y^2(\rho+3)^{1/3}}(t_f - t)^{1/3} - O(t_f - t)^{4/3}}, \tag{2.25}$$

where we used (2.22). The exponent of $s = 1/3$ is in reasonably good agreement with the range of values associated with AMR $s = 0.26 \pm 0.15$ as given above. This agreement was also noted by BEN-ZION and LYAKHOVSKY (2002).

As our final example we will consider a rod to which a constant uniaxial tensional strain $\epsilon_0 > \epsilon_y$ has been applied instantaneously at $t = 0$ and is held constant. The applicable equation for the rate of increase of damage with time is obtained from (2.2)–(2.4) with the result

$$\frac{d\alpha}{dt} = \frac{1}{t_d}\left(\frac{\epsilon_0}{\epsilon_y} - 1\right)^{\rho+2}[1 - \alpha(t)]^{\rho}. \tag{2.26}$$

Integrating with the initial condition $\alpha(0) = 0$, we find

$$\alpha(t) = 1 - \left[1 + (\rho - 1)\frac{t}{t_d}\left(\frac{\epsilon_0}{\epsilon_y} - 1\right)^{\rho+2}\right]^{-\frac{1}{\rho-1}}. \tag{2.27}$$

This result is valid as long as $\rho > 1$. The damage increases monotonically with time and as $t \to \infty$ the maximum damage is $\alpha(\infty) = 1$. Using (2.27) with (2.2) one derives the stress relaxation in the material as a function of time t

$$\frac{\sigma(t)}{\sigma_y} = 1 + \left(\frac{\epsilon_0}{\epsilon_y} - 1\right)\left[1 + (\rho - 1)\frac{t}{t_d}\left(\frac{\epsilon_0}{\epsilon_y} - 1\right)^{\rho+2}\right]^{-\frac{1}{\rho-1}}. \tag{2.28}$$

At $t = 0$ we have linear elasticity corresponding to $\alpha = 0$. In the limit $t \to \infty$ the stress relaxes to the yield stress $\sigma(\infty) = \sigma_y$ below which no further damage can occur. This behavior is illustrated in Figure 1 for the case $\epsilon_0/\epsilon_y = 2.0$. Initially, at $t = 0$, we have $\sigma/\sigma_y = 2.0$ at point A. The sample then moves along the constant strain path AD until the stress has relaxed to the yield stress σ_y.

This stress relaxation process has been applied to the understanding of the aftershock sequence that follows an earthquake. During an earthquake some regions in the vicinity of the fault rupture experience a rapid increase of strain (stress). This is in direct analogy to the instantaneous application of strain considered above. Just as the microcracks associated with damage relax stress in our model, aftershocks relax stresses applied during the main shock. We recognize that all aftershocks have secondary aftershocks and so forth but all aftershocks contribute to stress relaxation similarly.

A universal scaling law is applicable to the temporal decay of aftershock activity following an earthquake. This is known as the modified Omori's law and can be written in the form (UTSU et al., 1995, SCHOLZ, 2002)

$$\frac{dN}{dt} = \frac{1}{\tau[1 + t/c]^p},$$ (2.29)

where dN/dt is the rate of occurrence of aftershocks with magnitudes greater than m, t is the time that has elapsed since the main shock, τ and c are characteristic times, and the exponent p has a value near unity. The total number of aftershocks with magnitudes greater than m, $N_{tot}(\geq m)$, was obtained by integrating (2.29) with the result (SHCHERBAKOV et al., 2004)

$$N_{tot}(\geq m) = \int_0^\infty \frac{dN}{dt} dt = \int_0^\infty \frac{dt}{\tau(1 + t/c)^p} = \frac{c}{\tau(p-1)}.$$ (2.30)

If the modified Omori's law is assumed to be valid for large times (no cutoff) then the total number of aftershocks is finite only for $p > 1$. Combining (2.29) and (2.30) gives

$$\frac{1}{N_{tot}}\frac{dN}{dt} = \frac{p-1}{c}\frac{1}{(1 + t/c)^p},$$ (2.31)

We will next show that this result can be derived using continuum damage mechanics.

In order to relate our continuum damage mechanics model to aftershocks we determine the rate of energy release in the relaxation process considered above. In order to do this we use the approach applied to the AE experiments considered in the previous section. Since the strain is constant during the stress relaxation, no work is done on the sample. We hypothesize that if the applied strain (stress) is instantaneously removed during the relaxation process then the sample will return to a state of zero stress and strain following a linear stress-strain path (aY) with slope $E_0(1 - \alpha)$ to stress σ_y and following the path (YO) with slope E_0 to zero stress (Fig. 1). We assume that the difference between the energy added e_{YA} and the energy recovered e_{aY} is lost in aftershocks and find that this energy e_{as} is given by

$$e_{as} = e_{YA} - e_{aY} = \frac{1}{2}E_0(\epsilon_0 - \epsilon_y)^2\alpha(t).$$ (2.32)

The rate of energy release is obtained by substituting (2.27) into (2.32) and taking the time derivative with the result

$$\frac{de_{as}}{dt} = \frac{E_0 \epsilon_y^2}{2t_d} \left(\frac{\epsilon_0}{\epsilon_y} - 1\right)^{p+4} \left[1 + (\rho - 1)\left(\frac{\epsilon_0}{\epsilon_y} - 1\right)^{p+2} \left(\frac{t}{t_d}\right)\right]^{-\frac{\rho}{\rho-1}}. \tag{2.33}$$

The total energy of aftershocks e_{ast} is obtained by substituting $\alpha(\infty) = 1$ into (2.32) with the result

$$e_{ast} = \frac{1}{2} E_0 \left(\epsilon_0 - \epsilon_y\right)^2. \tag{2.34}$$

Combining (2.33) and (2.34) gives

$$\frac{1}{e_{ast}} \frac{de_{as}}{dt} = \frac{\frac{1}{t_d}\left(\epsilon_0/\epsilon_y - 1\right)^{p+2}}{\left[1 + (\rho - 1)\left(\epsilon_0/\epsilon_y - 1\right)^{p+2}\left(\frac{t}{t_d}\right)\right]^{\frac{\rho}{\rho-1}}}. \tag{2.35}$$

This result is clearly similar to the aftershock relation given in (2.31). To demonstrate this further let us make the substitutions

$$\rho = \frac{p}{p-1}, \tag{2.36}$$

$$c = \frac{t_d}{(\rho - 1)\left(\epsilon_0/\epsilon_y - 1\right)^{p+2}}. \tag{2.37}$$

Substitution of (2.36) and (2.37) into (2.35) gives

$$\frac{1}{e_{ast}} \frac{de_{as}}{dt} = \frac{p-1}{c} \frac{1}{(1+t/c)^p}. \tag{2.38}$$

This damage mechanics result is identical in form to the modified Omori's law for aftershocks given in (2.31).

We next consider a specific example. In Fig. 3 the rate of occurrence of aftershocks with $m \geq 2.5$ following the $m_{ms} = 7.3$ Landers (California) earthquake, June 28, 1992, is given as a function of the time after the earthquake. Also shown is the correlation with (2.31) taking $\tau = 2.22 \times 10^{-3}$ days, $c = 2.072$ days, and $p = 1.22$. From (2.36) we find that the corresponding power law exponent is $\rho = 5.55$. Further assuming that $\epsilon_0/\epsilon_y = 1.2$ we find from (2.37) that the damage time is $t_d = 4.3$ s.

3. Discussion

In this paper we have presented a damage mechanics model with a yield stress. We have also applied the solutions of the model to several problems in the fracture of materials and earthquakes. This model is derived from several damage mechanics models that have been applied to problems in mechanical and civil engineering (ANIFRANI et al., 1995; KRAJCINOVIC, 1996). The approach based on specifying the

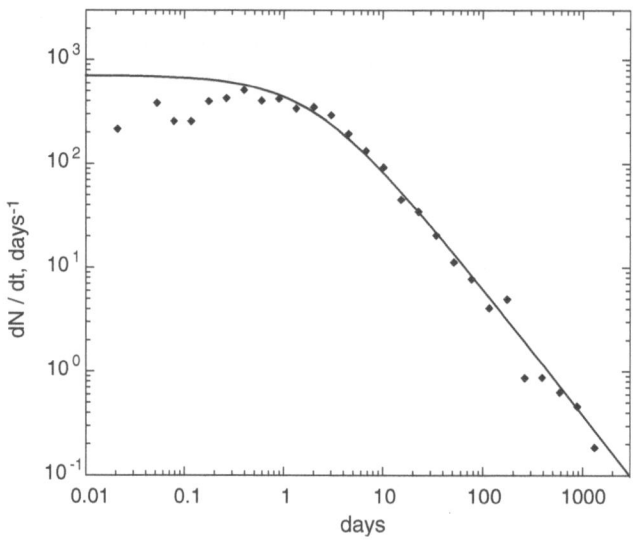

Figure 3

The rate of occurrence of aftershocks as a function of time following the June 28, 1992 Landers earthquake ($m_{ms} = 7.3$). The magnitude cutoff for aftershocks is $m = 2.5$. Also shown is the correlation with (29) taking $\tau = 2.22 \times 10^{-3}$ days, $c = 2.072$ days, and $p = 1.22$.

kinetic equation for damage evolution is empirical, however, it is justified by experimental observations. Although the physical mechanisms of time delays leading to fracture of materials remain unclear.

It should be pointed out that a similar time-to-failure model is routinely applied to composite materials. In the time-dependent fiber-bundle model the failure statistics of the individual fibers that make up the fiber bundle are specified (COLEMAN, 1957, 1958; SMITH and PHOENIX, 1981; NEWMAN and PHOENIX, 2001). It has been shown by KRAJCINOVIC (1996) and TURCOTTE *et al.* (2003) that the damage variable α can be introduced into the studies of fiber-bundle models where it defines the fraction of broken fibers. In the dynamic fiber-bundle models the time delays of failures of individual fibers are specified through an empirical distribution function.

The description of the damage evolution based on the empirical equation is valid for a gradual deterioration of ductile materials under creep (KACHANOV, 1986). This should be distinguished from the case of brittle-elastic behavior where damage can be defined as a density of microcracks and cavities and can be calculated exactly for certain geometries of these defects. It was noted by KACHANOV (1994) that the use of a kinetic equation for the scalar damage variable in the case of brittle-elastic solids can lead to inconsistencies. Our proposed kinetic equation is based on the studies of the fiber-bundle model where a similar power-law dependence on stress is observed (TURCOTTE *et al.*, 2003) and can be considered as an approximation. This model is also one-dimensional and extrapolations to higher dimensions should be done with care.

We have compared the predictions of our damage model with laboratory studies of the time delays associated with the rupture of chipboard panels and also with the time delays associated with the occurrence of aftershocks. In both cases the predicted power-law dependence of damage on time is consistent with the observations. The damage mechanics model basically has two parameters: the power-law exponent ρ and the characteristic time t_d. For the failure of the chipboard we find $\rho = 0.25$ and $t_d = 168$ s. For the aftershock sequence we have $\rho = 5.55$ and $t_d = 4.3$ s. Although both behave brittley, the chipboard is a much "softer" material than the rock in which aftershocks occur. We attribute the differences in the parameter values to this difference in material properties.

Some forms of damage are clearly thermally activated. The irreversible deformation of a solid by diffusion or dislocation creep is an example. The ability of vacancies and dislocations to move through a crystal is governed by an exponential dependence on absolute temperature with a well-defined activation energy. Time delays associated with fracture have been attributed to stress corrosion which is also a thermally activated process (DAS and SCHOLZ, 1981). However, GUARINO et al. (1998) varied the temperature in their experiments on the fracture of chipboard and found no effect. An alternative explanation for the time delay associated with microcracking has been given by SCORRETTI et al. (2001) and CILIBERTO et al. (2001). An effective "temperature" can be attributed to the spatial disorder (heterogeneity) of the solid. The spatial variability of stress in the solid is caused by the microcracking itself, not by thermal fluctuations.

Acknowledgment

This work has been supported by NSF Grant ATM-0327571.

REFERENCES

ANIFRANI, J.C., LEFLOCH, C., SORNETTE, D., and SOUILLARD, B. (1995), *Universal log-periodic correction to renormalization-group scaling for rupture stress prediction from acoustic emissions*, J. Phys. I *5*, 631–638.

BEN-ZION, Y. and LYAKHOVSKY, V. (2002), *Accelerated seismic release and related aspects of seismicity patterns on earthquake faults*, Pure Appl. Geophys. *159*, 2385–2412.

BOWMAN, D.D. and KING, G.C.P. (2001), *Accelerating seismicity and stress accumulation before large earthquakes*, Geophys. Res. Lett. *28*, 4039–4042.

BOWMAN, D.D., OUILLON, G., SAMMIS, C.G., SORNETTE, A., and SORNETTE, D. (1998), *An observational test of the critical earthquake concept*, J. Geophys. Res. *103*, 24359–24372.

BOWMAN, D.D. and SAMMIS, C.G. (2004), *Intermittent criticality and the Gutenberg-Richter distribution*, Pure Appl. Geophys. *161*, 1945–1956.

BREHM, D.J. and BRAILE, L.W. (1998), *Intermediate-term earthquake prediction using precursory events in the New Madrid seismic zone*, Bull. Seismol. Soc. Am. *88*, 564–580.

BREHM, D.J. and BRAILE, L.W. (1999a), *Intermediate-term earthquake prediction using the modified time-to-failure method in Southern California*, Bull. Seismol. Soc. Am. *89*, 275–293.

BREHM, D.J. and BRAILE, L.W. (1999b), *Refinement of the modified time-to-failure method for intermediate-term earthquake prediction*, J. Seismol. *3*, 121–138.

BUFE, C.G., NISHENKO, S.P., and VARNES, D.J. (1994), *Seismicity trends and potential for large earthquakes in the Alaska-Aleutian region*, Pure Appl. Geophys. *142*, 83–99.

BUFE, C.G. and VARNES, D.J. (1993), *Predictive modeling of the seismic cycle of the greater San Francisco Bay region*, J. Geophys. Res. *98*, 9871–9883.

CILIBERTO, S., GUARINO, A., and SCORRETTI, R. (2001), *The effect of disorder on the fracture nucleation process*, Physica D *158*, 83–104.

COLEMAN, B.D. (1957), *Time dependence of mechanical breakdown in bundles of fibers. I Constant total load*, J. Appl. Phys. *28*, 1058–1064.

COLEMAN, B.D. (1958), *Statistics and time dependence of mechanical breakdown in fibers*, J. Appl. Phys. *29*, 968–983.

DAS, S. and SCHOLZ, C.H. (1981), *Theory of time-dependent rupture in the Earth*, J. Geophys. Res. *86*, 6039–6051.

FREUND, L.B., *Dynamic Fracture Mechanics* (Cambridge University Press, Cambridge 1990).

GLUZMAN, S. and SORNETTE, D. (2001), *Self-consistent theory of rupture by progressive diffuse damage*, Phys. Rev. E *6306*, Art. No. 066129.

GUARINO, A., CILIBERTO, S., and GARCIMARTIN, A. (1999), *Failure time and microcrack nucleation*, Europhys. Lett. *47*, 456–461.

GUARINO, A., CILIBERTO, S., GARCIMARTIN, A., ZEI, M., and SCORRETTI, R. (2002), *Failure time and critical behaviour of fracture precursors in heterogeneous materials*, Eur. Phys. J. B *26*, 141–151.

GUARINO, A., GARCIMARTIN, A., and CILIBERTO, S. (1998), *An experimental test of the critical behaviour of fracture precursors*, Eur. Phys. J. B *6*, 13–24.

HILD, F., *Discrete versus continuum damage mechanics: A probabilistic perspective*. In *Continuum Damage Mechanics of Materials and Structures* (O. ALLIX, F. HILD eds.) (Elsevier, Amsterdam 2002), pp. 79–114.

HIRATA, T. (1987), *Omori's power law aftershock sequences of microfracturing in rock fracture experiment*, J. Geophys. Res. *92*, 6215–6221.

HIRATA, T., SATOH, T., and ITO, K. (1987), *Fractal structure of spatial-distribution of microfracturing in rock*, Geophys. J. R. Astr. Soc. *90*, 369–374.

JAUMÉ, S.C. and SYKES, L.R. (1999), *Evolving towards a critical point: A review of accelerating seismic moment/energy release prior to large and great earthquakes*, Pure Appl. Geophys. *155*, 279–305.

JOHANSEN, A. and SORNETTE, D. (2000), *Critical ruptures*, Eur. Phys. J. B *18*, 163–181.

KACHANOV, L.M. *Introduction to Continuum Damage Mechanics* (Martinus Nijhoff, Dordrecht 1986).

KACHANOV, M. (1994), *On the concept of damage in creep and in the brittle-elastic range*, Int. J. Damage Mech. *3*, 329–337.

KATTAN, P.I. and VOYIADJIS, G.Z. *Damage Mechanics with Finite Elements: Practical Applications with Computer Tools* (Springer, Berlin 2002).

KING, G.C.P. and BOWMAN, D.D. (2003), *The evolution of regional seismicity between large earthquakes*, J. Geophys. Res. *108*, 2096.

KNOPOFF, L., LEVSHINA, T., KEILIS-BOROK, V.I., and MATTONI, C. (1996), *Increased long-range intermediate-magnitude earthquake activity prior to strong earthquakes in California*, J. Geophys. Res. *101*, 5779–5796.

KRAJCINOVIC, D. (1989), *Damage mechanics*, Mech. Mater. *8*, 117–197.

KRAJCINOVIC, D. *Damage Mechanics* (Elsevier, Amsterdam 1996).

LOCKNER, D. (1993), *The role of acoustic-emission in the study of rock fracture*, Int. J. Rock Mech. Min. Sci. *30*, 883–899.

LOCKNER, D.A., BYERLEE, J.D., KUKSENKO, J.D., PONOMAREV, V., and SIDORIN, A. *Observations of quasi-static fault growth from acoustic emissions*. In *Fault Mechanics and Transport Properties of Rocks* (Academic Press, London 1992), pp. 3–31.

LYAKHOVSKY, V., BEN-ZION, Y., and AGNON, A. (2001), *Earthquake cycle, fault zones, and seismicity patterns in a rheologically layered lithosphere*, J. Geophys. Res. *106*, 4103–4120.

LYAKHOVSKY, V., BENZION, Y., and AGNON, A. (1997), *Distributed damage, faulting, and friction*, J. Geophys. Res. *102*, 27635–27649.

MAIN, I.G. (1999), *Applicability of time-to-failure analysis to accelerated strain before earthquakes and volcanic eruptions*, Geophys. J. Int. *139*, F1–F6.

MOGI, K. (1962), *Study of elastic shocks caused by the fracture of heterogeneous materials and its relations to earthquake phenomena*, Bull. Earthquake Res. Inst. *40*, 125–173.

NEWMAN, W.I. and PHOENIX, S.L. (2001), *Time-dependent fiber bundles with local load sharing*, Phys. Rev. E *6302*, Art. No. 021507.

ROBINSON, R. (2000), *A test of the precursory accelerating moment release model on some recent New Zealand earthquakes*, Geophys. J. Int. *140*, 568–576.

RUNDLE, J., KLEIN, W., TURCOTTE, D.L., and MALAMUD, B.D. (2000), *Precursory seismic activation and critical-point phenomena*, Pure Appl. Geophys. *157*, 2165–2182.

SAMMIS, C.G., BOWMAN, D.D., and KING, G. (2004), *Anomalous seismicity and accelerating moment release preceding the 2001 and 2002 earthquakes in northern Baja California, Mexico*, Pure Appl. Geophys. *161*, 2369–2378.

SCHOLZ, C.H., *The Mechanics of Earthquakes and Faulting* (Cambridge University Press, Cambridge 2002), 2nd ed.

SCORRETTI, R., CILIBERTO, S., and GUARINO, A. (2001), *Disorder enhances the effects of thermal noise in the fiber bundle model*, Europhys. Lett. *55*, 626–632.

SHCHERBAKOV, R. and TURCOTTE, D.L. (2003), *Damage and self-similarity in fracture*, Theor. Appl. Frac. Mech. *39*, 245–258.

SHCHERBAKOV, R., TURCOTTE, D.L., and RUNDLE, J.B. (2004), *A generalized Omori's law for earthquake aftershock decay*, Geophys. Res. Lett. *31*, Art. No. L11613.

SHCHERBAKOV, R., TURCOTTE, D.L., and RUNDLE, J.B. (2005), *Aftershock statistics*, Pure Appl. Geophys. *162*, 1051–1076.

SMITH, R.L. and PHOENIX, S.L. (1981), *Asymptotic distributions for the failure of fibrous materials under series-parallel structure and equal load-sharing*, J. Appl. Mech. *48*, 75–82.

SORNETTE, D. and ANDERSEN, J.V. (1998), *Scaling with respect to disorder in time-to-failure*, Eur. Phys. J. B *1*, 353–357.

SYKES, L.R. and JAUMÉ, S.C. (1990), *Seismic activity on neighboring faults as a long-term precursor to large earthquakes in the San Francisco Bay area*, Nature 348, 595–599.

TURCOTTE, D.L., NEWMAN, W.I., and SHCHERBAKOV, R. (2003), *Micro and macroscopic models of rock fracture*, Geophys. J. Int. *152*, 718–728.

UTSU, T., OGATA, Y., and MATSU'URA, R.S. (1995), *The centenary of the Omori formula for a decay law of aftershock activity*, J. Phys. Earth *43*, 1–33.

VARNES, D.J. and BUFE, C.G. (1996), *The cyclic and fractal seismic series preceding an $m_b = 4.8$ earthquake on 1980 February 14 near the Virgin Islands*, Geophys. J. Int. *124*, 149–158.

YANG, W.Z., VERE-JONES, D., and LI, M. (2001), *A proposed method for locating the critical region of a future earthquake using the critical earthquake concept*, J. Geophys. Res. *106*, 4121–4128.

ZÖLLER, G., HAINZL, S., and KURTHS, J. (2001), *Observation of growing correlation length as an indicator for critical point behavior prior to large earthquakes*, J. Geophys. Res. *106*, 2167–2175.

(Received March 10, 2005, revised October 19, 2005, accepted October 21, 2005)

To access this journal online:
http://www.birkhauser.ch

Pure appl. geophys. 163 (2006) 1047–1057
0033–4553/06/061047–11
DOI 10.1007/s00024-006-0057-7

© Birkhäuser Verlag, Basel, 2006

▌**Pure and Applied Geophysics**

An Update on the Fracture Toughness Testing Methods Related to the Cracked Chevron-notched Brazilian Disk (CCNBD) Specimen

R. J. Fowell,[1] C. Xu,[1] and P. A. Dowd[1]

Abstract—This paper reviews the use of the cracked Chevron-notched Brazilian disc (CCNBD) for fracture toughness testing. Theoretical and experimental backgrounds of the method are described. Some issues regarding the current development (i.e., recalibration) of the specimen geometry are presented and discussed. A number of geometries related to the CCNBD proposed recently for fracture toughness testing of rock are then introduced and commented on.

Key words: Fracture toughness testing, Brazilian disc, CCNBD, stress intensity factor.

1. Introduction

Due to the great popularity of using Brazilian disks in the rock mechanics community, the introduction of the CCNBD specimen for rock fracture toughness testing (FOWELL and XU, 1994; CHEN, 1990) did not encounter any difficulty in gaining wide acceptance. Compared with the Chevron bend (CB) and short rod (SR) specimens, the CCNBD has numerous advantages which include easier sample preparation, much higher failure load, simpler testing procedure and it is easily adaptable for mixed-mode fracture toughness testing. Some selected publications about the use of the specimen include CHANG *et al.* (2002), DWIVEDI *et al.* (2000), AL-SHAYEA *et al.* (2000) and KRISHNAN *et al.* (1998).

2. Background

The basic CCNBD testing configuration is given in Figure 1. The relations between the geometrical entities are expressed in Equation (1) below.

---

[1]Department of Mining and Mineral Engineering, The University of Leeds, Leeds, UK.

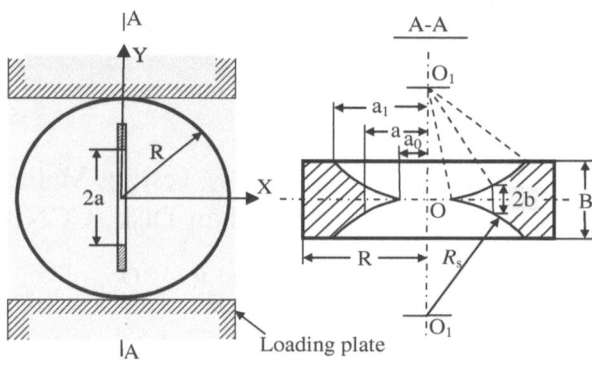

Figure 1

Basic configuration of CCNBD fracture toughness testing.

$$
\begin{cases}
\alpha_s = R_s/R = \sqrt{\alpha_0^2 + \left(\alpha_1^2 - \alpha_0^2 + \alpha_B^2/4\right)^2 \div \alpha_B^2} \\[2mm]
h_c = \left(\alpha_s - \sqrt{\alpha_s^2 - \alpha_1^2}\right) \cdot R = \left(\alpha_s - \sqrt{\alpha_s^2 - \alpha_0^2}\right) \cdot R + B/2 \\[2mm]
\alpha_0 = \sqrt{\alpha_s^2 - \left(\sqrt{\alpha_s^2 - \alpha_1^2} + \alpha_B/2\right)^2} \\[2mm]
\alpha_1 = \sqrt{\alpha_s^2 - \left(\sqrt{\alpha_s^2 - \alpha_0^2} - \alpha_B/2\right)^2} \\[2mm]
\alpha_B = 2 \cdot \left(\sqrt{\alpha_s^2 - \alpha_0^2} - \sqrt{\alpha_s^2 - \alpha_1^2}\right)
\end{cases} \tag{1}
$$

where α_0, α_1 and α_B are dimensionless and are calculated as: $\alpha_0 = a_0/R$, $\alpha_1 = a_1/R$ and $\alpha_B = B/R$. The ISRM suggested that standard specimen dimensions are given in Table 1 (ISRM, 1995).

In practical experiments, the dimensions of CCNBD specimens obtained will deviate from the standard figures. To obtain the stress intensity factor (SIF) for

Table 1

Standard CCNBD geometrical dimensions (Fig. 1)

Descriptions	Values	Dimensionless Expression
Diameter **D** (mm)	75.0	
Thickness **B** (mm)	30.0	$\alpha_B = B/R = 0.80$
Initial chevron-notched crack length $\mathbf{a_0}$ (mm)	9.89	$\alpha_0 = a_0/R = 0.2637$
Final chevron-notched crack length $\mathbf{a_1}$ (mm)	24.37	$\alpha_1 = a_1/R = 0.65$
Saw diameter $\mathbf{D_s}$ (mm)	52.0	$\alpha_s = D_s/R = 0.6933$
Cutting depth $\mathbf{h_c}$ (mm)	16.95	
Y^*_{min} (dimensionless)	0.84	
Critical crack length $\mathbf{a_m}$ (mm)	19.31	$\alpha_m = a_m/R = 0.5149$

different dimensions, the specimen geometry is analyzed theoretically using the combination of compliance method, dislocation method and superimposition technique. The SIF for a CCNBD specimen with crack length α can be calculated as (XU and FOWELL, 1993):

$$Y^*(\alpha) = \left[\frac{\alpha_B^4 \cdot g_3(\alpha)}{8 \cdot g_1(\alpha) \cdot g_2(\alpha)} \right]^{\frac{1}{2}}, \tag{2}$$

where

$$\begin{cases} g_1(\alpha) = (\alpha_1^2 - \alpha_0^2) + \dfrac{\alpha_B^2}{4} - \sqrt{(\alpha_1^2 - \alpha^2) \cdot \alpha_B^2 + \left(\alpha_1^2 - \alpha_0^2 - \dfrac{\alpha_B^2}{4} \right)^2} \\[4mm] g_2(\alpha) = \left[\dfrac{1}{B'E} \cdot \dfrac{g_1(\alpha)}{C(\alpha)} + \dfrac{1}{B'E} \cdot \displaystyle\int_\alpha^{\alpha_1} \dfrac{g_4(\xi)}{C(\xi)} \cdot d\xi \right]^2 \\[4mm] g_3(\alpha) = \dfrac{2}{(B'E)^2} \cdot \dfrac{Y(\alpha) \cdot g_1(\alpha)}{C^2(\alpha)} - \dfrac{(1 - c_k)}{B'E} \cdot \dfrac{g_4(\alpha)}{C(\alpha)} \\[4mm] g_4(\alpha) = \dfrac{\alpha_B^2 \cdot \alpha}{\sqrt{(\alpha_1^2 - \alpha^2) \cdot \alpha_B^2 + (\alpha_1^2 - \alpha_0^2 - \alpha_B^2/4)^2}} \end{cases} \tag{3}$$

The results are then validated using the finite element method (FEM) and the boundary element method (BEM) (XU and FOWELL, 1993). For fracture toughness testing, the important SIF value for the specimen is the minimum SIF, denoted as Y^*, as it corresponds to the failure load recorded during the testing. From them the fracture toughness value can then be calculated as:

$$K_C = \frac{P_{\max}}{B \cdot \sqrt{R}} \cdot Y_m^*. \tag{4}$$

Note the above calculation only depends on the compliance $C(\alpha)$ of the corresponding cracked straight-through Brazilian disc (CSTBD) and is independent of fracture mode, i.e., if the correct compliance is supplied the above fracture toughness is equally applicable to mode I, II, III or mixed mode fracture testing.

However, due to the plane strain constraint, not all geometries of the CCNBD are valid to be used for fracture testing. Studies (XU and FOWELL, 1993, 1994) showed that CCNBD geometries must fall within the range outlined in Figure 2 to yield valid fracture toughness results. These ranges can also be expressed below:

$$\begin{cases} \alpha_1 \geq 0.4, & \text{Line 0} \\ \alpha_1 \geq \alpha_B/2, & \text{Line 1} \\ \alpha_B \leq 1.04, & \text{Line 2} \\ \alpha_1 \leq 0.8, & \text{Line 3} \\ \alpha_B \geq 1.1729 \cdot (\alpha_1)^{5/3}, & \text{Line 4} \\ \alpha_B \geq 0.44, & \text{Line 5} \end{cases}$$

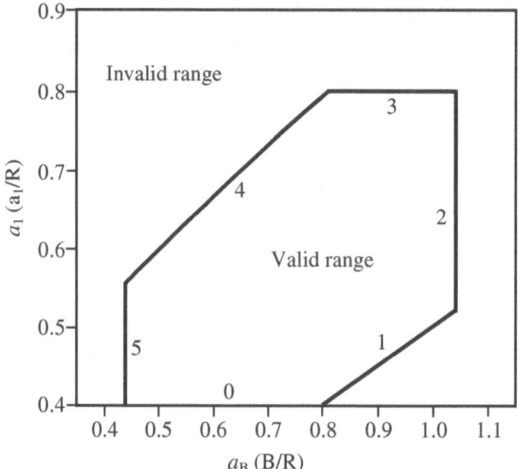

Figure 2
Valid geometrical ranges for fracture toughness testing.

For CCNBD specimens within these ranges, an easier version (compared to Equation 3) to calculate the minimum dimensionless SIF values is given in (ISRM, 1995):

$$Y^*_{min} = u \cdot e^{v \cdot \alpha_1}, \tag{6}$$

where constants u and v are listed in ISRM (1995). A selected portion for some common configurations is given in Table 2 below:

WANG (1998) introduced a correction factor to account for the compliance of uncracked disc in the SIF evaluation of CCNBD specimens and the results of full-scale calibration using FEM are presented in WANG *et al.* (2004).

3. Geometry Related to CCNBD: 1 – Flattened Brazilian Disc (FBD)

The uncracked Brazilian disc was used by GUO *et al.* (1993) directly for mode I fracture toughness testing. This configuration was revisited by WANG *et al.* (2004) by introducing two parallel flat loading planes as shown in Figure 3(a). The geometry is analyzed by WANG *et al.* (2004) using the FEM and Figure 3(b) shows the dimensionless SIF for the configuration for different stages of the crack propagation. The experiment using this testing method can only be performed in a displacement controlled loading system and a typical load-displacement curve is given in Figure 3(c). The mode I fracture toughness K_{IC} can then be calculated as:

$$K_{IC} = \frac{P_{min}}{\sqrt{R} \cdot t} \cdot \Phi^*_{max}, \tag{7}$$

Table 2

Selected valuesof u and v

α_0	0.200	0.250	0.275	0.300	0.325	0.350	0.375	0.400
				u				
α_B								
0.680	0.2667	0.2704	0.2718	0.2744	0.2774	0.2807	0.2848	0.2888
0.720	0.2650	0.2683	0.2705	0.2727	0.2763	0.2794	0.2831	0.2871
0.760	0.2637	0.2668	0.2693	0.2719	0.2744	0.2781	0.2819	0.2860
0.800	0.2625	0.2657	0.2680	0.2706	0.2736	0.2772	0.2811	0.2845
0.840	0.2612	0.2649	0.2672	0.2699	0.2727	0.2763	0.2801	0.2831
0.880	0.2602	0.2642	0.2668	0.2691	0.2723	0.2754	0.2793	0.2816
0.920	0.2598	0.2634	0.2658	0.2684	0.2716	0.2747	0.2782	0.2811
0.960	0.2593	0.2633	0.2655	0.2685	0.2710	0.2746	0.2767	0.2799
1.000	0.2591	0.2630	0.2653	0.2679	0.2709	0.2738	0.2768	0.2786
				v				
0.680	1.7676	1.7711	1.7757	1.7759	1.7754	1.7741	1.7700	1.7666
0.720	1.7647	1.7698	1.7708	1.7722	1.7693	1.7683	1.7652	1.7617
0.760	1.7600	1.7656	1.7649	1.7652	1.7662	1.7624	1.7593	1.7554
0.800	1.7557	1.7611	1.7613	1.7603	1.7596	1.7561	1.7525	1.7512
0.840	1.7522	1.7547	1.7551	1.7548	1.7535	1.7499	1.7469	1.7473
0.880	1.7487	1.7492	1.7478	1.7487	1.7463	1.7452	1.7403	1.7434
0.920	1.7423	1.7446	1.7443	1.7432	1.7411	1.7389	1.7360	1.7363
0.960	1.7370	1.7373	1.7372	1.7346	1.7344	1.7309	1.7343	1.7331
1.000	1.7308	1.7307	1.7306	1.7297	1.7273	1.7270	1.7258	1.7302

where P_{min} is the local minimum load reading as shown in Figure 3(c), Φ^*_{max} is the maximum dimensionless SIF as shown in Figure 3(b). The main attraction of this method is the even simpler sample preparation as no slot needs to be cut at the center of the disc. However extensive analytical, numerical and experimental validation of the method is needed. A sensitive displacement loading requirement and low success rate for tests are also some of the disadvantages of this specimen geometry.

A geometry closely related to the flattened Brazilian disc is the modified ring (MR) configuration reported in FISCHER *et al.* (1996). The geometry is depicted in Figure 4 and the mode I fracture toughness value is obtained from the test graph shown in Figure 5. Finite-element analysis of the geometry is also presented by FISCHER *et al.* (1996).

4. Geometry Related to CCNBD: 2 – Semi-circular Specimen under Three-point Bend (SCB)

This geometry was proposed by CHONG (1987) and received extensive study by LIM *et al.* (1994 a,b,c) and experimental attention from KRISHNAN *et al.* (1998) and FUNATSU *et al.* (2004). The specimen and configuration for testing are shown in

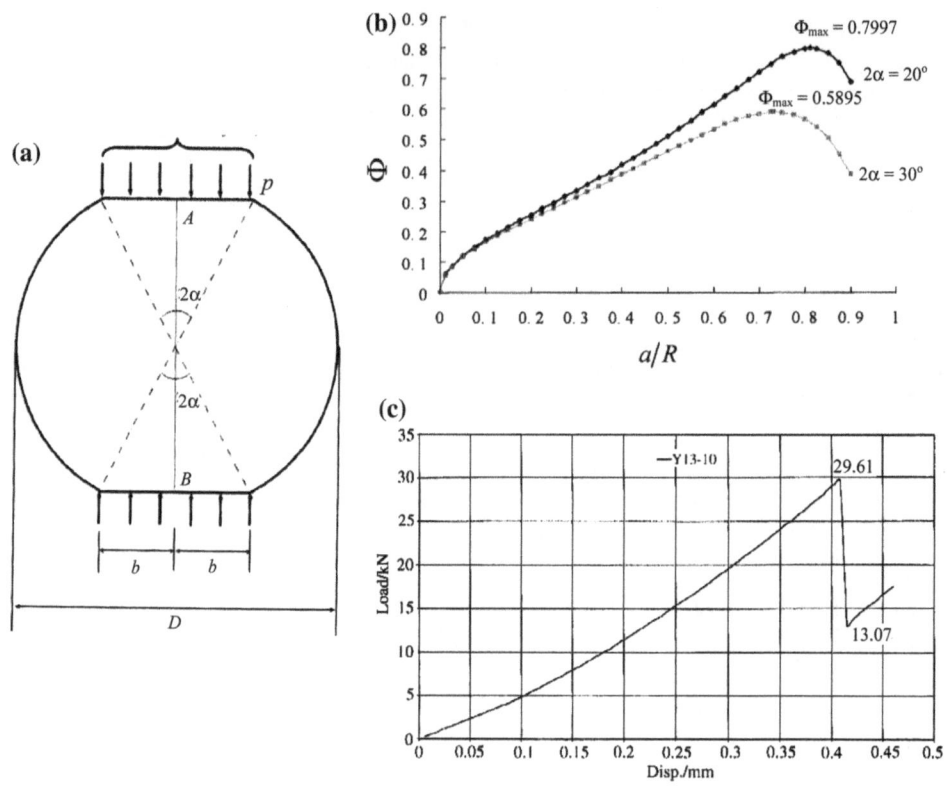

Figure 3
Flattened Brazilian disk, a) Geometrical dimensions, b) SIF c) L-D curve (WANG *et al.*, 2004).

Figure 6(a). The fracture toughness for mode I and II is then obtained by Equation (8) below (LIM *et al.*, 1995).

$$
\begin{cases}
K_{IC} = \dfrac{P}{2rt} \cdot \sqrt{\pi a} \cdot Y_I^* \\[2mm]
K_{IIC} = \dfrac{P}{2rt} \cdot \sqrt{\pi a} \cdot Y_{II}^*
\end{cases}
\tag{8}
$$

where Y_I^* and Y_{II}^* are dimensionless SIF values for the specimen with crack length a, t is the thickness, r is the radius and P is the recorded failure load. Figure 6 (b) gives the dimensionless SIF Y_I for pure mode I loading fixture and the dimensionless SIF Y_I and Y_{II} for mixed conditions are given in Figure 6(c). A typical load-displacement curve for the testing is shown in Figure 6(d).

This geometry, although retaining some merit such as being easily adaptable for mixed mode testing, loses some advantages of the CCNBD specimen. The most notable one is the low failure load which can sometimes be difficult to implement in practical experiments and is error prone, as it will be more difficult to obtain accurate

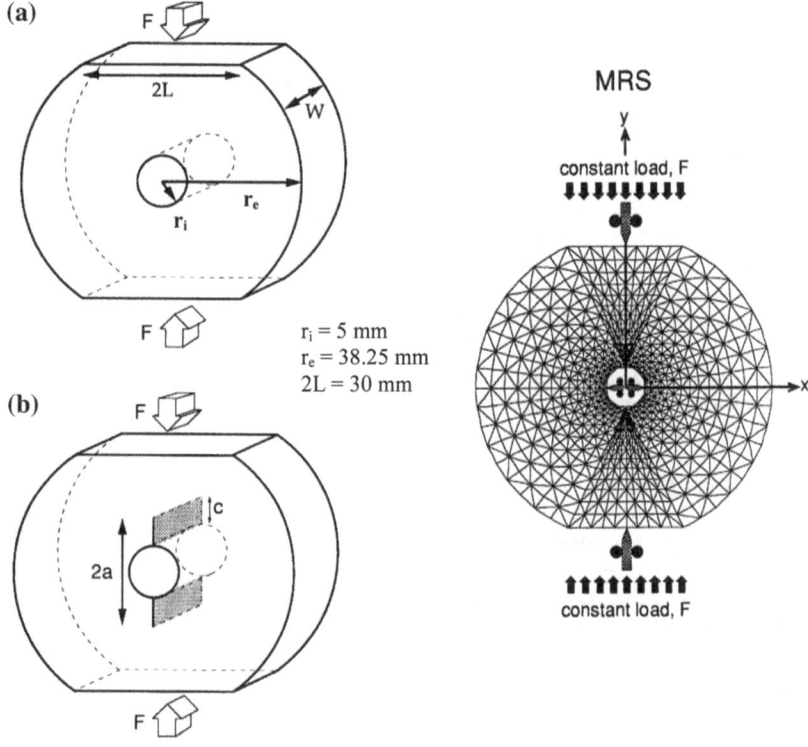

Figure 4
Modified ring (MR) specimen geometry (FISHER *et al.*, 1996).

load-displacement recordings. Another disadvantage will be the possible need for a pre-cracking process prior to testing although this can be easily overcome by introducing a Chevron-notch instead of a straight-through crack.

5. Geometry Related to CCNBD: 3–Double-edge Cracked Brazilian Disc (DECBD)

This geometry has been recently proposed by CHEN *et al.* (2001) for mode I fracture toughness testing. The specimen and loading configuration is as shown in Figure 7(a). The specimen is studied in CHEN *et al.* (2001) using a weight function and the FEM method. The dimensionless mode I SIF for the specimen for different crack inclination angles is given in Figure 7 (b).

In the authors' opinion, this geometry has great potential. An initial impression of the specimen is that it will retain most of the merit of the CCNBD specimen, with the improvement of an easier sample preparation. The specimen, if configured properly (i.e., certain crack inclination angle), will act like a shear box which will

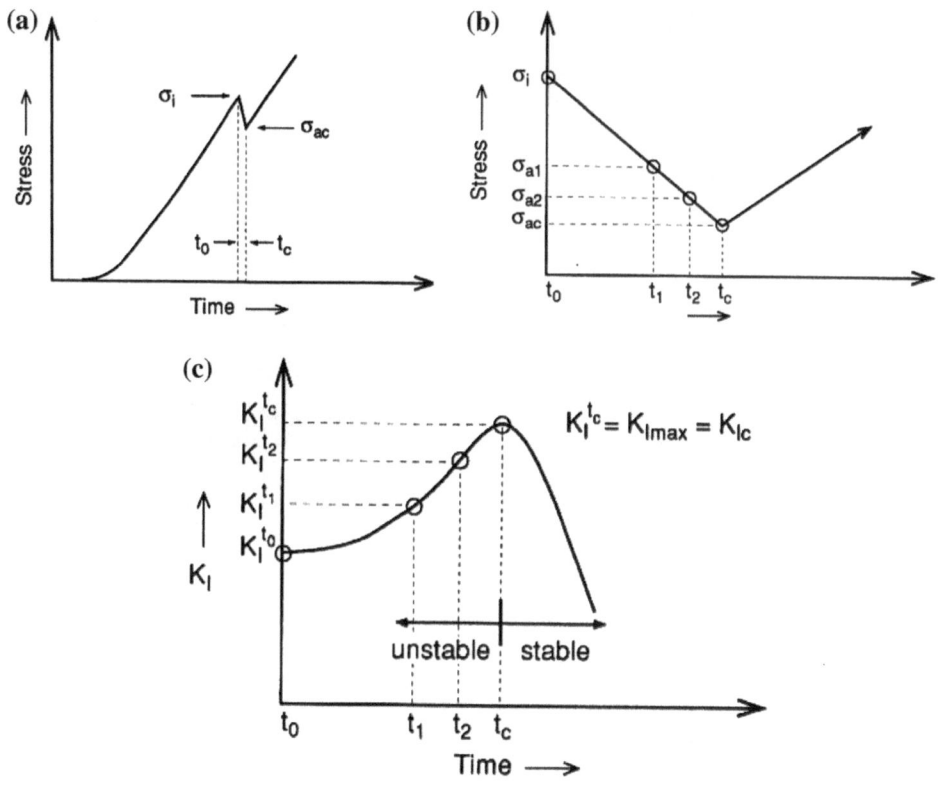

Figure 5
Fracture toughness determination using MR testing (FISHER *et al.*, 1996).

make it extremely suitable for mixed modes I and II fracture toughness testing. Certainly the geometry is still in its early stages of development and we cannot be sure at this stage if these claims will be correct. Some work for this geometry is imminent, for example, mode II SIF evaluation and extensive experimental testing validation.

6. Conclusion

Since its introduction, fracture toughness testing using the CCNBD specimen has attracted considerable attention in the rock fracture research community. Advantages of using the method have started to be realized, which include, easier sample preparation, much higher failure load, and hence simpler and less error-prone testing procedure. Another superb advantage of the specimen is that it can easily be adapted for pure Model II and mixed Modes I and II fracture toughness testing, which has

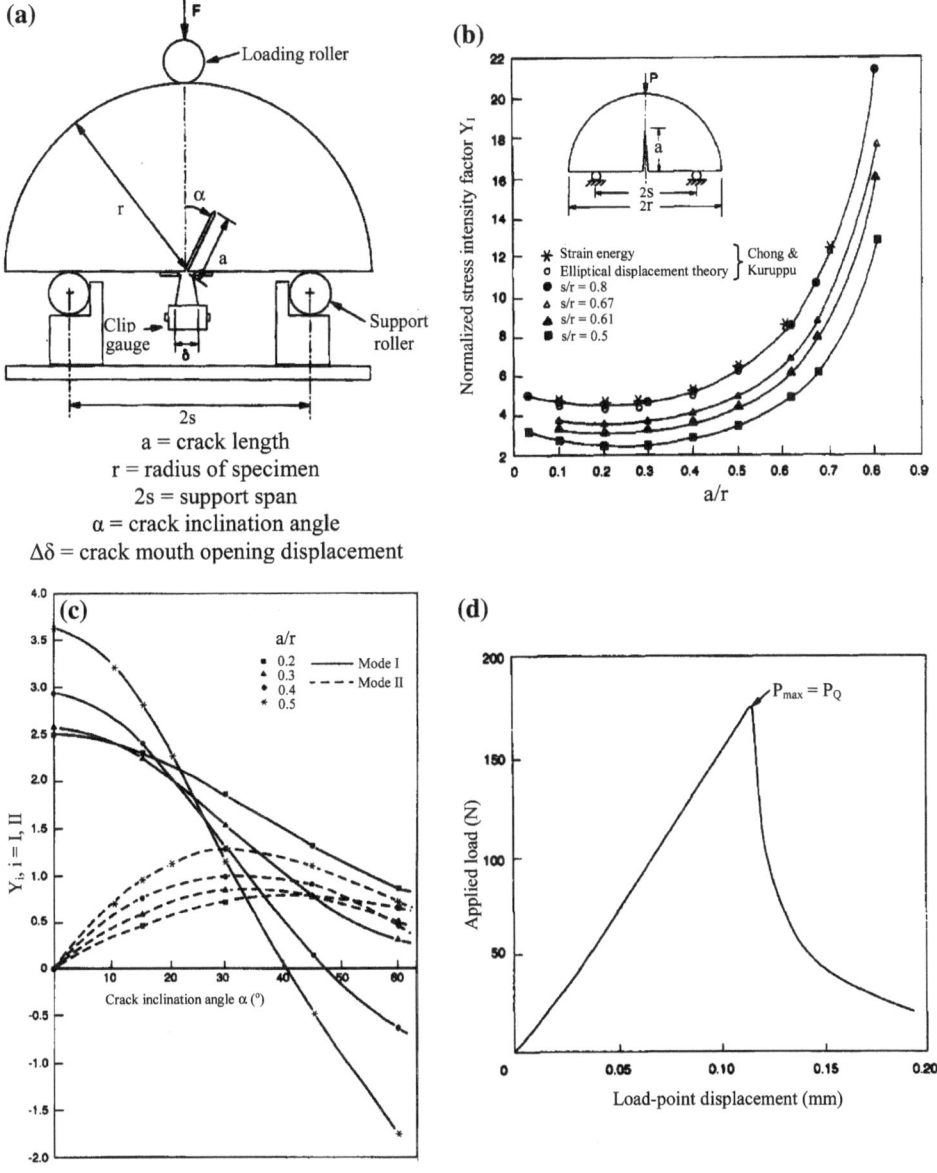

Figure 6
SCB a) Configuration, b) Mode I SIF c) Mixed-mode SIF d) L-D curve (LIM et al., 1994a,b).

attracted extensive research in the past decade and is expected to remain as a very active research topic for the foreseeable future. It may be necessary to revise the dimensionless SIF values for a future release of the suggested method to incorporate some recent developments. More research and input from different sources need to be coordinated.

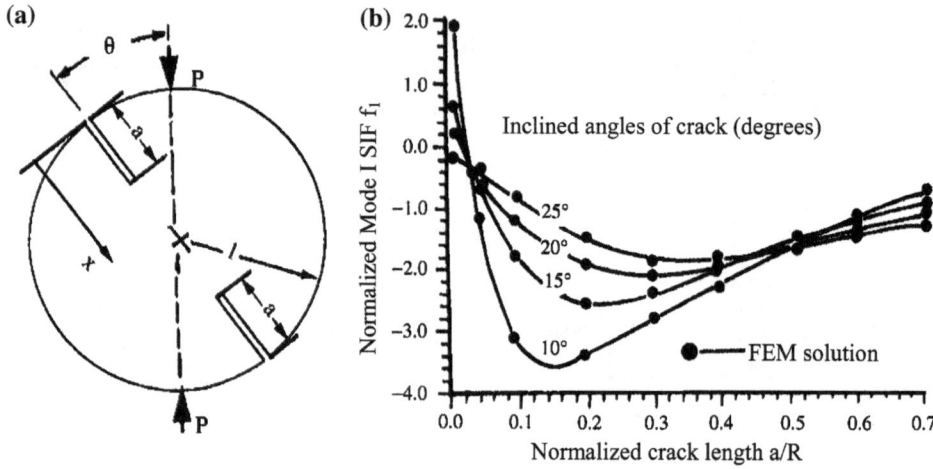

Figure 7
Double edge Brazilian disk - DECBD a) Geometry b) Mode I SIF (CHEN *et al.*, 2001).

Several specimen geometries closely related to the CCNBD have also been developed and used, however in the authors' opinion, none of them have the same unique desirable combinations of features and advantages as the CCNBD geometry. The DECBD specimen, however, is believed to have great potential for further development.

REFERENCES

AL-SHAYEA, N.A., KHAN, K., and ABDULJAUWAD, S.N. (2000), *Effects of confining pressure and temperature on mixed-mode (I-II) fracture toughness of a limestone rock*, Int. J. Rock. Mech. and Min. Sci. *37*, 629–643.

CHANG, S.H., LEE, C.I., and JEON, S. (2002), *Measurement of rock fracture toughness under modes I and II and mixed-mode conditions by using disc-type specimens*, Engin. Geology *66*, 79–97.

CHEN, F., SUN, Z., and XU, J. (2001), *Mode I fracture analysis of the double edge cracked Brazilian disk using a weight function method*, Int. J. Rock. Mech. and Min. Sci. *38*, 475–479.

CHEN, J.F. (1990), *The development of chevron-notched Brazilian disk method for rock fracture toughness measurement*, Proceeding of 1990 SEM Spring Conf. on Experimental Mechanics, Albuquerque, USA, pp. 18–23.

CHONG, K.P., KURUPPU, M.D., and KUSZMAUL, J.S. (1987), *Fracture toughness determination of rocks with core-based specimens*, Proceedings of SEM/RILEM Internat. Conf. on Fracture of Concrete and Rocks, Texas, (Shah S.P. and Swartz, S.E. eds.), pp. 13–25.

DWIVEDI, R.D., SONI, A.K., GOELL, R.K., and DUBE, A.k. (2000), *Fracture toughness of rock under sub-zero temperature conditions*, Int. J. Rock. Mech. and Min. Sci. *37*, 1267–1275.

FISCHER, M.P., ELSWORTH, D., ALLEY, R.B., and ENGELDER, T. (1996), *Finite element analysis of the modified ring test for determining mode I fracture toughness*, Int. J. Rock. Mech. and Min. Sci. *33*, 1–15.

FOWELL, R.J. and XU, C. (1994), *The use of the cracked Brazilian disc geometry for rock fracture investigation*, Int. J. Rock. Mech. and Min. Sci. *31*, 571–579.

FUNATSU, T., SETO, M., SHIMADA, H., MATSUI, K., and KURUPPU, M. (2004), *Combined effects of increasing temperature and confining pressure on the fracture toughness of clay bearing rocks*, Int. J. Rock. Mech. and Min. Sci., *41*, 927–938.

GUO, H., AZIZ, N.I., and SCHMIDT, L.C. (1993), *Rock fracture toughness determination by Brazilian test*, Engin. Geology *33*, 177–188.

KRISHNAN, G.R., ZHAO, X.L., ZAMAN, M., and ROEGIERS, J.C. (1998), *Fracture toughness of a soft sandstone*, Int. J. Rock. Mech. and Min. Sci. *35*, 695–710.

ISRM COMMISSION ON TESTING METHODS (1995), *Suggested method for determining Mode I fracture toughness using cracked Chevron-notched Brazilian disc (CCNBD) specimens*, R.J. FOWELL (co-ordinator), Int. J. Rock. Mech. and Min. Sci. *32*, 57–64.

LIM, I.L., JOHNSTON, I.W., CHOI, S.K., and BOLAND, J.N. (1994a), *Fracture testing of a soft rock with semi-circular specimens under three-point bending part 1–mode I*, Int. J. Rock. Mech. and Min. Sci. *31*, 185–197.

LIM, I.L., JOHNSTON, I.W., CHOI, S.K., and BOLAND, J.N. (1994b), *Fracture testing of a oft rock with semi-circular specimens under three-point bending part 2–mixed modes*, Int. J. Rock. Mech. and Min. Sci., *31*, 199–212.

LIM, I.L., JOHNSTON, I.W., and CHOI, S.K. (1994c), *Assessment of mixed-mode fracture toughness testing methods for rock*, Int. J. Rock. Mech. and Min. Sci. *31*, 265–272.

XU, C. and FOWELL, R.J. (1994), *The crack chevron-notched Brazilian dist test–geometrical considerations for practical rock fracture toughness measurement*, Int. J. Rock. Mech. and Min. Sci. *30*, 821–824.

XU, C. and FOWELL, R.J. (1993), *Experimental validation of the cracked chevron-notched Brazilian disc geometry specimen for rock fracture toughness testing*, LUMA Magazine, Leeds University Mining Association, pp. 57–68.

WANG, Q.Z. (1998), *Stress intensity factors of the ISRM suggested CCNBD specimen used for mode I fracture toughness determination*, Int. J. Rock. Mech. and Min. Sci. *35*, 977–982.

WANG, Q.Z., JIA, X.M., and WU, L.Z. (2004), *Wide range stress intensity factors for the ISRM suggested method using CCNBD specimens for the rock fracture toughness tests*, Int. J. Rock. Mech. and Min. Sci. *41*, 709–716.

WANG, Q.Z., JIA, X.M., KOU, S.Q., ZHANG, Z.X., and LINDQVIST, P.-A. (2004), *The flattened Brazilian disc specimen used for elastic modulus, tensile strength and fracture toughness of brittle rocks: analytical and numerical results*, Int. J. Rock. Mech. and Min. Sci. *41*, 245–253.

(Received March 18, 2005, revised September 19, 2005, accepted October 21, 2005)
Published Online First: May 12, 2006

To access this journal online:
http://www.birkhauser.ch

Pure appl. geophys. 163 (2006) 1059–1072
0033–4553/06/061059–14
DOI 10.1007/s00024-006-0060-2

❙ Pure and Applied Geophysics

Cohesive Crack Analysis of Toughness Increase Due to Confining Pressure

KAZUSHI SATO,[1] and TOSHIYUKI HASHIDA[2]

Abstract:—Apparent fracture toughness in Mode I of microcracking materials such as rocks under confining pressure is analyzed based on a cohesive crack model. In rocks, the apparent fracture toughness for crack propagation varies with the confining pressure. This study provides analytical solutions for the apparent fracture toughness using a cohesive crack model, which is a model for the fracture process zone. The problem analyzed in this study is a fluid-driven fracture of a two-dimensional crack with a cohesive zone under confining pressure. The size of the cohesive zone is assumed to be negligibly small in comparison to the crack length. The analyses are performed for two types of cohesive stress distribution, namely the constant cohesive stress (Dugdale model) and the linearly decreasing cohesive stress. Furthermore, the problem for a more general cohesive stress distribution is analyzed based on the fracture energy concept. The analytical solutions are confirmed by comparing them with the results of numerical computations performed using the body force method. The analytical solution suggests a substantial increase in the apparent fracture toughness due to increased confining pressures, even if the size of the fracture process zone is small.

Key words: Rock, fracture toughness, cohesive crack model, confining pressure, stress intensity factor.

1. Introduction

It has been reported that fracture toughness of rocks increases with increasing confining pressure (SCHMIDT and HUDDLE, 1977; ABOU-SAYED, 1978; FUNATSU *et al.*, 2004). Fracture toughness is often expressed in terms of the stress intensity factor required for crack propagation. In the fracture of a linear elastic solid, the fracture toughness is assumed to be an inherent material property and independent of loading configurations. In rocks, however, the stress intensity factor for crack propagation varies with confining pressure. Therefore, the fracture toughness of rocks should be viewed as an apparent material property.

[1]Miyagi National College of Technology, 48 Aza–Nodayama, Shiote, Medeshima, Natori, 981–1239, Japan. E-mail: kazushi@miyagi-ct.ac.jp
[2]Fracture and Reliability Research Institute, Tohoku University, 01 Aoba, Aramaki, Aoba-ku, Sendai, 980–8579, Japan.

It is well known that the fracture process zone that is formed prior to crack growth is significantly large in size compared to the specimen dimensions for many rock types (LABUZ et al., 1987). Therefore, a suitable model for the fracture process zone is necessary to evaluate the crack growth on the basis of laboratory experiments. A Barenblatt-type cohesive crack model has been applied to express the formation of the fracture process zone in rocks (HASHIDA, 1990; SATO et al., 1995). HASHIDA et al. (1993) have reported that the crack growth behavior of Iidate granite under confining pressure up to 26.5 MPa can be predicted using the cohesive crack model (tension-softening model). In the tension-softening model, the fracture process zone is represented as a crack subjected to a cohesive stress which depends on the crack opening displacement. The relation between the cohesive stress and the crack opening displacement is referred to as the tension–softening curve. The critical fracture energy is provided by the area under the tension-softening curve (HILLER-BORG, 1983). HASHIDA et al. (1993) have concluded that the tension-softening curve is independent of confining pressure. Therefore, it can be seen that the tension-softening curve is the inherent material property to be used to evaluate the crack growth under confining pressure. The cohesive crack model is expected to provide a useful tool for predicting the apparent fracture toughness.

In this study, the apparent fracture toughness under confining pressure is examined using a cohesive crack model. A two-dimensional fluid-driven fracture with cohesive zone under confining pressure is analyzed. Two types of cohesive stress distributions are studied. The size of cohesive zone is assumed to be negligibly small. The apparent fracture toughness is measured in terms of the stress intensity factor induced by the fluid pressure. The theoretical model is verified by comparing it with numerical results obtained by the body force method.

2. Problem Formulation

The problem to be analyzed is a fluid-driven fracture of a two-dimensional crack with cohesive zones under confining pressure in an infinite body, as shown in Figure. 1. Tension is taken to be positive. The plane strain condition is assumed.

The fracture process zone is represented by the cohesive zone. The cohesive zone length is R. The fluid pressure P is applied to the crack surface except within the cohesive zone. The pressurized crack length is $2a$, and the total crack length including the cohesive zone is $2c$. The cracked body is subjected to the confining pressure S at a far distance.

Two types of cohesive stress distributions were analyzed, as shown in Figure. 2. Case 1 represents a model of constant cohesive stress, while case 2 corresponds to a linearly decreasing stress model.

When the fracture process zone is negligibly small, the critical stress intensity factor K_c for the case of constant cohesive stress is

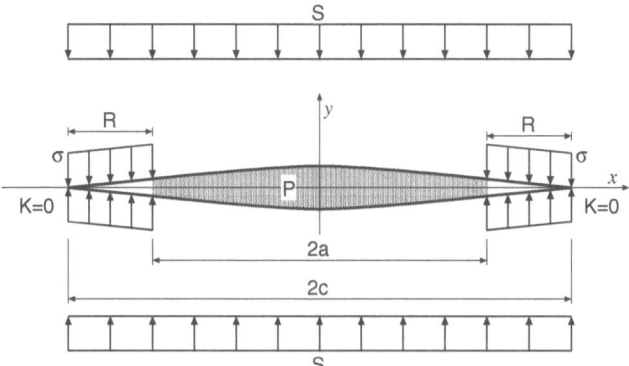

Figure 1
Fluid-driven crack with cohesive zone under confining stress.

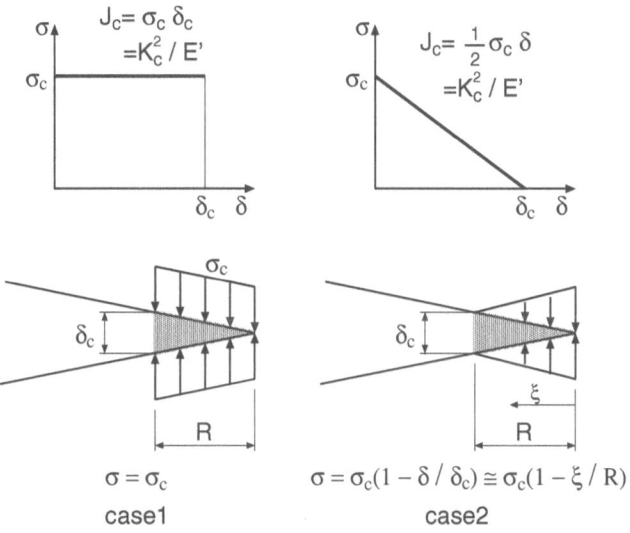

Figure 2
Cohesive stress distributions analyzed in this study: (a) constant cohesive stress and (b) linearly decreasing stress. The cohesive zone length is determined by the critical crack opening displacement δ_c.

$$\frac{K_c^2}{E'} = \sigma_c \delta_c, \tag{1}$$

where σ_c is the peak stress of the cohesive stress, δ_c is the critical crack opening displacement and E' is the effective Young's modulus. For the case of linearly decreasing stress, the critical stress intensity factor is given by

$$\frac{K_c^2}{E'} = \frac{1}{2}\sigma_c \delta_c. \tag{2}$$

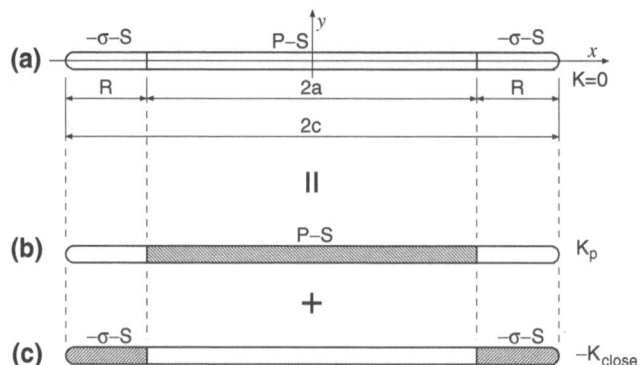

Figure 3
Problem analyzed by superimposing two subproblems with respect to applied pressure and cohesive stress:
(a) original problem, (b) subproblem of applied pressure and (c) subproblem of cohesive stress.

It is assumed that the σ–δ relation is maintained under confined conditions as well as atmospheric (no confining pressure) condition. Therefore, K_c is the inherent material property and should be viewed as the fracture toughness of the rock.

This study analyzes the stress intensity factor required for crack growth when the fracture process zone is assumed to be negligibly small. This stress intensity factor corresponds to the apparent fracture toughness.

3. Theoretical Analysis

3.1. Stress Intensity Factor Based Analysis

This section presents the analytical results based on the stress intensity factor for both the cases of constant and linearly decreasing cohesive stress.

In order to derive a predictive model for the apparent fracture toughness, we considered subproblems with respect to the applied pressure and to the closing stress, as shown in Figure. 3. The stresses acting on the crack in the problem to be analyzed, as shown in Figure. 1, are the fluid pressure P, the cohesive stress σ and the confining stress S. The applied pressure that is the opening stress is the excess pressure $(P - S)$ acting on the pressurized region $2a$. The closing stress is the composition of the cohesive stress and the confining stress $(-\sigma - S)$ acting on the cohesive zone R.

The first subproblem is that of the crack, length $2c$, subjected to the applied pressure acting on the pressurized interval $2a$. The second subproblem is that of the crack subjected to the closing stress acting on the cohesive zone R. The original problem can be analyzed by superimposing the solution of the two subproblems. In the original problem, the stresses at the crack tip $(x = \pm c)$ have to be finite. This requires that the stress intensity factor at the crack tip equals to zero. Based on this condition, the apparent fracture toughness that is produced by the applied pressure can be determined.

When the length of the cohesive zone is sufficiently small in comparison to the crack length, the stress intensity factor K_P produced by the applied pressure (see Figure. 3b) is

$$K_P = \Delta P \sqrt{\pi c} = \Delta P \sqrt{\pi(a + R)} \tag{3}$$

$$= \Delta P \sqrt{\pi a}\left(1 - \frac{R}{2a} + \cdots\right) \tag{4}$$

where $\Delta P = P - S$ is the excess pressure. For $R/a \ll 1$,

$$K_P = \Delta P \sqrt{\pi a}. \tag{5}$$

K_P corresponds to the apparent fracture toughness, as mentioned above.

Considering a constant cohesive stress σ_c acting over the length R at the tip of a semi–infinite crack, the stress intensity factor K_{close} produced by the closing stress (TADA et al., 1973a) is

$$K_{close} = \sqrt{\frac{8}{\pi}}(\sigma_c + S)\sqrt{R}. \tag{6}$$

The condition that the stress at the tip of the cohesive zone has to be finite requires

$$K_P - K_{close} = 0. \tag{7}$$

Substituting equation (6) into equation (7) yields

$$R = \frac{\pi}{8}\left(\frac{K_P}{\sigma_c + S}\right)^2. \tag{8}$$

The crack opening displacement at $x = \pm a$ due to the excess pressure ΔP (TADA et al., 1973a) is

$$\delta_P = \frac{2(1 - v)}{G}\Delta P\sqrt{c^2 - a^2} \tag{9}$$

$$= \frac{2(1 - v)}{G}\Delta P\sqrt{2aR}\left(1 - \frac{R}{4a} + \cdots\right). \tag{10}$$

For $R/a \ll 1$,

$$\delta_P = \frac{2(1 - v)}{G}\Delta P\sqrt{2aR}, \tag{11}$$

where v is Poisson's ratio and G is the shear modulus. For a semi-infinite crack, the crack opening displacement due to the closing stress at the edge of the cohesive zone is

$$\delta_{close} = \frac{2(1-v)}{G} \cdot \frac{2}{\pi}(\sigma_c + S)R. \tag{12}$$

The overall crack opening displacement at $x = \pm a$ is

$$\delta_{|\pm a} = \delta_P - \delta_{close}. \tag{13}$$

The crack growth takes place when the crack opening displacement at $x = \pm a$ reaches the critical crack opening displacement δ_c. Considering the crack growth condition, $\delta_{|\pm a}$ in equation (13) is replaced by δ_c. Then, substituting equations (1), (8), (11) and (12) into equation (13), yields

$$\frac{K_P^2}{K_c^2} = 1 + \frac{S}{\sigma_c}, \tag{14}$$

where $E' = E/(1 - v)$ and $G = E/2(1 + v)$ were used.

Equation (14) suggests that the stress intensity factor required for crack growth under confining pressure should exceed the inherent material property K_c due to the presence of the cohesive zone. Equation (14) provides a predictive model for the apparent fracture toughness when the cohesive stress is assumed to be constant.

Next, the case where the cohesive stress decreases linearly from the crack tip is considered.

In this case, in order to evaluate the stress intensity factor due to the cohesive stress, the distribution of cohesive stress along the crack should be given. However, the cohesive stress distribution is not linear since the crack opening displacement is not proportional to the distance from the crack tip. Therefore, the profile of crack opening displacement is needed for a rigorous analysis. Nonetheless, in this study, a linear distribution of cohesive stress along the crack is assumed for simplicity as

$$\sigma = \sigma_c\left(1 - \frac{\xi}{R}\right), \qquad (0 \leq \xi \leq R), \tag{15}$$

where ξ is the distance from the crack tip. This approximation results in an error of about 10%, as described below. For this case, the stress intensity factor produced by the cohesive stress is obtained by integrating the stress intensity factor for the problem of the semi-infinite crack with a point load (TADA et al., 1973b)

$$K_{close} = \frac{4}{3\pi} \sigma_c\sqrt{2\pi R}. \tag{16}$$

Therefore, the cohesive zone length is

$$R = \frac{\pi}{8} \frac{K_P^2}{\left(\frac{2}{3}\sigma_c + S\right)^2}, \tag{17}$$

where K_P is given by equation (5).

The crack opening displacement due to the cohesive stress is obtained by integrating the crack opening displacement for the semi–infinite crack with a point load (TADA et al., 1973b)

$$\delta_{close} = \frac{2(1-v)}{G} \cdot \frac{2}{3\pi} \sigma_c R. \tag{18}$$

K_c is given by equation (2), then the result is

$$\frac{K_P^2}{K_c^2} = \frac{2\left(\frac{2}{3} + \frac{S}{\sigma_c}\right)^2}{1 + \frac{S}{\sigma_c}}. \tag{19}$$

Equation (19) indicates that K_P^2/K_c^2 becomes 8/9 when the confining pressure is zero ($S = 0$). This deviation from unity may be due to the approximation of the cohesive stress distribution along the crack.

3.2. Fracture Energy Based Analysis

This section describes an analysis based on the fracture energy concept. The fundamental idea is that the energy required for opening the crack under confining pressure is the sum of the two components needed to overcome the confining pressure as well as the cohesive stress.

When the cohesive zone length is small, for the constant cohesive stress case, the total fracture energy is

$$G_{total} = \sigma_c \delta_c + S\delta_c = \frac{K_P^2}{E'}. \tag{20}$$

When the confining pressure is zero, the total fracture energy is given by

$$G_{total} = \sigma_c \delta_c = \frac{K_c^2}{E'}. \tag{21}$$

Dividing (20) by (21), yields

$$\frac{K_P^2}{K_c^2} = 1 + \frac{S}{\sigma_c}. \tag{22}$$

This equation is in agreement with equation (14), which is derived based on the stress intensity factor. On the basis of the fracture energy approach, a solution for a general stress distribution (nonlinear) also can be obtained.

When the cohesive stress is provided by

$$\sigma = \sigma_c \left(1 - \frac{\delta}{\delta_c}\right)^n, \tag{23}$$

the fracture energy can be calculated for unconfined conditions by the J–integral

$$\frac{K_c^2}{E'} = \int_0^{\delta_c} \sigma d\delta = \frac{\sigma_c \delta_c}{n+1}. \tag{24}$$

Therefore, under the confining pressure, the total fracture energy is

$$\frac{K_P^2}{E'} = \frac{\sigma_c \delta_c}{n+1} + S\delta_c. \tag{25}$$

Then, the apparent fracture toughness can be obtained as

$$\frac{K_P^2}{K_c^2} = 1 + (n+1)\frac{S}{\sigma_c}. \tag{26}$$

When n is zero (constant cohesive stress), equation (26) coincides with equation (14). When n is unity, equation (26) corresponds to the linear decreasing cohesive stress case.

The predictive model of equation (26) holds true independent of the loading configuration. The only requirement is that the fracture process zone is small enough in comparison to the crack length.

Next, the result from the fracture energy approach is compared with that obtained based on the stress intensity factor. Thus, three equations (14), (19) and (26) deduced in this study are compared in Figure. 4. In this figure, equation (26) with $n = 1$ is compared with equation (19). The increasing tendency of equation (26) is consistent with equation (19), although equation (19) shows slightly smaller values due to the

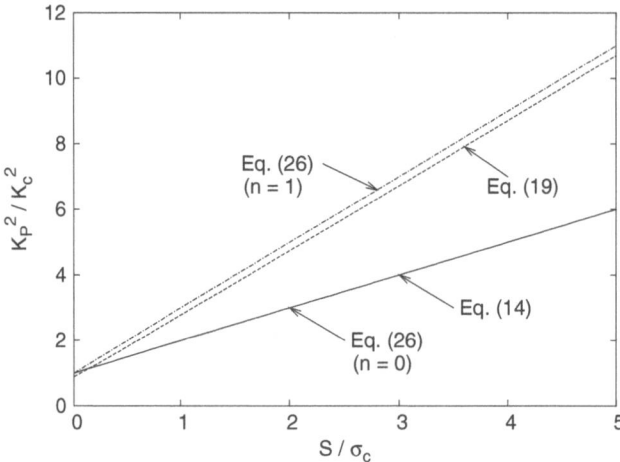

Figure 4
Comparison of derived equations. Equation (14) and equation (26) with $n = 0$ corresponds to the case of the constant cohesive stress. equation (19) and equation (26) with $n = 1$ corresponds to the case of the linear cohesive stress.

approximation of the stress distribution along the crack. The formulation obtained from the fracture energy approach is expected to provide a general model for predicting the increase of the apparent fracture toughness under confining pressure.

The apparent fracture toughness significantly increases with the confining pressure. This result demonstrates that the fracture toughness increases when the fracture process zone exists at the crack tip, even if the size of the fracture process zone is negligibly small.

4. Numerical Analysis

4.1. Numerical Procedure

In order to examine the validity of the derived equations, numerical analyses were performed in this study.

Two-dimensional numerical computations were also carried out, based on the cohesive crack model, in addition to the theoretical analysis described in the previous sections. The body force method (NISHITANI, 1968; NISHITANI et al., 1990) was used to determine the stress intensity factor and the inner pressure. The body force method allows us to examine the effect of confining pressure on the apparent fracture toughness, under the condition in which the extent of the cohesive zone is not negligibly small. The problem configuration is shown in Figure. 1. In the numerical computations, the size of the cohesive zone and the cohesive stress were predetermined and the inner pressure was adjusted to maintain the stress intensity factor zero at the crack tip ($x = \pm c$). Therefore, the fracture toughness K_c was varied by changing the size of the cohesive zone and cohesive stress in the computations. Here, K_c is evaluated by equations (1) or (2). The size parameter of the cohesive zone, R/c in the numerical computations was in the range of $0.001 \cong 0.5$. As seen in the size parameter, the numerical computations were conducted to include the condition in which the size of the cohesive zone is nonnegligible with respect to the crack length.

4.2. Numerical Results and Discussion

Figures 5 and 6 show the dependency of K_c on the cohesive zone size obtained by numerical calculations. Figure 5 is obtained for the constant cohesive stress condition and Figure. 6 is for the linearly decreasing stress condition. The fracture toughness K_c increases with the increase of the cohesive zone size. For the same cohesive zone length, K_c increases with increasing confining pressure.

The numerical results are compared with the analytical solutions. The dependency of apparent fracture toughness on the confining pressure is shown in Figures. 7(b) and 8(b), for the constant cohesive stress condition and for the linearly decreasing stress condition, respectively. These figures were constructed by selecting

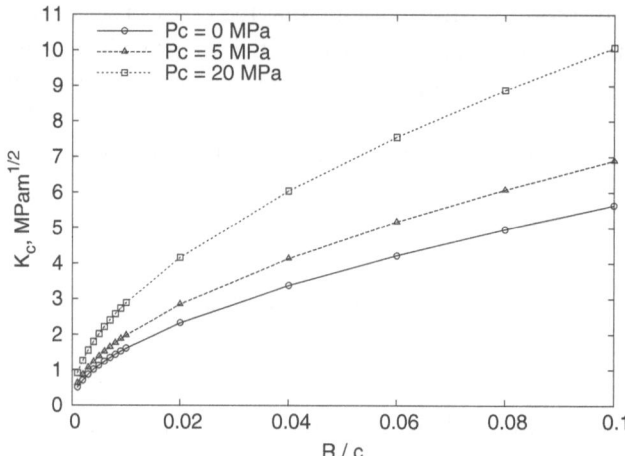

Figure 5
Dependency of K_c on the cohesive zone size R for the case of constant cohesive stress.

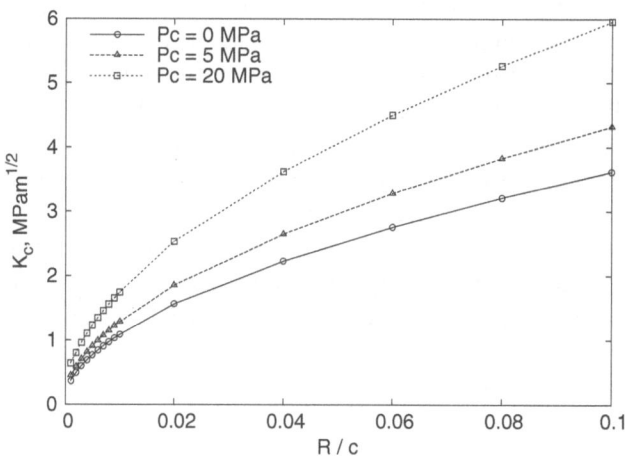

Figure 6
Dependency of K_c on the cohesive zone size R for the case of linear cohesive stress.

the set of data presented in Figures. 5 and 6 which provided the similer same K_c value. The analytical results given by equations (14) and (26) are indicated as solid lines in Figures. 7(b) and 8(b), respectively. In the case of the linearly decreasing cohesive stress, the parameter n in equation (26) is taken to be 1.0. The apparent fracture toughness K_P is evaluated by equation (5).

Figures 7(a) and 8(a) show the cohesive zone size for different K_c values and confining pressure conditions. It can be seen that the numerical result obtained for the lower K_c value and smaller R/c is very close to the trend predicted by the

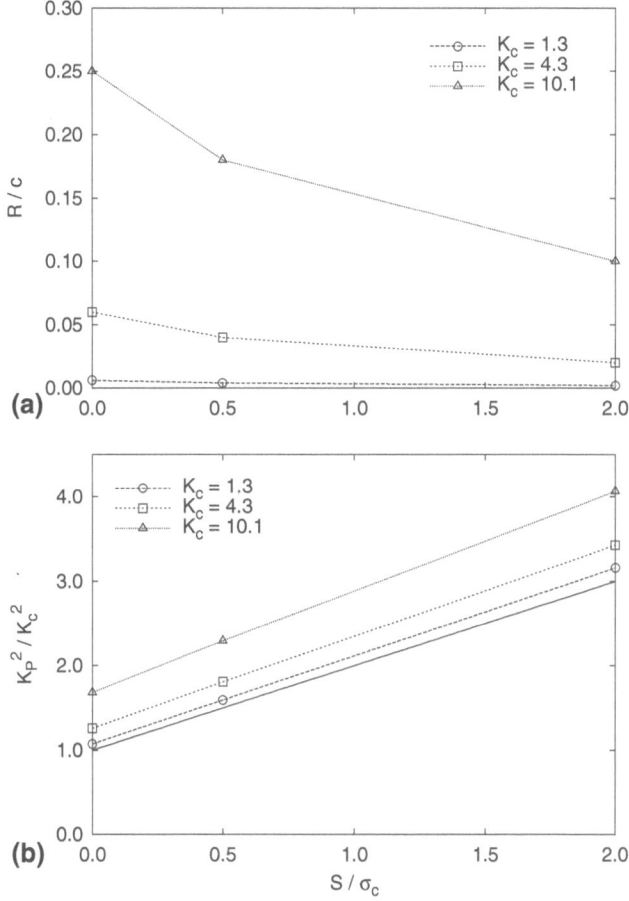

Figure 7
Comparison of numerical results and the analytical solution for the case of constant cohesive stress: (a) Size of cohesive zone and (b) dependency of apparent fracture toughness on confining pressure.

theoretical analysis. This evidence demonstrates the validity of the theoretical analysis for the condition of the negligible cohesive zone size. When the K_c value increases and the cohesive zone size becomes larger with respect to the crack length, the numerical result tends to deviate from the theoretical result. It is very interesting to note that the slope in the plot of $K_P^2/K_c^2-S/\sigma_c$ shows no drastic change even when the cohesive zone size becomes larger.

The apparent fracture toughness significantly increases with increasing confining pressure as shown in Figs. 7 and 8. MATSUKI et al. (1995) have stated that the reason for the increase of apparent fracture toughness is the closure of pre-existing cracks involved in the rock specimen. This causes the rock specimen to become less damaged material. RUBIN (1993), and FIALKO and RUBIN (1997) have reported that

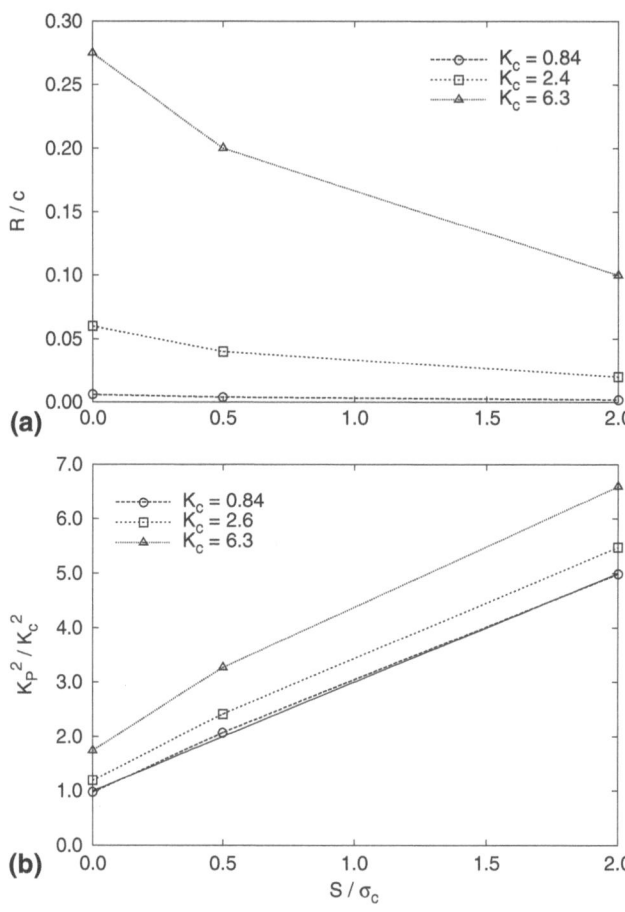

Figure 8
Comparison of numerical results and the analytical solution for the case of linear cohesive stress: (a) Size of cohesive zone and (b) dependency of apparent fracture toughness on confining pressure.

the high tensile stress field appears off crack plane when the confining pressure becomes sufficiently high. The magnitude of the tensile stress can be higher than the peak cohesive stress, namely the tensile strength of rocks. They have concluded that the increase of the apparent fracture toughness is the result of inelastic deformation near the crack tip region due to this high tensile stress. However, no quantitative model has been proposed for predicting the increase of apparent fracture toughness under higher confining pressures. In contrast with the above-mentioned previous studies, the present study points out the different mechanism for the increase of apparent fracture toughness. Based on our theoretical analysis, we propose that the apparent toughness increase may be due to the additional energy required to open up the crack frank within the cohesive zone against the confining pressure. Furthermore, a quantitative model for the apparent toughness increase has been derived based on the theoretical analysis.

5. Concluding Remarks

The present study provides a predictive model for the apparent fracture toughness, which was derived on the basis of a cohesive crack model. A two–dimensional fluid-driven fracture with a cohesive zone under confining pressure was analyzed. Two types of cohesive stress distributions were studied. Furthermore, the problem for more general cohesive stress distribution was analyzed based on the fracture energy concept. The size of the cohesive zone was assumed to be negligibly small. The apparent fracture toughness was measured as the stress intensity factor induced by the fluid pressure. The derived results were compared with numerical results.

The analytical solution suggests a substantial increase in the apparent fracture toughness due to the increased confining pressures, even if the size of the fracture process zone is small. The apparent fracture toughness is shown to be a function of the tensile strength (the peak value of the cohesive stress) and confining pressure. Their functional form depends on the cohesive stress distribution. Consequently, the evaluation of the fracture process zone in rocks is of importance to fracture problems.

REFERENCES

ABOU-SAYED, A.S. (1978), *An experimental technique for measuring the fracture toughness of rock under downhole stress condition*, VDI-Berichte *313*, 819–824.

FIALKO, Y.A. and RUBIN, A.M., (1997), *Numerical simulation of high-pressure rock tensile fracture experiments: Evidence of an increase in fracture energy with pressure?* Int. J. of Geophys. Res. *102*-B3, 5231–5242.

FUNATSU, T., SETO, T., SHIMADA, H., MATSUI, K. and KURUPPU, M. (2004), *Combined effects of increasing temperature and confining pressure on the fracture toughness of clay bearing rocks,* Int. J. of Rock. Mech. and Min. Sci. *41*, 927–938.

HASHIDA, T. *Evaluation of fracture processes in granites based on the tension-softening model. In Micromechanics of Failure of Quasi-Brittle Materials* (eds. Mihashi, H., Takahashi, H. and Wittmann, F. H.) (Elsevier Applied Science, 1990) pp. 233–243.

HASHIDA, T., OGHIKUBO, H., TAKAHASHI, H. and SHOJI, T. (1993), *Numerical simulation with experimental verification of the fracture behavior in granite under confining pressures based on the tension-softening model,* Int. J. of Fracture 59, 227–244.

HILLERBORG, A. *Analysis of one single crack, In Fracture Mechanics of Concrete* (ed. Wittmann F.H.) (Elsevier Science Publishers B. V., Amsterdam, 1983) pp. 223–250.

LABUZ, J. F., SHAH, S. P. and DOWDING, C. H. (1987), *The fracture process zone in granite: Evidence and effect,* Int. J. Rock. Mech. Min. Sci. and Geomech. Abstr. *24*-4, 235–246.

MATSUKI, K., KANEKO, T. and SATO, T. (1995), *K-resistance curve of rock under confining pressure,* Shigen-to-Sozai (in Japanese) *111*, 755–760.

NISHITANI, H. (1968), *The two-dimensional stress problem solved using an electronic digital computer,* Bull. Japan Soc. Mech. Engrs. *11*-43, 15–23.

NISHITANI, H., SAIMOTO, A. and NOGUCHI, H. (1990), *Versatile method of analysis of two-dimensional elastic problem by body force method* (1st Report, Basic Theory), Trans. JSME(A) (in Japanese) *56*-530, 2123–2129.

RUBIN, A.M. (1993), *Tensile fracture of rock at high confining pressure: Implications for dike propagation,* Int. J. of Geophys. Res. *98*-B9, 15, 919–15, 935.

SATO, K., HASHIDA, T. and TAKAHASHI, H. *Acoustic emission study of process zone in granite and tension-softening model.* In AE/MS *Activity in Geological Structures and Materials* (ed. Hardy, JR., H.R.) (Trans Tech Publications, 1995) pp. 81–95

SCHMIDT, R.A. and HUDDLE, C.W. (1977), *Effect of confining pressure on fracture toughness of Indiana limestone,* Int. J. Rock. Mech. Min. Sci. and Geomech. Abstr. *14*, 289–293.

TADA, H, PARIS, P.C., and IRWIN, G.R. (1973a), *The Stress Analysis of Cracks Handbook*, Del Research Corporation, 3.7

TADA, H, PARIS, P.C., and IRWIN, G.R. (1973b), *The Stress Analysis of Cracks Handbook,* Del Research Corporation, 3.6

(Received February 28, 2005, revised October 28, 2005, accepted November 4, 2005)
Published Online First: May 12, 2006

To access this journal online:
http://www.birkhauser.ch

Pure appl. geophys. 163 (2006) 1073–1089
0033–4553/06/061073–17
DOI 10.1007/s00024-006-0066-6

| Pure and Applied Geophysics |

Fracture Toughness Evaluation Based on Tension-softening Model and its Application to Hydraulic Fracturing

KAZUSHI SATO,[1] and TOSHIYUKI HASHIDA[2]

Abstract—This paper discusses the applicability of the tension-softening model in the determination of the fracture toughness of rocks, where the fracture toughness evaluated based on the tension-softening model is compared with the crack growth resistance deduced from laboratory-scale hydraulic fracturing tests. It is generally accepted that the fracture process is dominated by the growth of a fracture process zone for most types of rocks. In this study, the J-integral based technique is employed to determine the fracture toughness of Iidate granite on the basis of the tension-softening model, where compact tension specimens of different dimensions were tested in order to examine the specimen size effect on the measured fracture toughness. It was shown that the tension-softening relation deduced from the J-integral based technique allowed us to determine the specimen size independent fracture toughness K_c of Iidate granite. Laboratory-scale hydraulic fracturing tests were performed on cubic specimens (up to a 10 m sized specimen), where cyclic pressurization was conducted using a rubber-made straddle packer to observe the extent of the hydraulically induced crack. The experimental results of pressure and crack length were then used to construct the crack growth resistance curve based on the stress intensity factor K. The crack growth resistance obtained from the hydraulic fracturing tests was observed to initially increase and then level off, giving a constant K value for a long crack extension stage. The plateau K value in the crack growth resistance curve was found to be in reasonable agreement with the fracture toughness K_c deduced from the tension-softening relation. It was demonstrated that the tension-softening model provides a useful tool to determine the appropriate fracture toughness of rocks, which may be applicable for the analysis of the process of large-scale crack extension in rock masses.

Key words: Rock, fracture toughness, tension-softening model, hydraulic fracturing, J-based technique.

1. Introduction

The hydraulic fracturing is one of the most promising methods for enhancing the recovery of geothermal energy from a deep rock mass. Because the size of the subsurface cracks induced hydraulically is important to the heat exchange process, a fracture mechanics approach is expected to provide a useful tool in the development

[1]Miyagi National College of Technology, 48 Aza Nodayama, Shiote, Medeshima, Natori, 981–1239, Japan. E-mail: kazushi@miyagi-ct.ac.jp
[2]Fracture and Reliability Research Institute, Tohoku University, 01 Aoba, Aramaki, Aoba-ku, Sendai, 980–8579, Japan

of the artificial geothermal energy extraction system. The prediction of the crack size calls for the determination of appropriate fracture toughness of rocks, initially.

It is well known that the fracture process of rocks is dominated by the growth of a fracture process zone (LABUTZ et al., 1987). The crack growth resistance of rocks increases with the growth of the fracture process zone, as is often observed in rising crack growth resistance curves for many types of rocks. After the fracture process zone is fully formed and the steady-state of crack growth is achieved, then the crack growth resistance curve levels off and saturates, giving the constant value of crack growth resistance. In order to analyze the growth of hydraulically induced cracks on a field scale, the saturated and constant crack growth resistance must be determined for the rock of concern. It has been shown that apparent fracture toughness values determined on the basis of linear elastic fracture mechanics (LEFM) increases with respect to specimen size, and that a valid K_{Ic} test for rocks generally requires an impracticably large specimen size (SCHMIDT and LUTZ, 1979: INGRAFFEA, 1987: HASHIDA and TAKAHASHI, 1993). The difficulty can be generally attributed to the non-negligible size of the fracture process zone, which is often comparable with the specimen dimension used for the laboratory-scale fracture toughness tests. Therefore, the above-mentioned results indicate that it may be difficult to employ LEFM-based testing methods to evaluate the appropriate crack growth resistance and then the fracture toughness for the analysis of the field scale crack extension. Hence, a suitable model for characterizing the formation of the fracture process zone is necessary in order to evaluate the fracture toughness on the basis of laboratory experiments.

The use of a Barenblatt-type cohesive crack model has been proposed by HILLERBORG (1983) for concrete. This cohesive crack model is often referred to as the tension-softening model for the crack opening mode. As illustrated in Figure 1, the tension-softening model represents the fracture process zone as an extension of the crack subjected to a closing stress σ which depends on the crack opening displacement δ of the fictitiously extended crack. The relation between the cohesive stress σ and the crack opening displacement δ is called the tension-softening curve. By definition, the tension-softening curve can be obtained from a direct tension test. Therefore, the peak of the cohesive stress corresponds to the uniaxial tensile strength. Furthermore, the area under the tension-softening curve provides the critical J-integral value which corresponds to the crack growth resistance for the steady-state of crack growth after the fracture process zone is fully formed. Consequently, the fracture toughness for the analysis of the hydraulically induced crack may be evaluated by the measurement of the tension-softening curve in laboratory tests.

In this paper, we investigate the usefulness of the fracture toughness based on the tension softening model for the prediction of the growth of hydraulically induced cracks. By employing the J-integral based testing method, fracture tests were conducted to determine the tension softening relation of a granite using fracture toughness specimens of various sizes. The appropriate fracture toughness of the

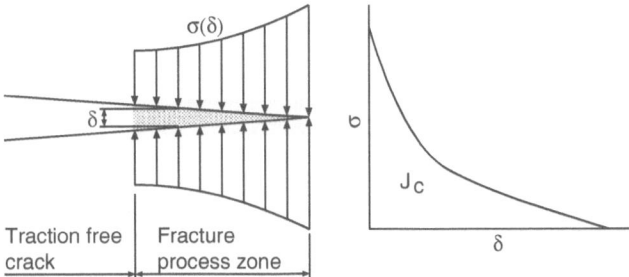

Figure 1

Schematic drawing of the tension–softening model. Fracture process zone is represented by a fictitious crack with cohesive stress. The relation between the cohesive stress and the crack opening displacement of the fictitious crack is the tension-softening curve.

granite was determined by examining the effect of the specimen size on the measured tension softening relation. Laboratory-scale hydraulic fracturing tests were performed to evaluate the crack growth resistance curve of the granite. The crack growth resistance curve was evaluated using the stress intensity factor. The crack growth resistance curve obtained from the hydraulic fracturing tests was compared with the fracture toughness value in order to examine the validity of the tension softening model for rocks.

2. Fracture Toughness Test

This section presents the results of fracture toughness tests of a granite based on the tension-softening model. The J-based technique is used to measure the tension-softening curve. The fracture toughness deduced from the tension-softening curve is compared with the fracture mechanics parameters estimated from the fracture energy method and the crack growth resistance approach.

In principle, the tension-softening curve can be evaluated by conducting direct tensile tests. However, it is generally very difficult to perform a stable direct tensile test and to obtain the complete stress-displacement curve. LI et al., (1987) have introduced a J-based technique for determining the tension-softening curve for concrete. In this technique, a set of fracture toughness tests are conducted instead of the direct tension test, and the load vs. load-line displacement relation registered is used to produce the tension-softening curve. The technique allows us to perform a stable fracture test using a conventional testing machine. In the following, a brief description of the J-based technique is presented.

The J-based technique is based on the nonlinear fracture mechanics parameter, J-integral. Taking advantage of a cohesive crack model, the area under the tension-softening curve is related to the J-integral

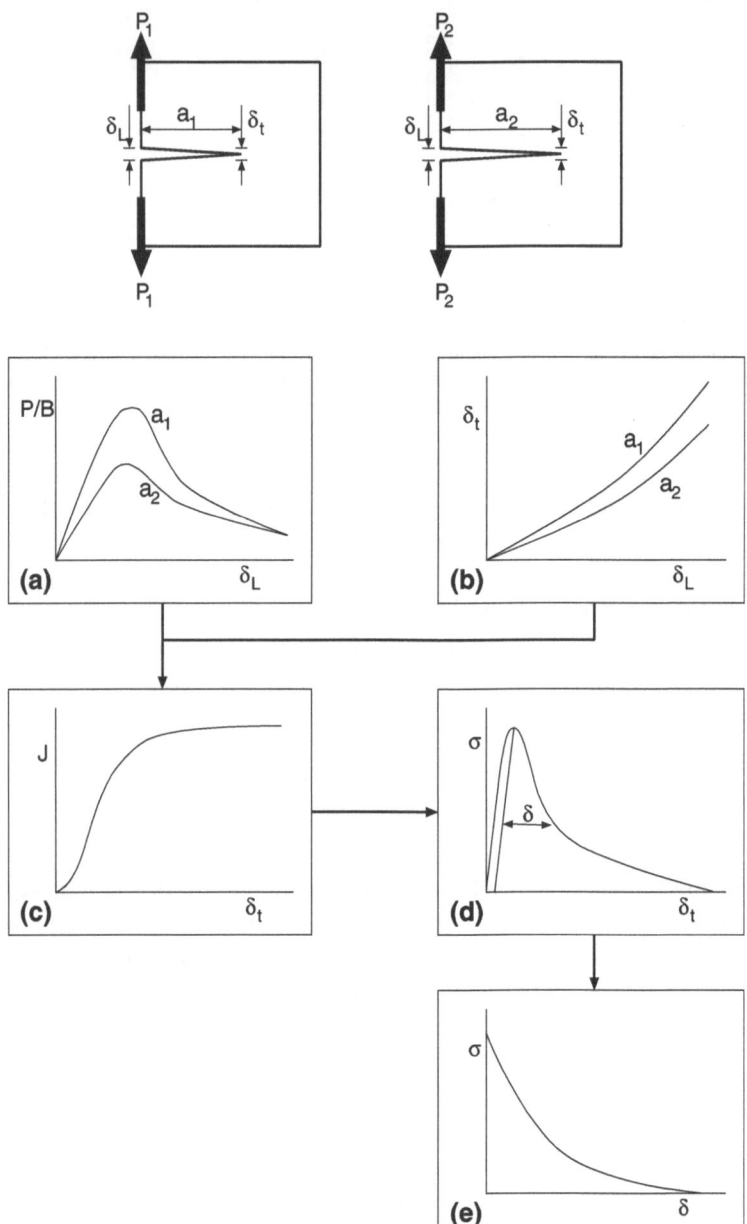

Figure 2
Flow chart of testing procedure by the J-based technique.

$$J = \int_{0}^{\delta_t} \sigma \, d\delta, \tag{1}$$

where δ_t is the crack opening displacement at the initial crack tip. Differentiating Equation (1) with respect to δ_t, the cohesive stress σ may be determined from

$$\sigma = \frac{\partial J}{\partial \delta_t}. \tag{2}$$

Thus, if the relation between J and δ_t can be obtained experimentally, the tension-softening curve can be derived from Equation (2).

Figure 2 illustrates the flow of the testing procedure of the J-based technique. The J-integral value can be evaluated experimentally by conducting tests on two specimens with slightly different crack lengths. During the tests, load P, load line displacement δ_L and crack tip opening displacement δ_t are measured simultaneously. The value of the J-integral for a given value of δ_L is calculated using the following equation

$$J(\delta_L) = \frac{1}{a_2 - a_1} \int_{0}^{\delta_L} \left(\frac{P_1}{B_1} - \frac{P_2}{B_2} \right) d\delta_L, \tag{3}$$

where B is the specimen thickness. The subscripts, 1 and 2, refer to the different crack lengths, a_1 and a_2, respectively. Based on a set of P/B-δ_L and δ_t-δ_L relations for each specimen with different crack lengths, the relation between J and δ_t is obtained as in Figure 2(c). In this procedure, the average of the crack tip opening displacement for each specimen is used as δ_t. Figure 2(d) indicates the σ-δ_t relation deduced from the slope of the J-δ_t data. The initial rising portion of the σ-δ_t curve is due to an elastic deformation between the measuring points of δ_t, and therefore should not be regarded as part of the tension-softening curve. Assuming the unloading path from the peak of the curve which is parallel to the initial slope of the curve, and subtracting the elastic deformation, the true crack opening displacement δ is extracted from the measured δ_t as shown in Figure 2(d). Figure 2(e) shows the corrected tension-softening curve.

The above-mentioned procedure was employed to measure the tension-softening curve using compact tension (CT) specimens of various sizes and to determine the valid fracture toughness based on the tension-softening model. The rock tested was granite from a quarry in Iidate, Fukushima prefecture, Japan. Iidate granite is a light gray biotite granite, containing a rift plane typical of granites. In this study, four blocks of Iidate granite were used to conduct fracture toughness tests and hydraulic fracturing tests. As shown in Table 1, the fracture property of the individual blocks was found to be slightly different, depending on the sampling location of the quarry. Hereafter, in order to distinguish the sampling locations, the four rock blocks used are labeled as A, D, F and H, respectively. The geometry of the CT specimen used is shown in Figure 3. The specimen dimensions are given in Table 2. The CT specimens

Table 1

Elastic and fracture properties measured for four blocks of Iidate granite.

Rock	E' (GPa)	J_c (N/m)	$K_c(=\sqrt{E'J_c})$ (MPa$\sqrt{m}$)	σ_f (MPa)	σ_{uts} (MPa)
A	41.9	163	2.71	5.78	5.38
D	29.6	386	3.38	5.61	4.83
F	59.5	166	3.14	7.28	—
H	36.3	112	2.00	3.28	—

of several different sizes were tested for Rock A to examine the specimen size effect on the measured tension-softening curve (HASHIDA, 1989). Side grooves, width 4.3 mm and depth 7.0 mm, were machined into the specimen to enforce the crack growth to remain in the median plane. An artificial notch of width of 0.15 mm was introduced in each specimen. Hashida and Takahashi (1993) verified that the notch width of 0.15 mm is sufficiently narrow to simulate a sharp crack for Iidate granite. The CT specimens were loaded to total failure using a screw driven testing machine. In order to calculate the J-integral value by Equation (3), different initial crack lengths were machined into each set of the specimen size.

Typical sets of load-load line displacement (P-δ_L) curve for Rock A are shown in Figure 4. Figures 4(a) and (b) indicate the P-δ_L curves obtained from the CT specimens with different initial crack lengths ($a_0/W = 0.5$ and 0.6) for specimen size 6 T and 2.5 T, respectively. It is noted in the 6 T size specimen (Fig. 4(a)) that the P-δ_L curve for the longer crack length specimen ($a_0/W = 0.6$) joins with that for the shorter crack length specimen ($a_0/W = 0.5$) at the displacement about 300 μm, demonstrating that the fracture process zone is fully formed in this specimen size. From the viewpoint of the deformation behavior, the test results of 6 T size specimen can be taken as valid for measuring the tension-softening curve of Rock A. On the other hand, the test records of 1.5 T, 2.5 T and 4 T size specimens indicate that the

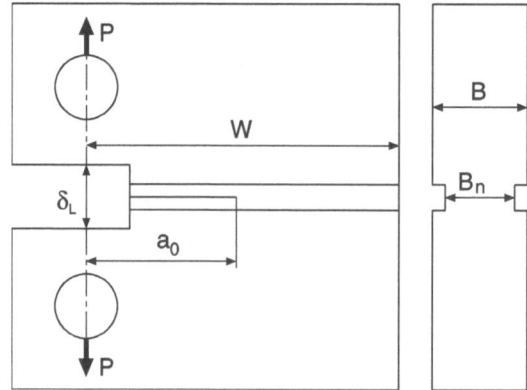

Figure 3
Geometry of the compact tension specimen used.

Table 2

Dimensions of the compact tension specimens used.

Rock	Nominal specimen size	a_0 (mm)	W (mm)	B (mm)	B_n (mm)
	1.5T (8)[*]	36–46	75	40	25
	2.5T (7)	50–74	125	50	33
	4Ta (2)	110, 133	222	60	40
A	6T (2)	150, 180	300	60	50
	1T[**] (1)	16.5	49.7	20.4	—
	3T[**] (1)	47.4	151	47.4	—
	4Tb[**] (1)	114	210	47.0	—
D	4T (2)	81, 90	180	60	41
F	2T (3)	45, 50	100	40	32
H	4T (2)	100, 120	200	61	47

[*]: Number of specimens used.
[**]: Crack growth resistance curve was measured.
Side grooves were not introduced.

specimen sizes are insufficiently large to allow the fracture process zone to extend entirely during the tests.

Following the procedure of the J-based technique, the tension-softening curves were deduced for each of specimen pairs from the experimentally obtained records of P-δ_L and δ_t-δ_L. For the 2.5 T size specimen of Rock A, several pairs of the specimens were tested to examine the reproducibility for the obtained result, after which the tension-softening curve was obtained for each pair of the specimens. Figure 5(a) gives the experimental scatter band for the tension-softening curve determined for the 2.5 T size specimen. An upper and lower bound are drawn to indicate the scatter of the resulting tension-softening curves. An averaged tension-softening curve shown in Figure 5(a) was determined by means of the least-square method from the tension-softening curves obtained. The averaged tension-softening curve is compared with the tension-softening curves obtained for 1.5 T, 4 T and 6 T size specimens.

Deduced tension-softening curves of Rock A for four specimen sizes by means of the J-based technique are indicated in Figure 5(b). The tension-softening curves shown in Figure 5(b) is the averaged data for each specimen size. Taking into account the scatter band of the tension-softening curve as observed for the 2.5 T size specimen, it can be seen that the tension-softening curves obtained for the specimen sizes greater than the 2.5 T size are approximately consistent with each other. However, the tension-softening curve of the 1.5 T size specimen slightly deviates from the general trend obtained for the larger specimens. This comparison demonstrates that the specimen size independent tension-softening curve can be determined by the J-based technique using subsized specimens.

The measured fracture parameters obtained by the J-based technique are summarized in Table 1 for Rock's A, D, F and H. Averaged values are given in Table 1. E' is the effective Young's modulus measured from the initial tangent of the load-load line displacement curve of the fracture toughness specimens tested using the J-based technique (HASHIDA and TAKAHASHI, 1985). J_c is the critical J-integral value obtained from the area under the tension-softening curve. K_c is the critical stress intensity factor converted from the J_c value using the equation $K_c = \sqrt{E'J_c}$. σ_f is the peak value of the deduced tension-softening curve which corresponds to the uniaxial tensile strength by the definition of the tension-softening model. σ_{uts} is the tensile strength measured by a direct tensile test of Rock A and D (HASHIDA, 1990) in which only peak load was recorded. It can be seen that the σ_f value compares well with the σ_{uts} value for the two kinds of granites. This result supports the use of the J-based technique, which is an indirect method to measure the tension-softening curve of rock.

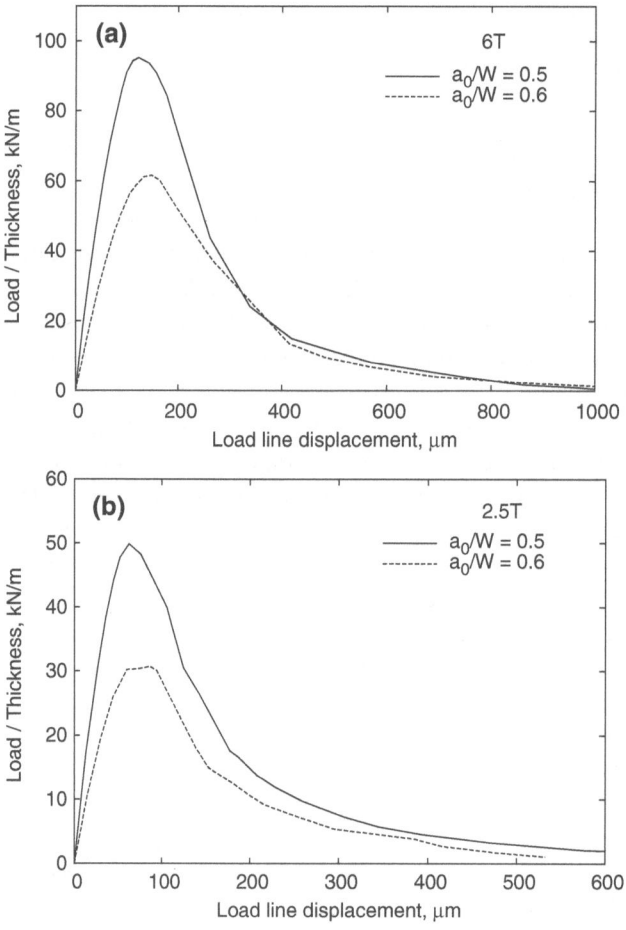

Figure 4
Load-load line displacement curves measured for two specimens with slightly different crack length. (a) is measured curves for 6 T size specimens and (b) is for 2.5 T size specimens.

Next, we examine the size dependency of the J-based technique from the viewpoint of fracture toughness. Figure 6 shows the critical J-integral values J_c determined for four specimen sizes of Rock A, which are computed from the deduced tension-softening curves. The J_c is plotted against the ligament size $W - a_0$ as the representative specimen size, where as the ligament size is the averaged value of the specimens used for the J-based tests. The fracture energy G_F (RILEM50-FMC, 1987) is also shown for comparison. The fracture energy G_F is calculated from the load-load line displacement record using the following equation

$$G_F = \frac{A}{B(W - a_0)},$$ (4)

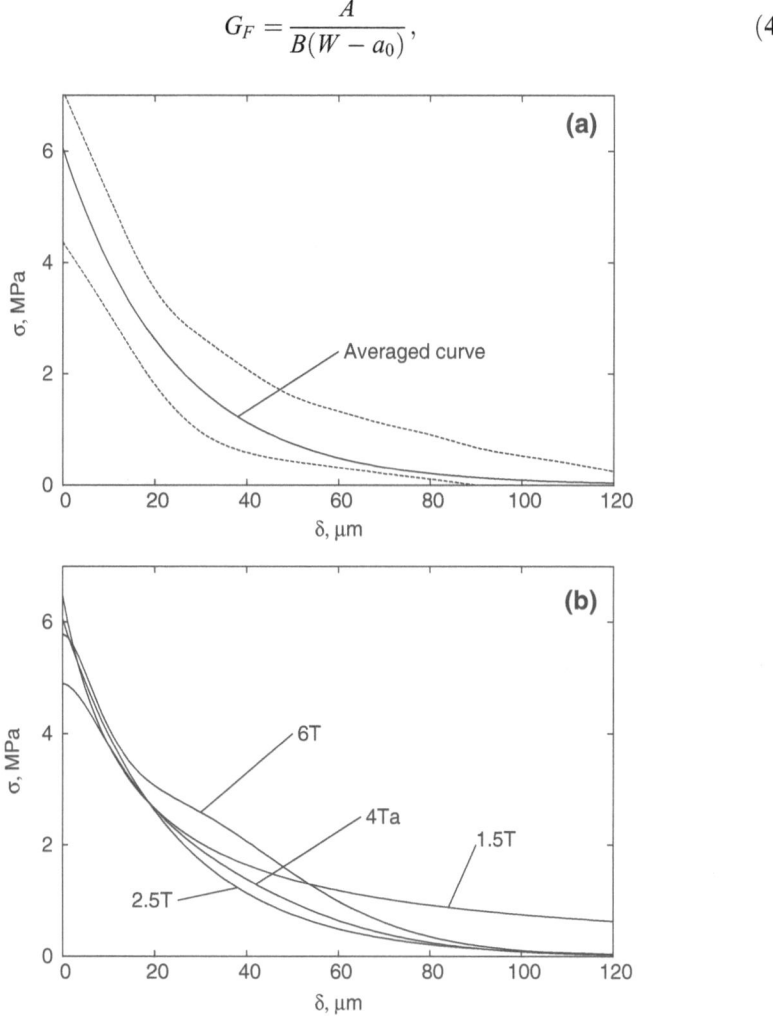

Figure 5

Tension-softening curves Deduced by the J-based technique. (a) deduced tension-softening curve for 2.5 T size specimen. Dashed line indicates the scatter band of the test results. (b) Summary of the deduced tension-softening curves for four specimen sizes.

where A is the area under the load-load line displacement curve, and B, W and a_0 are the specimen dimensions as shown in Figure 3, respectively. Since the side grove was introduced into the specimen used, B_n was used for B in Equation (4). The G_F values increase with the increase of the specimen size, and then tend to approach a constant value. It also has been shown that the critical stress intensity factor computed from the peak load of the fracture toughness specimens exhibits a significant specimen size dependency and no constant value is obtained with the specimen dimensions used (HASHIDA, 1990). In contrast, the J_c determined from the tension-softening relation provides a reasonably constant value irrespective of the different specimen sizes used. It is also noted that the J_c value is relatively close to the constant G_F value for the larger specimen sizes. The agreement of the two fracture parameters for the larger specimen sizes also supports the usefulness of the J_C, which may be applicable for the analysis of large-scale crack growth behavior in rocks. In the J-based technique, two specimens with slightly different crack lengths are used to evaluate the J-integral as a function of the deformation, whereas the G_F value is evaluated from the load-load line displacement data of one specimen. Even though the reason for the specimen size effects is still unclear, the above comparison may suggest that the differentiation method used in the J-based technique is more accurate for the determination of the true facture energy, compared with the G_F approach. However, a more detailed examination is needed to clarify the reason for the results shown in Figure 6. It might be considered that using two specimens in the J-based technique cancels out a size-dependent effect in rock specimens. The above-mentioned result indicates that the size-independent fracture toughness can be obtained using subsized specimens on the basis of the tension-softening model.

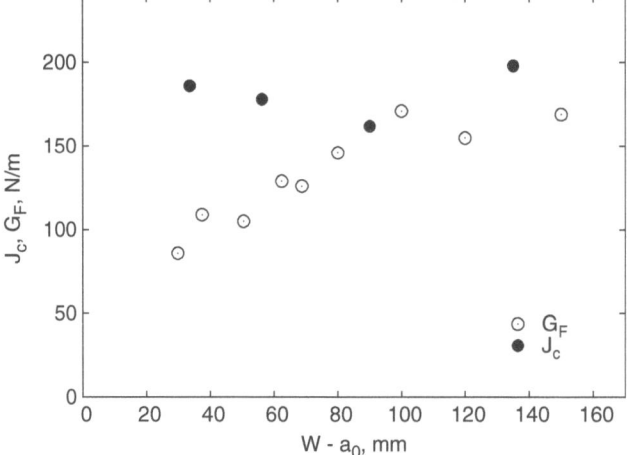

Figure 6

Comparison of the fracture energy G_F and the critical J-integral value J_c based on the tension-softening model.

Furthermore, we compare the fracture toughness K_c determined based on the tension-softening model with K-resistance curves. During the fracture toughness tests using CT specimens of Rock A, unloading and reloading cycles were frequently performed. Based on the unloading compliance data, the amount of crack growth Δa was determined (ISRM, 1988). The crack growth resistance is measured by the stress intensity factor for the unloading point. The measured K-resistance curves are summarized in Figure 7 for various CT specimens of Rock A. The K_c value computed from the J_c (see Table 1) is also indicated in Figure 7 for Rock A. The shaded region in Figure 7 shows the error band of the K_c values obtained from the determined tension-softening curves. The measured K-resistance is shown to initially increase, and then tends to saturate at the longer crack growth stage. In some test results (1 T, 1.5 T, and 6 T size specimens), the K-resistance shows a decreasing trend when the extending crack tip approaches the back surface of the specimen, probably due to the specimen end effect. Nonetheless, the general trend observed in Figure 7 suggests that the saturated K-resistance may be approximately in the range of 2.2–2.5 MPam$^{1/2}$. The range of K-resistance value is relatively close to the K_c value computed from the J_c. In order to predict the process of the crack growth in large-scale rock masses, it is necessary to evaluate the upper bound of the crack growth resistance corresponding to the steady-state crack growth condition. The above-mentioned result demonstrates that the crack growth resistance under the steady-state crack growth condition can be evaluated from the tension-softening relation using subsized specimens.

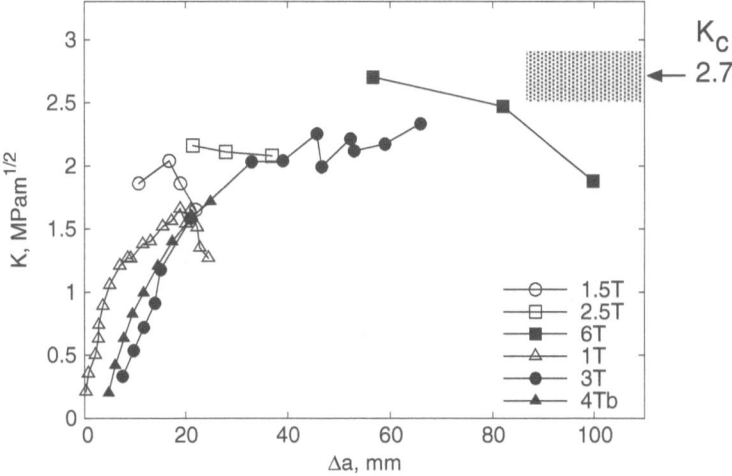

Figure 7

K-resistance curve of Rock A. K_c is the critical stress intensity factor calculated from the tension-softening curve. The shaded region indicates the scatter band of the K_c values.

This section discussed the validity of the fracture toughness evaluated on the basis of the tension-softening model. The next section describes the applicability of the tension-softening model to the analysis of the crack growth in hydraulic fracturing.

3. Hydraulic Fracturing

In this section, the results of hydraulic fracturing experiments are analyzed to deduce the crack growth resistance curve for Iidate granite, and then compared with the fracture toughness, by evaluation based on the tension-softening model. The loading configuration and crack geometry in the hydraulic fracturing experiment are quite different from those for the fracture toughness testing of compact tension specimens. Thus, the comparison may be useful to verify the validity of the fracture toughness evaluated based on the tension-softening model.

HASHIDA et al., (1993) conducted a series of laboratory scale hydraulic fracturing tests using several rectangular specimens prepared from Rock F, as shown in Figure 8. As shown in Table 3, the sizes of the rectangular specimens were in the range of 15 cm-1m. The dimensions of the specimens are nominally designated by 15 cm, 20 cm, 30 cm, 50 cm, and 1 m. A simulated vertical borehole of 15 mm in diameter was drilled in the center of all specimens, and a bilobed pre-notch was introduced on the borehole using a specially designed devise in order to facilitate the analysis of crack propagation resistance. The depth of the pre-notches a_i/λ was approximately 0.5. In addition, larger-scale hydraulic fracturing tests have also been performed using a 10 m sized specimen, which was quarried at the quarry in Iidate as illustrated in Figure 9. The rock type used for the 10m sized specimen is designated as Rock H. Vertical boreholes, 48 mm in diameter were drilled into the 10 m sized specimen.

The hydraulic fracturing experiments were performed by pressurizing the borehole using rubber packers. A set of rubber packers with different pressurization intervals was utilized to extend the crack step by step in such a way that the subsequent injection interval enclosed the crack region created at the preceding stage. Dyed fluid was used as the pressurization fluid to delineate the hydraulically induced crack. After the tests, the specimen was broken open along the dyed crack plane to examine the crack growth process. In the following, we briefly describe a method for determining the crack propagation resistance, based on the results of the hydraulic fracturing tests. Schematics of crack propagation process and corresponding borehole pressure vs. crack length (P-Δa) are shown in Figure 10. We consider the pressure variation with crack growth during the stepwise pressurization. The locations A and B represent the positions of the rubber packer used for the hydraulic fracturing test. The pressurization interval for the first cycle is A-A, and B-B for the subsequent cycle, respectively. The point 1 on the P-Δa curve denotes the crack initiation from the pre-notch tip. When the crack tip along the borehole reaches the location A, the pressure decreases due to the leakage of the injection fluid without

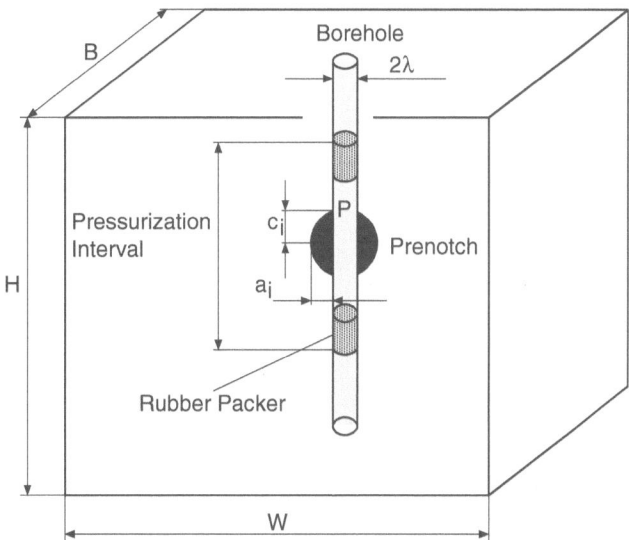

Figure 8
Specimen geometry used for laboratory scale hydraulic fracturing tests.

significant crack growth. At the subsequent cycle, where the borehole interval B-B is subjected to fluid injection, the crack starts to propagate after the pressure corresponding to the point 2 is reached. When the fluid gets over the straddle packer at the location B, the pressure decreases again. Thus, the borehole pressure, which corresponds to the crack length induced at the pressure $P(\Delta a)$ can be found by comparing the magnitude of the maximum pressure achieved at the cycle P_{pre} and that at the subsequent cycle P_{sub}. Namely, $P(\Delta a) = P_{pre}$, if $P_{pre} \leq P_{sub}$, and $P(\Delta a) = P_{sub}$, if $P_{pre} > P_{sub}$. This discussion enables us to determine the crack dimension corresponding to the borehole pressure, and then to construct crack propagation resistance curves obtained from hydraulic fracturing tests. For the 10 m sized specimen, in addition to the above-mentioned simple procedure, a detailed analytical investigation (ABÉ et al., 1989) has been undertaken, in which the crack

Table 3

Specimen dimensions used for small-scale hydraulic fracturing tests.

Nominal specimen size	W (cm)	B (cm)	H (cm)	Borehole diameter 2λ (mm)
15 cm	15	15	15	15
20 cm	20	20	20	15
30 cm	30	30	30	15
50 cm	50	50	50	15
1 m	100	100	100	15

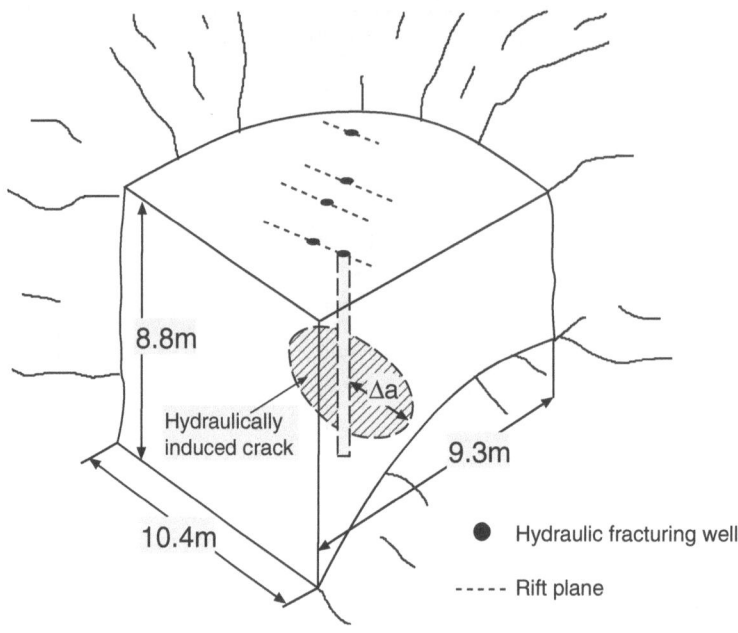

Figure 9
Geometry of the 10 m sized rock specimen for hydraulic fracturing test.

growth resistance data were obtained assuming that the stress intensity factor is
uniform along the periphery of the hydraulically induced crack. In the analysis, the
process of crack growth observed on the dyed fracture surface and borehole pressure
vs. time records were used to evaluate the crack growth resistance as a function of the

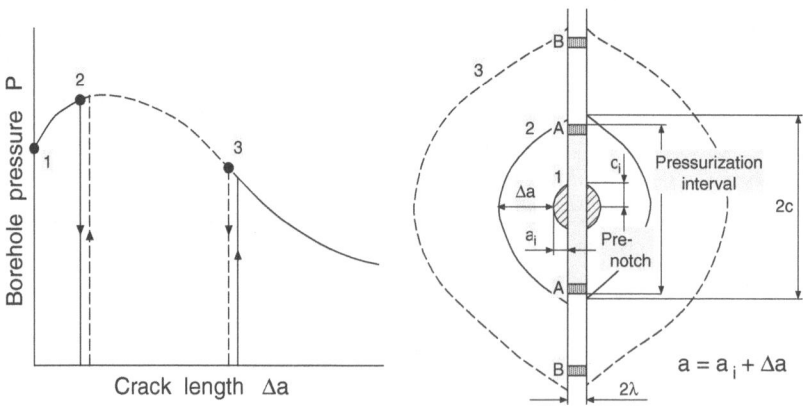

Figure 10
Procedure for the analysis of crack growth resistance in hydraulic fracturing test.

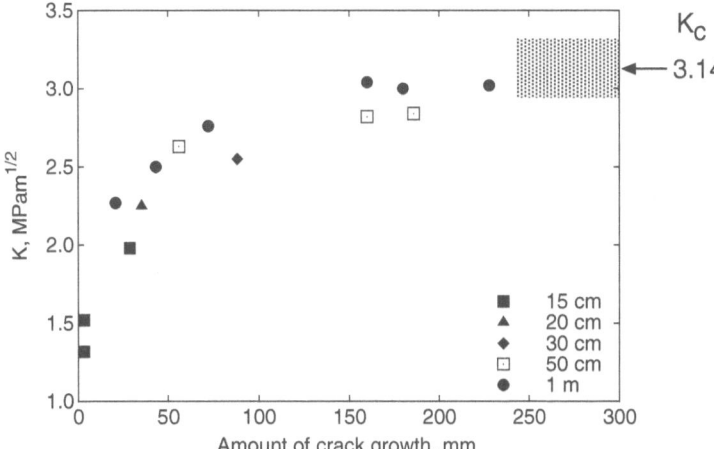

Figure 11
Crack growth resistance curve measured by hydraulic fracturing tests. K_c is the critical stress intensity factor calculated from the tension-softening curve. The shaded region indicates the scatter band of the K_c values.

crack length. For detailed information of the experimental procedures used for the hydraulic fracturing experiments and the analytical methods, the reader is referred to the literature (HASHIDA et al., 1993: ABÉ et al. 1989).

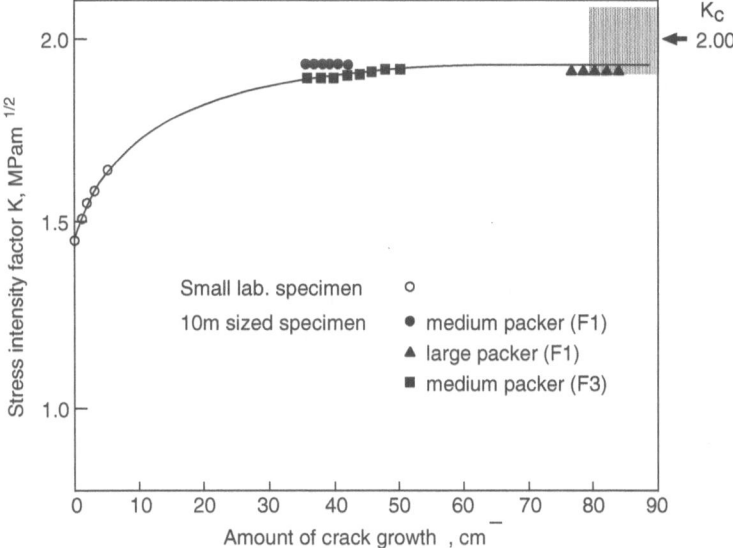

Figure 12
Crack growth resistance curve measured by the 10 m sized rock specimen. K_c is the critical stress intensity factor calculated from the tension-softening curve. The shaded region indicates the scatter band of the K_c values.

The crack propagation resistance curves determined are summarized for the hydraulic fracturing tests conducted on Rock F and H in Figures 11 and 12, respectively. Figure 12 gives the experimental data from the hydraulic fracturing tests conducted on the two boreholes (F1 and F3) that were drilled into the 10 m sized specimen. It has been shown that the aspect ratio of the induced cracks a/c was in proximity to 1.0, particularly for the longer crack growth stage, indicating that the induced cracks in the hydraulic fracturing experiments were approximately of circular shape. It is seen that the crack propagation resistance initially increases and gradually levels off as the crack extension progresses. It is particularly noted that the curve effects a nearly constant crack propagation resistance, and a steady-state crack growth condition is achieved at the larger crack growth stage. The so-called rising R-curve behavior observed in Figures 11 and 12 is consistent with the growth of fracture process zone. The fracture toughness value determined by the J-based testing method K_c is also indicated in those figures. Note that the constant crack propagation resistance at the larger crack growth stage is close to the K_c. In order to assess the crack growth process in large-scale rock masses, it is necessary to evaluate an upper limit of fracture toughness values. These results suggest that the K_c determined by the J-based method is judged to be an appropriate fracture toughness value which may be applicable to the analysis of crack growth in hydraulic fracturing.

4. Concluding Remarks

This paper discusses the usefulness of the tension-softening model in the determination of the appropriate fracture toughness of rocks. The validity of the tension-softening model was discussed by comparing the experimental results of fracture toughness tests and laboratory scale hydraulic fracturing tests conducted on Iidate granite. The fracture toughness test results demonstrated that the evaluation method based on the tension-softening model allowed us to measure the specimen size-independent fracture toughness of the rock using subsized specimens. It was also established from the fracture toughness test results that the size-independent fracture toughness corresponded to the constant value of the crack growth resistance in the steady-state condition. The fracture toughness evaluated using the tension-softening model was compared with the crack propagation resistance obtained from the hydraulic fracturing tests. The results suggested that the tension-softening model may be able to provide an appropriate fracture toughness for predicting the large-scale crack growth induced by hydraulic fracturing.

REFERENCES

ABÉ, H., HAYASHI, K., and HASHIDA, T. *Studies on crack propagation resistance of rocks based on hydrofrac data of large specimens*. In Proc. SEM-RILEM *Int. Conf. on Fracture of Concrete and Rock* (S. P. Shah and S. E. Swartz eds.) (Houston 1989), pp. 354–361.

HASHIDA, T. *Tension-softening curve measurements for fracture toughness determination in granite.* In *Fracture Toughness and Fracture Energy* (Mihashi, H., Takahashi, H., and Wittmann, F. H. eds.) (A. A. Balkema 1989) pp. 47–55.

HASHIDA, T. (1990) *Evaluation of fracture processes in granites based on the tension-softening model.* In *Micromechanics of Failure of Quasi-Brittle Materials* (Mihashi, H., Takahashi, H., and Wittmann, F. H. eds.) (Elsevier Applied Science, 1990) pp. 233–243.

HASHIDA, T, SATO, K., and TAKAHASHI, H. (1993), *Significance of crack opening monitoring for determining the growth behavior of hydrofractures.* In Proc. 18th *Workshop on Geothermal Reservoir Eng.* (Stanford University, Stanford, 1993) pp. 241–246.

HASHIDA, T. and TAKAHASHI, H. (1985), *Simple determination of the effective Young's modulus of rock by the compliance method*, J. Test. Evaluat. *13*, 77–84.

HASHIDA, T. and TAKAHASHI, H. (1993), *Significance of AE Crack Monitoring in Fracture Toughness Evaluation and Non-linear Rock Fracture Mechanics*, Int. J. Rock Mech. Min. Sci. and Geomech. Abstr. *30*–1, 47–60.

HILLERBORG A. *Analysis of one single crack*, In *Fracture Mechanics of Concrete* (F. H. Wittmann ed.) (Elsevier Science Publishers B. V., Amsterdam 1983) pp. 223–250.

INGRAFFEA, A. R. *Theory of crack initiation and propagation in rock. In Fracture Mechanics of Rock* (Atkinson, B. K. eds.) (Academic Press 1987) pp. 71–110.

ISRM COMMISSION ON TESTING METHOD (1988), *Suggested method for determining fracture toughness of rock*, Int. J. Rock Mech. Min. Sci.and Geomech. Abstr. *25*–1, 71–96.

LABUZ, J. F., SHAH, S. P., and DOWDING C. H. (1987), *The fracture process zone in granite: Evidence and effect*, Int. J. Rock. Mech. Min. Sci. and Geomech. Abstr. *24*–4, 235–246.

LI, V. C., CHAN, C. M., and LEUNG, C. K. Y. (1987), *Experimental determination of the tension-softening relations for cementitious composites*, Cem. and Conc. Res. *17*, 441–452.

RILEM50-FMC (1987) *Determination of the fracture energy of mortar and concrete by means of three-point bend tests on notched beams*, Mate. Struct. *18*–106, 45–48.

SCHMIDT, R. A. and LUTZ, T. J. (1979), K_{Ic} and J_{Ic} of Westerly granite — effects of thickness and in-plane dimensions. In *Fracture Mechanics Applied to Brittle Materials* (Freiman, S. W. ed.) (ASTM STP 678, 1979) pp. 166–182.

(Received February 28, 2005, Revised/accepted November 4, 2005)

Pure appl. geophys. 163 (2006) 1091–1100
0033–4553/06/061091–10
DOI 10.1007/s00024-006-0056-8

© Birkhäuser Verlag, Basel, 2006

❙Pure and Applied Geophysics

A Method for Testing Dynamic Tensile Strength and Elastic Modulus of Rock Materials Using SHPB

Q. Z. Wang,[1] W. Li,[1] and X. L. Song[1]

Abstract—An experimental procedure for testing dynamic tensile strength and elastic modulus of rock materials at high strain rate loading is presented in this paper. In our test the split Hopkinson pressure bar (SHPB) was used to diametrally impact the Brazilian disc (BD) and flattened Brazilian disc (FBD) specimens of marble. A tensile strain rate of about 45 1/s was achieved at the center of the specimen. In order to improve the accuracy of the analysis, the initiation time difference between the strain waves acting on the two flat ends of the FBD specimen was treated properly. Typical failure modes corresponding to different loading conditions were observed. It was verified with a finite-element simulation that the equilibrium condition was established in the specimen before its failure. This numerical simulation validates the experimental procedure and also proves the suitability of formulation for the basic equations.

Key words: Flattened Brazilian disc, dynamic splitting test, dynamic tensile strength, dynamic elastic modulus, split Hopkinson pressure bar (SHPB), rock.

1. Introduction

Drilling, excavation, blasting and crashing of rocks are often encountered in civil and mining engineering, where dynamic tensile strength and elastic modulus of rock materials are needed, however, the research on this topic is relatively limited. Dynamic tension test at high strain rate is very difficult to conduct for rocks, since "in most cases, existing experimental techniques will not allow for reaching the maximum strain rate of $\sim$ 30 1/s" (Brara and Klepaczko 2004). The international Society for Rock Mechanics (ISRM) issued the suggested method for determining tensile strength of rock materials by using the Brazilian disc (BD) specimen (ISRM, 1978). This splitting test method has been widely used in static indirect tension for rocks, concrete and many other materials. Wang and Xing (1999) proposed introducing two parallel flat ends to the disc specimen for loading, in this way a BD becomes the flattened Brazilian disc (FBD). The FBD is favorable for reducing stress concentration effect, thus ensuring crack initiation from the center of the disc instead of from the loading

[1]Department of Civil Engineering and Applied Mechanics, Sichuan University, Chengdu, Sichuan 610065, China. E-mail: qzwang2004@163.com

point. RODRIGUEZ *et al.* (1994) extended the static Brazilian test into dynamic testing of ceramics using the split Hopkinson pressure bar (SHPB) setup.

In this paper, dynamic splitting test for dynamic tensile strength and elastic modulus of marble using BD and FBD specimens is reported, also presented are basic equations for analysis and numerical simulation. The dynamic load can be calculated with the dynamic strain recordings on the incident bar and transmission bar. RODRIGUEZ *et al.* (1994) pointed out: "A complete assessment of the splitting tests as a good method of determining the dynamic tensile strength should consider at least three critical aspects: the elastic behavior assumption, the time evolution of the stress distribution in the specimen and the failure pattern." These aspects are observed in our test and analysis, special attention is paid to the nonuniformity of stress waves in the specimen. A finite element simulation for the dynamic stress evolution in the specimen was performed to testify the validity of the experimental procedure.

2. Experimental Procedure

2.1 Specimens, SHPB Setup and Basic Formulas

The SHPB setup with the BD or FBD specimen is shown in Figure 1, the specimen is subjected to diametral impact. The diameter of the bars is 100 mm. In order to reduce the high frequency fluctuation, a soft media acting as wave shaper is glued to the left end of the incident bar (Fig. 1), the wave shaper filters out the high-frequency oscillation and makes the rising front of the pulse exerted on the specimen less steep, which facilitates to satisfying the equilibrium condition before specimen failure occurs. The stress wave in the specimen could be uniform after some time, as demonstrated by both the experimental recordings and the numerical simulation.

According to the theory of elasticity (TIMOSHENKO and GOODIER *et al.*, 1970), under static diametral compression of BD, the principal stress conditions on the loading diameter (Fig.2, excluding the near region of load application) are as follows:

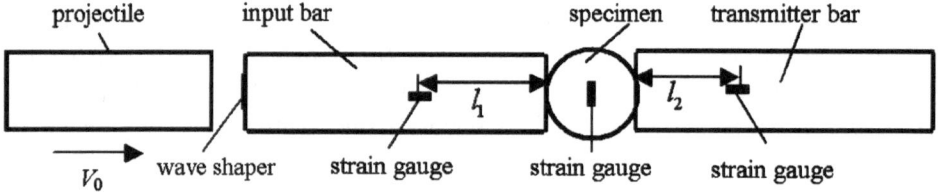

projectile   input bar   specimen   transmitter bar

V_0 wave shaper   strain gauge   strain gauge   strain gauge

Figure 1
The SHPB with specimen under diametral compression.

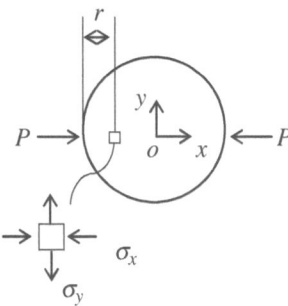

Figure 2
BD with point load.

$$\sigma_x = \frac{2P}{\pi DB} \frac{D^2}{r(D-r)},$$ (1)

$$\sigma_y = -\frac{2P}{\pi DB}.$$ (2)

Compression is considered positive in this paper, so the minus sign belongs to tensile stress and tensile strain. D is the diameter, and B is the thickness of the specimen, r is the distance measured from the loading point. The tensile strength measured with the BD specimen can be calculated from eq. (2) using the maximum load recorded in the test.

BD is adapted into FBD as shown in Figure 3 (WANG and XING 1999), thus the original stress concentration on the loading point can be reduced, also crack initiation at the center of the disc can be more likely guaranteed. However, the stress distribution on the loading diameter of FBD cannot be described with eq. (1) and eq. (2), so that WANG et al. (2004) did the calibration for FBD with the loading angle $2\alpha = 20°$ and obtained the formulas for σ_x and σ_y on the center of FBD, respectively. Furthermore the formula for the tensile strength σ_t based on the Griffith strength criterion was derived. These three formulas are listed below:

$$\sigma_x = 2.973 \frac{2P}{\pi DB}, \quad \sigma_y = -0.964 \frac{2P}{\pi DB}$$ (3)

$$\sigma_t = 0.95 \frac{2P}{\pi DB}.$$ (4)

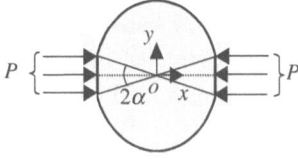

Figure 3
FBD with distributed load.

For the dynamic splitting test using SHPB, the following formula for calculating the load is based on recorded dynamic strains on the bars. It is assumed that the equilibrium condition is satisfied before the specimen breaks, so that $\varepsilon_i(t) + \varepsilon_r(t) = \varepsilon_t(t)$. The average value of the two pressures P_1 and P_2, where P_1 corresponds to $\varepsilon_i(t) + \varepsilon_r(t)$, and P_2 to $\varepsilon_t(t)$, exerted on the two sides of the specimen, is taken as the load (P) applied to the specimen, thus the average load is as follows (TEDESCO *et al.*, 1989; HUGHES *et al.*, 1993):

$$P = \frac{P_1 + P_2}{2} = \frac{E_0 A_0}{2} (\varepsilon_i(t) + \varepsilon_r(t) + \varepsilon_t^*(t)), \tag{5}$$

where A_0 and E_0 are the cross-sectional area and elastic modulus of the bar, respectively, $\varepsilon_i(t)$, $\varepsilon_r(t)$, $\varepsilon_t^*(t)$ are the incident strain, reflected strain and adapted transmitted strain on the bars, respectively. $\varepsilon_t^*(t)$ is adapted from the $\varepsilon_t(t)$ in order to reduce the effect of nonuniformity during loading. The adaptation is a shift of the transmitted strain-wave form along the time axis for reasonable superposition (ZHOU *et al.*, 1992). As the initiation time for the transmitted wave on the right end of the specimen is different from that of the incident wave and reflected wave on the left end of the specimen, it takes a time interval of τ_0 for the elastic wave to travel through the specimen, so that we have $\varepsilon_t^*(t) = \varepsilon_t(t + \tau_0)$.

For the determination of the dynamic elastic modulus, noting that the stress condition in the specimen is two-dimensional, and compressive normal stress is taken as positive, then we have (SU and WANG, 2004):

$$E = \frac{\sigma_y(t) - \mu\sigma_x(t)}{\varepsilon_y(t)}. \tag{6}$$

Thus, from eqs. (3) and (6) and taking Poisson's ratio μ as 0.3, it is easy to see that the elastic modulus can be calculated using the average load $P(t)$ from eq. (5) and the tensile strain $\varepsilon_y(t)$ on the center of the specimen as follows

$$E(t) = -1.856 \frac{2P(t)}{\pi D B \varepsilon_y(t)}. \tag{7}$$

2.2 *Experimental Records*

A white marble taken from Ya'an, Sichuan province of China was tested; its Poisson's ratio is 0.3, Young's modulus is 16 GPa, and density is 2527 kg/m³. 15 BDs and 5 FBDs with $2\alpha = 20°$ (see Fig. 3) were prepared. The disc diameter is 75 mm, and the thickness is 30 mm. The diameter of SHPB is 100 mm, the incident bar length 450 cm, the output bar length 250 cm, their elastic modulus 210 GPa, Poisson's ratio 0.25~0.30, the density 7850 kg/m³, and the elastic wave speed is 5172 m/s. The length of the projectile is 500 mm. The wave shape of the dynamic load can be adjusted by changing the speed of the projectile punching the incident

Table 1

Specimen specifications, loading conditions and wave forms

Spec. No.	Diam. /mm	Thickness /mm	Flat ends	Gas gun pressure /MPa	Projectile speed /m/s	Incident rising time/μs	Incident height /$\mu\varepsilon$	Transmitted height /$\mu\varepsilon$
1	74.53	30.52	no	0.125	2.01	184	312.10	39.05
2	75.40	30.35	no	0.125	2.07	188	321.80	36.60
3	75.65	29.85	yes	0.4	6.83	84	1302.20	62.15
4	75.42	31.30	yes	1.0	12.47	88	2009.50	147.25
5	75.52	29.70	yes	0.2	3.75	372	469.85	56.05

bar and also by modifying the configuration of the projectile. Strain gauges were stuck at the center of two sides of the disc respectively, and the average of the two measured values was taken for analysis. Also strain gauges were put at l_1 = 149.5 cm of the incident bar, and l_2 = 20 cm of the transmission bar, as shown in Figure 1, the average values at the two lateral sides were taken for the needed strain in order to avoid the bending effect of the bars. Typical data of our test are presented in Table 1, where the first three columns are specifications for the five representative specimens, the following two columns are loading conditions, and the last three columns correspond to characteristics for wave forms.

3. Analysis of Experimental results

The corresponding failure mode of the specimen is shown in Figure 4. The impact speed for specimen No. 4 is so high that it broke into three pieces. The relatively ideal failure mode (two equal halves divided along a loading diameter) belongs to No. 3 and No. 5, they are all FBDs. Taking specimen No. 5 for example, its incident, reflected and transmitted wave forms are shown in Figure 5. The signal recorded on the transmitted bar is small and is also interfered by the noise as shown in Figure 5(b). From Figures 5(a) and 5(b) it can be seen that the transmitted strain signal is only about 1/10 of that on the incident bar.

Figure 6(a) shows the tensile strain record on the center of specimen No. 5. The lower horizontal line was caused by the signal exceeding the range of the instrument, as the maximum strain is too large, however we did not use this maximum strain, we chose the steepest point on the curve as the critical point using a differentiation operation on the curve, done with the software Origin. Figure 6(b) is a comparison of the strain signal on the two ends of the specimen, which shows that the signal on the right end lags behind that on the left end with a time lag of 97.5 μs. The two waves are not of the same shape, implying that nonuniformity exists both in time and space, which should be handled properly (ZHOU *et al.*, 1992). The adaptation of the transmitted strain wave in eq. (5) is based on this consideration.

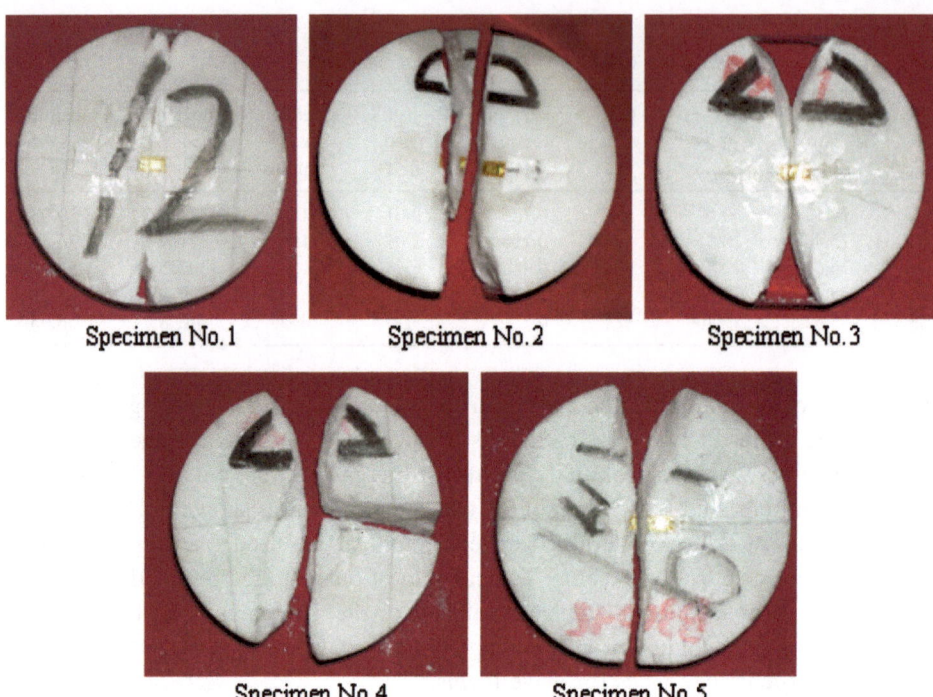

Figure 4
Failure modes of five representative specimens.

Su and Wang (2004) used an Instron 1342 type servohydraulic testing machine to test marble with a strain rate of 0.11 1/s. They derived a tensile strength of 4.47 MPa and an elastic modulus of 20.9 GPa. Now using SHPB, the strain rate for specimen No. 5 is 44.86 1/s, which can be estimated with Figure 6 using the strain difference divided by the corresponding time interval. At such a high strain rate, the tensile strength is 20.55 MPa and the maximum elastic modulus is 114.2 Gpa. These quantities are 4.60 times and 5.46 times as large as their static values, respectively.

4. Finite Element Simulation

To prove the validity of the present experiment, the stress history in the specimen was analyzed using the finite element software ANSYS. The mesh consists of 519 plane 8-node elements and 1648 nodes in total. A ramp load with a peak value of 99.23 kN and a duration of 288 μs, was applied. These data were taken from experiments. Figures 7(b) to (f) shows the time evolution of the dynamic tensile stress σ_y in the specimen. The isolines of the tensile stress in FBD under dynamic split

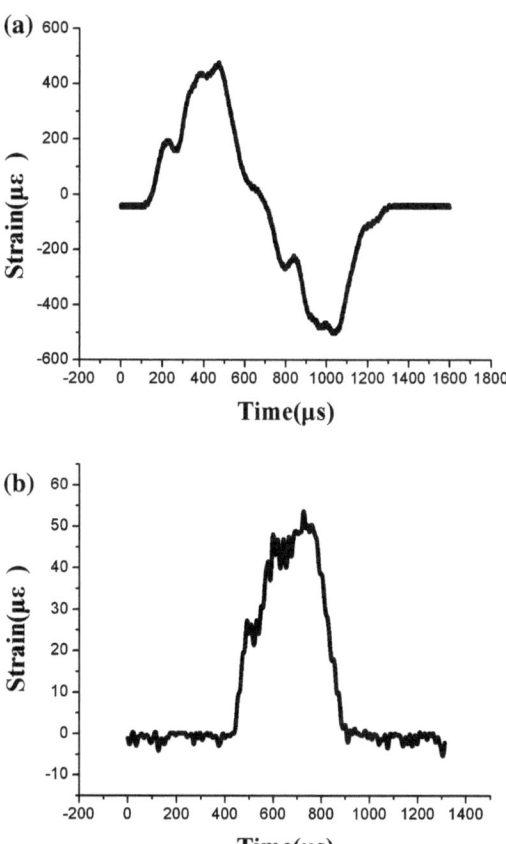

Figure 5
Wave forms on bars for testing specimen No. 5. (a) Incident and reflected waves. (b) Transmitted wave.

loading are given. Shown in Figure 7(a) is the counterpart static loading case, which is given as a comparison base. Figure 7(d) basically resembles Figure 7(a), and this identity proves that under dynamic loading and after a time interval of 75 μs the equilibrium is reached, which occurs before the specimen breaks. Figures 7(d), (e) and (f) show that after a time interval of 75 μs, σ_y distribution in the specimen is uniform, thereafter it does not change with time. For the corresponding BD specimen, about 150 μs is needed to arrive at such uniformity.

5. Conclusion

1. An experimental procedure for measuring dynamic tensile strength σ_t and elastic modulus E using SHPB and BD and FBD specimens is proposed, the basic

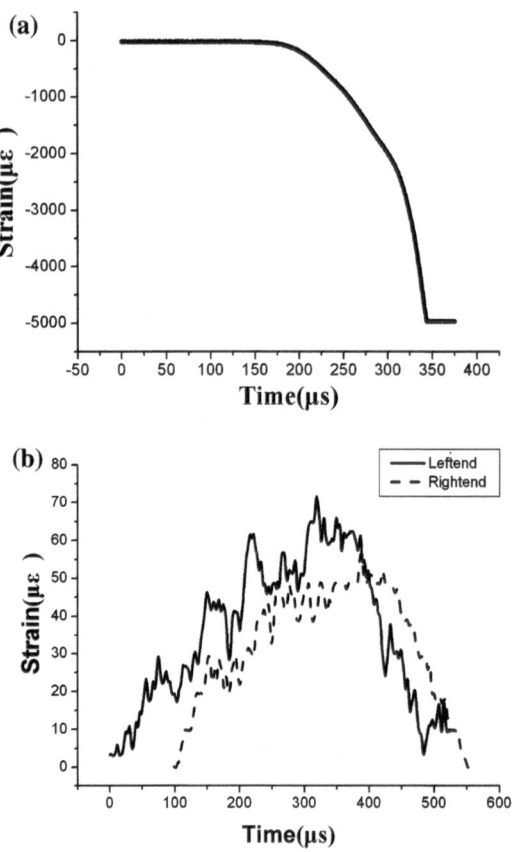

Figure 6
(a) ε_y strain at the specimen center. (b) Strains on the flat ends of the specimen.

formulas for σ_t and E are give in eqs. (4) and (7), respectively, while the average load is derived from dynamic strain recordings on the incident bar and transmission bar using eq. (5), the critical point is illustrated.

2. Dynamic strain recording and its analysis are key issues to the experimental procedure. The nonuniformity of strain on the two sides of the specimen, demonstrated in Figure 6(b), was considered for proper superposition of different stains. A method for the adaptation of the transmitted strain was used.

3. A tensile strain rate of almost 45 1/s was achieved at the center of the specimen as shown in Figure 6(a). At such high strain rate loading, the dynamic tensile strength and elastic modulus of a marble are several times higher than their static values.

4. It is proved that the equilibrium was established before the failure of the specimen occurred. This explains why the formulas for static loading can be used for

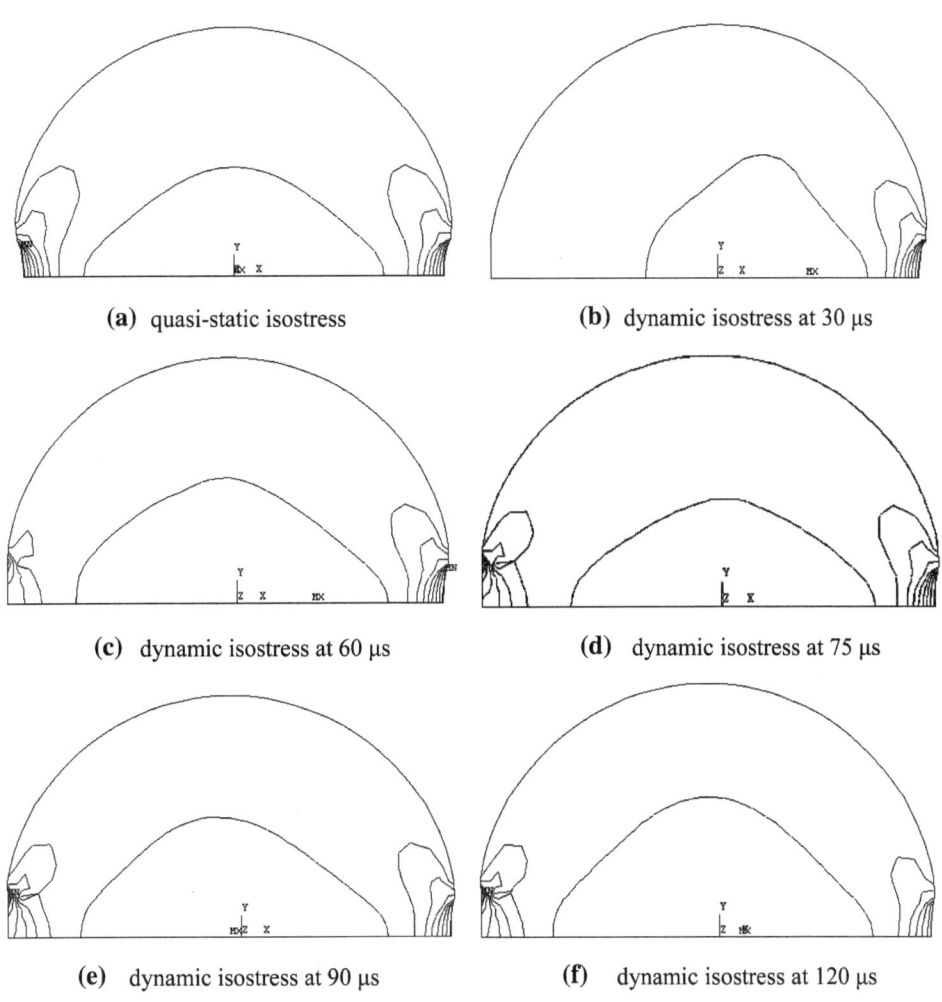

(**a**) quasi-static isostress

(**b**) dynamic isostress at 30 μs

(**c**) dynamic isostress at 60 μs

(**d**) dynamic isostress at 75 μs

(**e**) dynamic isostress at 90 μs

(**f**) dynamic isostress at 120 μs

Figure 7
Comparison of isostress contour for FBD under quasi-static and dynamic loading.

corresponding dynamic splitting test. The result of the numerical simulation shows that for reaching equilibrium in the specimen FBD needs about half the time of BD, hence FBD is superior to BD for dynamic splitting test.

Acknowledgement

This work was supported by the National Natural Science Foundation of China (Project No. 40172094). The authors are very grateful to Professor Ove Stephansson and another reviewer for their insightful comments.

REFERENCES

BRARA, A. and KLEPACZKO, J.R. (2004), *Dynamic tensile behavior of concrete: Experiment and numerical analysis*, ACI Material J. *101*, 162–167.

GOMEZ, J.T., SHUKLA, A., and SHARMA, A. (2001), *Static and dynamic behavior of concrete and granite in tension with damage*, Theoretical and Appl. Fract. Mech. *36*, 37–49.

HUGHES, M.L., TEDESCO, J.W., and ROSS, C.A. (1993), *Numerical analysis of high strain rate splitting-tensile tests*, Computers and Structures *47*, 653–671.

ISRM (1978), *Suggested methods for determining tensile strength of rock materials*, Int. J. Rock Mech. Min. Sci. and Geomech. Abstr. *15*, 99–103.

RODRIGUEZ, J., NAVARRO, C., and SANCHEZ-GALVEZ, V. (1994), *Splitting tests: An alternative to determine the dynamic tensile strength of ceramic materials*, Journal de Physique IV, *4*, c8-101-c8-106.

SU, B.J. and WANG, Q.Z. (2004), *Experimental study of flattened Brazilian disc specimen under dynamic loading*, Journal of Yangtze River Academy (in Chinese) *21*, 22–24.

TEDESCO, J.W., ROSS, C.A., and BRUNAIR, R.M. (1989), *Numerical analysis of dynamic split cylinder test*, Computers and Structures *32*, 609–624.

TIMOSHENKO, S.P. and GOODIER, J.N., *Theory of elasticity* (McGraw-Hill, New York 1970).

WANG, Q.Z. and XING, L. (1999), *Determination of fracture toughness K_{IC} by using the flattened brazilian disk specimen for rocks*, Eng. Fracture Mech. *64*, 193–201.

WANG, Q.Z., JIA, X.M., KOU, S.Q., ZHANG, Z.X., and LINDQVIST, P.-A. (2004), *The flattened Brazilian disc specimen used for determining elastic modulus, tensile strength and fracture toughness of brittle rocks: Theoretical and numerical results*, Int. J. Rock Mech. Min. Sci. *41*, 245–253.

ZHOU, F.H., WANG, L.L., and HU, S.S. (1992), *On the effect of stress nonuniformness in polymer specimen of SHPB test*, Experim. Mech. (in Chinese) *7*, 23–29.

(Received February 17, 2005, revised September 20, 2005, accepted October 26, 2005)
Published Online First: May 12, 2006

To access this journal online:
http://www.birkhauser.ch

Pure appl. geophys. 163 (2006) 1101–1130
0033–4553/06/061101–30
DOI 10.1007/s00024-006-0065-7

∎Pure and Applied Geophysics

True Triaxial Stresses and the Brittle Fracture of Rock

Bezalel Haimson[1]

Abstract—This paper reviews the efforts made in the last 100 years to characterize the effect of the intermediate principal stress σ_2 on brittle fracture of rocks, and on their strength criteria. The most common theories of failure in geomechanics, such as those of Coulomb, and Mohr, disregard σ_2 and are typically based on triaxial testing of cylindrical rock samples subjected to equal minimum and intermediate principal stresses ($\sigma_3 = \sigma_2$). However, as early as 1915 Böker conducted conventional triaxial extension tests ($\sigma_1 = \sigma_2$) on the same Carrara marble tested earlier in conventional triaxial compression by von Kármán that showed a different strength behavior. Efforts to incorporate the effect of σ_2 on rock strength continued in the second half of the last century through the work of Nadai, Drucker and Prager, Murrell, Handin, Wiebols and Cook, and others. In 1971 Mogi designed a high-capacity true triaxial testing machine, and was the first to obtain complete true triaxial strength criteria for several rocks based on experimental data. Following his pioneering work, several other laboratories developed equipment and conducted true triaxial tests revealing the extent of σ_2 effect on rock strength (e.g., Takahashi and Koide, Michelis, Smart, Wawersik). Testing equipment emulating Mogi's but considerably more compact was developed at the University of Wisconsin and used for true triaxial testing of some very strong crystalline rocks. Test results revealed three distinct compressive failure mechanisms, depending on loading mode and rock type: shear faulting resulting from extensile microcrack localization, multiple splitting along the σ_1 axis, and nondilatant shear failure. The true triaxial strength criterion for the KTB amphibolite derived from such tests was used in conjunction with logged breakout dimensions to estimate the maximum horizontal *in situ* stress in the KTB ultra deep scientific hole.

Key words: Rock mechanics, brittle fracture, true triaxial stress, failure criterion.

Introduction

In this paper I review some of the most important contributions of the last century dealing with rock brittle fracture criteria, with particular emphasis on the effect of the intermediate principal stress σ_2. The influence of stress condition on brittle fracture of rock has been studied at various levels since the beginning of civilization. In the early days, making stone tools and mining minerals required understanding of how rock breaks. The spectacular Egyptian pyramids, and the

[1] Department of Materials Science and Engineering and Geological Engineering Program, University of Wisconsin, 1509 University Avenue, Madison, WI 53706-1595, U.S.A. E-mail: bhaimson@wisc.edu

Greco-Roman temples, aqueducts, and sculptures, are just a few examples of sophisticated quarrying and carving techniques developed from an intuitive knowledge of the process of fracturing. In recent times numerous theoretical predictions and highly specialized experimental techniques have been introduced in attempts to formulate a uniform criterion of brittle fracture, relating the acting stresses and the rock properties necessary to induce rock failure. Although a uniform criterion has not been established yet, considerable progress has been made in understanding the effect of the intermediate principal stress on rock strength. This paper summarizes the advances made, and submits substantial evidence that the most commonly used failure criteria, which neglect the effect of σ_2, do not reflect the mechanical behavior of rock under a general state of stress.

The nomenclature used in this paper follows that of JAEGER and COOK (1979). "Dilatancy" is defined as the formation and extension of open microcracks aligned with the direction of the largest principal stress throughout the volume of a rock specimen. The term "yield point", i.e., the stress level at which the transition between elastic and ductile behavior occurs, is only mentioned in conjunction with a special case in the second section of the chapter "True triaxial testing at the University of Wisconsin." Most work on brittle rocks has not been concerned with the yield point because it is difficult to discern accurately. The terms "failure," "fracture," and "strength" are used interchangeably in this paper to describe the same phenomenon, namely "brittle fracture," i.e., the complete lack of cohesion along a surface in a rock body, assumed to occur at the peak point of the stress-strain curve in experiments conducted in conventional testing machines (JAEGER and COOK, 1979, p. 81).

Brittle Fracture Theories

In terms of the three principal stresses in rock (σ_1, σ_2, and σ_3, where compression is positive and σ_1 is the maximum stress), the condition at failure can be expressed as:

$$\sigma_1 = f_1(\sigma_2, \sigma_3) \text{ or } f_2(\sigma_1, \sigma_2, \sigma_3) = 0, \tag{1}$$

where the nature of the functions f_1 or f_2 depends on the specific rock properties. For ductile materials there exists a single well-established triaxial criterion for plastic yield, which originated with von Mises (NADAI, 1950):

$$\tau_{oct} \equiv \frac{1}{3}\sqrt{(\sigma_1 - \sigma_2)^2 + (\sigma_2 - \sigma_3)^2 + (\sigma_3 - \sigma_1)^2} = c. \tag{2}$$

Equation (2) states that the yield point is reached when the distortional energy, represented by the octahedral shear stress τ_{oct}, is equal to a constant c, which is material-dependent.

This is not the case, however, in brittle rock. One main reason is that rock strength varies considerably with confining pressure, σ_3. A number of brittle fracture criteria have been proposed over the years, of which the most known ones present the condition of failure in the form of:

$$\sigma_1 = F_1(\sigma_3) \text{ or } F_2(\sigma_1, \sigma_3) = 0, \tag{3}$$

i.e., by neglecting the effect of the intermediate principal stress σ_2. Here, F_1 and F_2 are functions controlled by the material properties. Equation (3) characterizes criteria such as those due to Coulomb, Mohr, Griffith, and McClintock and Walsh (JAEGER and COOK, 1979). For example, the Coulomb criterion can be expressed as:

$$\sigma_1 = C_0 + \sigma_3 \left(\sqrt{\mu^2 + 1} + \mu \right)^2, \tag{4}$$

where the two constants, C_0 and μ, the uniaxial compressive strength and the coefficient of internal friction, respectively, are rock properties that can be determined in the laboratory (JAEGER and COOK, 1979, p. 97).

Another important criterion of the type shown in equation (3) was obtained by Griffith (JAEGER and COOK, 1979), who assumed that rock is pervaded by randomly oriented microcracks that give rise to stress concentration at their tips. Brittle fracture occurs when the maximum stress at the tip of the most favorably oriented microcrack reaches a critical value characteristic of the rock. With this assumption, the criterion developed by Griffith for a plane stress condition led to a parabolic relationship between the maximum shear stress and the mean stress:

$$(\sigma_1 - \sigma_3)^2 = 8T_0(\sigma_1 + \sigma_3) \quad \text{for} \quad \sigma_1 + 3\sigma_3 \geq 0, \tag{5}$$

where T_0 is the uniaxial tensile strength of the material, a property that can be determined in the laboratory.

MURRELL (1963), who was an early proponent of the need to include all three principal stresses in brittle fracture criteria, extended Griffith's theory to three dimensions, and obtained a criterion that can be described in terms of stress invariants:

$$\tau_{oct}^2 = 8T_0\sigma_{oct}, \tag{6}$$

where τ_{oct} is the octahedral shear stress, defined in equation (2), $\sigma_{oct} = \frac{1}{3}(\sigma_1 + \sigma_2 + \sigma_3)$ is the mean stress, also called the octahedral normal stress. MURRELL (1963) found this criterion to be in reasonable agreement with the limited experimental results available to him.

The exclusion of σ_2 from brittle fracture theories was also challenged by NADAI (1950), FREUDENTHAL (1951), DRUCKER and PRAGER (1952), BRESLER and PISTER (1957), WIEBOLS and COOK (1968), and others, all of whom included all three principal stresses in their respective criteria. NADAI (1950, p. 231) recommended that the von Mises yield criterion for ductile metals be adapted to pressure-dependent polycrystalline materials such as rocks by replacing the constant c in equation (2) with a monotonically rising function f_N of the octahedral normal stress σ_{oct}:

$$\tau_{oct} = f_N(\sigma_{oct}). \tag{7}$$

Equation (7) indicates that in general the shear stress (τ_{oct}) increases with the mean pressure (σ_{oct}) upon brittle fracture. A specific form of this criterion in the form of a linear relationship between the octahedral shear and octahedral normal stresses was suggested by FREUDENTHAL (1951), DRUCKER and PRAGER (1952), and BRESLER and PISTER (1957) for materials such as concrete and soil:

$$\tau_{oct} = c_1 + c_2\sigma_{oct}, \tag{8}$$

where c_1 and c_2 are positive material constants. FREUDENTHAL (1951) also proposed an alternative simplification of equation (7) in the form of a parabolic relationship, which may better fit the fracture strength of some materials:

$$\tau_{oct}^2 = c_3 + c_4\sigma_{oct}. \tag{9}$$

Equation (9) is very similar to Murrell's extended Griffith criterion (equation 6).

An energy-based criterion for assessing rock strength under combined or true-triaxial (also called polyaxial) stresses was propounded by WIEBOLS and COOK (1968). The criterion postulates that rock can be considered an elastic material containing a multitude of uniformly distributed and randomly oriented closed microcracks. When stress is applied to a certain volume of rock, the strain energy in the material has two components: energy stored in the absence of cracks, and energy resulting from frictional sliding that occurs along some of the cracks. The cracks remain closed when the rock is subjected to three compressive principal stresses, and sliding within a crack occurs only if the shear and normal stress conditions are such that $|\tau| - \mu_s\sigma > 0$ (following Amonton's Law, see JAEGER and COOK, 1979), where μ_s is the coefficient of sliding friction between the surfaces of a crack. Wiebols and Cook call this inequality the effective shear stress τ_{eff}, and define the strain energy stored per unit volume around cracks as a result of frictional sliding as the effective shear strain energy W_{eff}. The effective shear stress on a given crack depends on the principal stress magnitudes and the orientation of the crack with respect to the principal axes. The amount of W_{eff} per unit volume is a function of τ_{eff} on each crack, the number of cracks and their size and shape. The assumption that the rock strength is reached when the effective strain energy attains a critical value implies that it is a function of both the material properties and all three principal stresses. Assessing W_{eff} for a general state of stress can only be done numerically. However, for a particular simple condition such as when $\sigma_2 = \sigma_3$, the relationship degenerates to:

$$W_{eff} = 2\pi N \int_{\theta_1}^{\theta_2} \tau_{eff}^2 \sin\theta d\theta, \tag{10}$$

where θ is the angle between the normal to the crack plane and σ_1, and N is the number of cracks aligned with a narrow zone around θ.

The numerically derived Wiebols and Cook effective strain energy failure criterion has three significant characteristics: under conventional triaxial compression ($\sigma_1 > \sigma_2 = \sigma_3$) the strength increases linearly with the confining pressure, as in the Coulomb criterion (equation 4); under triaxial extension ($\sigma_1 = \sigma_2 > \sigma_3$) the strength increases linearly with σ_3 at the same rate as when $\sigma_2 = \sigma_3$, but the biaxial compressive strength ($\sigma_3 = 0$) is always greater than the uniaxial compressive strength ($\sigma_2 = 0$); for a constant value of σ_3, the strength as σ_2 is raised from its initial value of $\sigma_2 = \sigma_3$, first rises gradually to where it reaches a plateau and then declines in such a way that when $\sigma_2 = \sigma_1$ it is still higher than when $\sigma_2 = \sigma_3$. WIEBOLS and COOK (1968) presented their criterion in separate plots for selected coefficients of sliding friction μ_s in the form of families of curves corresponding to constant σ_3 and varying σ_2. A typical plot representing the failure criterion for $\mu_s = 0.25$ is shown in Figure 1. One limitation of this criterion is that there are no known laboratory techniques of measuring the coefficient of sliding friction on microcracks (μ_s should not be confused with μ). On the other hand experimental results described below generally confirm the qualitative adequacy of this theoretical criterion.

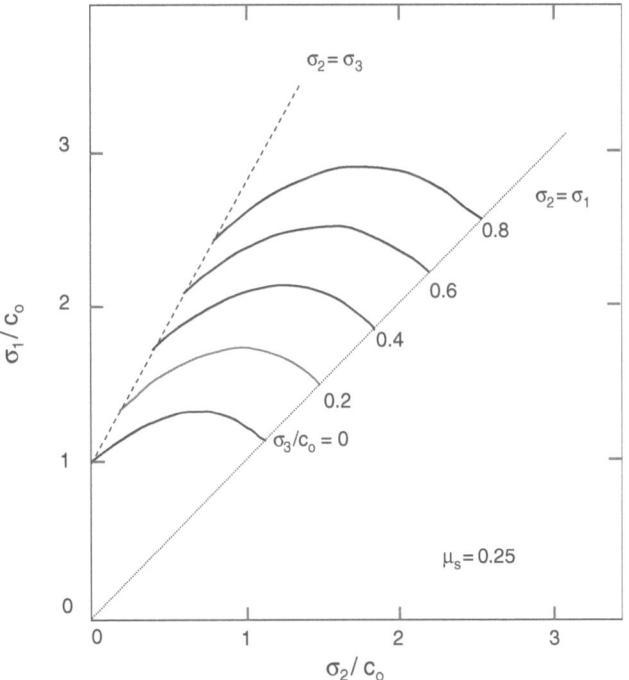

Figure 1

True triaxial failure criterion for a rock with a sliding coefficient of friction $\mu_s = 0.25$, as predicted by WIEBOLS and COOK (1968) theory. All principal stresses are normalized by the uniaxial compressive strength C_0.

Early Observations of the Effect of σ₂ on Brittle Fracture

Experiments that revealed the potential effect of the intermediate principal stress σ_2 on brittle fracture were performed as early as the turn of the last century by VON KÁRMÁN (1911) and BÖKER (1915). VON KÁRMÁN conducted so-called 'triaxial tests' on Carrara marble in which a confining fluid pressure around the cylindrical surface of a rock specimen was kept constant while the axial compressive load was raised until brittle fracture occurred. In these tests the maximum principal stress, σ_1, is aligned with the specimen axis, while the least and intermediate principal stresses, σ_3 and σ_2 respectively, are both equal to the confining pressure. Böker carried out similar triaxial tests on Carrara marble, except that he first applied an axial stress and kept it constant (σ_3), and then raised the confining pressure ($\sigma_1 = \sigma_2$) until failure. The compressive strength of the marble for the same σ_3 in one set of experiments was different than that in the other (Fig. 2), with the difference attributed to the variation of σ_2 from one extreme ($\sigma_3 = \sigma_2$) to the other ($\sigma_1 = \sigma_2$).

Although these results became widely known, they were not seen as a challenge to commonly used failure criteria such as those attributed to Coulomb, Mohr, and Griffith (JAEGER and COOK, 1979), which neglected the effect of σ_2 on rock strength. MURRELL (1963), however, expressed concern about the lack of a generalized

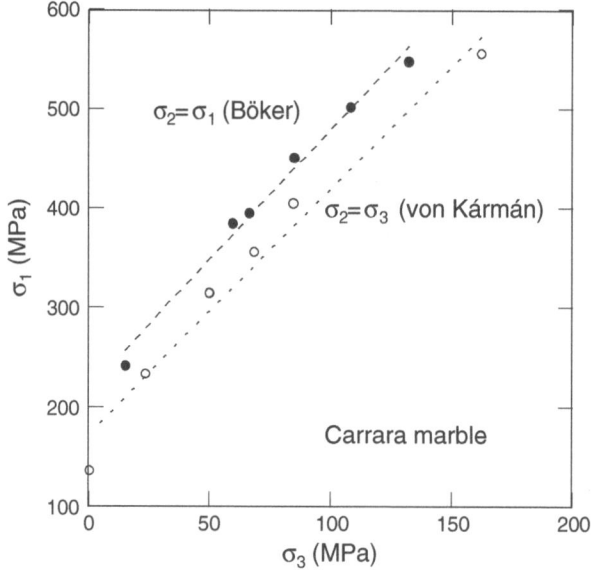

Figure 2

Compressive strength of Carrara marble under conventional triaxial compression ($\sigma_2 = \sigma_3$) and under triaxial extension ($\sigma_2 = \sigma_1$) based on experiments conducted by VON KÁRMÁN (1911) and BÖKER (1915), respectively, as reported by MURRELL (1963).

criterion for brittle fracture of rocks under a truly three-dimensional state of stress, similar to the von Mises criterion for plastic yield. He was among the first to point out that commonly conducted conventional triaxial tests in compression, like Von Kármán's, in which the intermediate principal stress σ_2 is equal to the least principal stress σ_3, do not lead to a general failure criterion. To demonstrate this point, he reproduced Von Kármán's and Böker's results, which are also recaptured in Figure 2 in the form of σ_1 at failure as a function of σ_3. The effect of the intermediate principal stress is obvious: Carrara marble is stronger when σ_2 equals σ_1, at any level of σ_3 tested. At $\sigma_3 = 100$ MPa, for example, the ultimate strength increases by some 15% as the stress condition changes from one in which $\sigma_2 = \sigma_3$ to one where $\sigma_2 = \sigma_1$.

HANDIN *et al.* (1967) conducted similar conventional triaxial compression $(\sigma_2 = \sigma_3)$ and triaxial extension $(\sigma_2 = \sigma_1)$ tests on a limestone, a dolomite and glass. Their results mirrored those of Von Kármán and Böker. Figure 3, showing the results obtained in specimens cored from the same Solnhofen limestone block, is representative of their findings. The ultimate strength in extension tests was clearly higher than that under compression conditions at any level of σ_3. Taking for example the case of $\sigma_3 = 100$ MPa, compressive strength when $\sigma_2 = \sigma_1$ was about 15% larger than that under $\sigma_2 = \sigma_3$. Handin *et al.* experiments on Blair dolomite, as well as on pyrex glass mimicked those on Solnhofen limestone. Since the mechanical properties

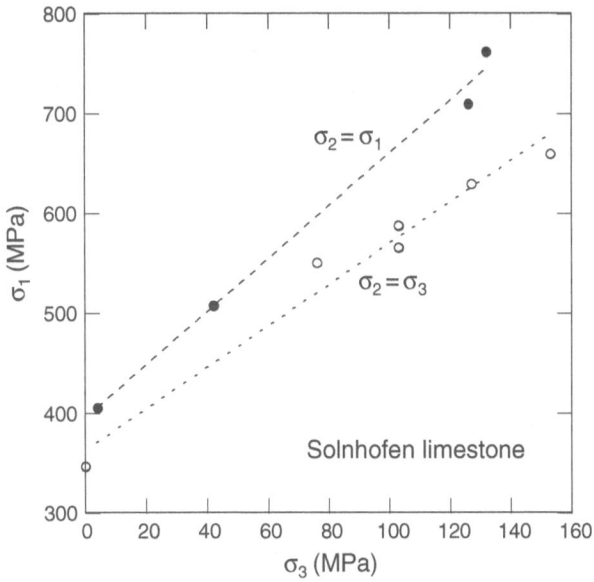

Figure 3

Compressive strength of Solenhofen limestone based on conventional triaxial compression $(\sigma_2 = \sigma_3)$ and extension $(\sigma_2 = \sigma_1)$ results (HANDIN *et al.*, 1967).

of these materials differ widely, the authors' reasonable conclusion was that the discrepancy in strength could only be the result of the different σ_2 between the two types of tests. Handin *et al.* also found that σ_2 caused the angle between the fault plane at brittle fracture and the direction of σ_1, to decrease between triaxial compression and triaxial extension. Similarly, the brittle-ductile transition stress in the limestone increased considerably between triaxial compression and triaxial extension.

Concurrent with Handin *et al.*'s work, JAEGER and HOSKINS (1966) conducted a series of innovative experiments they termed 'confined Brazilian tests', in which the typical Brazilian test consisting of a rock disc subjected to an increasing line load along the diametral plane until failure, was extended by jacketing the disc and applying a constant confining pressure to the specimen. This created a state of triaxial stress acting on the disc with three unequal principal stresses. However, these stresses were not independent of each other. Nevertheless, tests in Bowral trachyte, Gosford sandstone, and Carrara marble, supported previous findings suggesting that the magnitude of the intermediate principal stress was a factor influencing the conditions for failure in all three rocks.

Perhaps the greatest contribution to the study of brittle fracture and its dependence on the intermediate principal stress was made by MOGI (1967, 1971). He embarked on a long-term project of studying rock failure under the most general state of stress by first conducting experiments in a conventional triaxial cell. By nearly eliminating stress concentration due to end effects and other test refinements, he was able to demonstrate even more convincingly than previous workers the indisputable influence that σ_2 has on brittle fracture (MOGI, 1967). Figures 4 and 5 summarize his results in Westerly granite and Dunham dolomite. The substantial discrepancy between the strength of these two rocks in triaxial compression and triaxial extension is convincingly demonstrated, and the only variable that could have led to the difference in strength was σ_2.

True Triaxial Experiments and Strength Criteria

After conducting precision conventional triaxial compression and extension experiments, MOGI (1971) concluded that in order to observe more closely the role played by the intermediate principal stress on brittle fracture, he had to design and utilize a high-pressure true triaxial stress apparatus for testing hard rocks under three independently applied principal stresses ($\sigma_1 > \sigma_2 > \sigma_3$). His apparatus consisted of a pressure vessel that accommodated a rectangular prismatic rock sample (size $15 \times 15 \times 30$ mm) and two sets of pistons independently actuated to apply two of the principal stresses. The minimum principal stress, which has a major effect on fracture strength, was always provided by the confining pressure in the vessel, thus ensuring a uniform distribution of σ_3. Mogi minimized friction on the sample faces subjected to

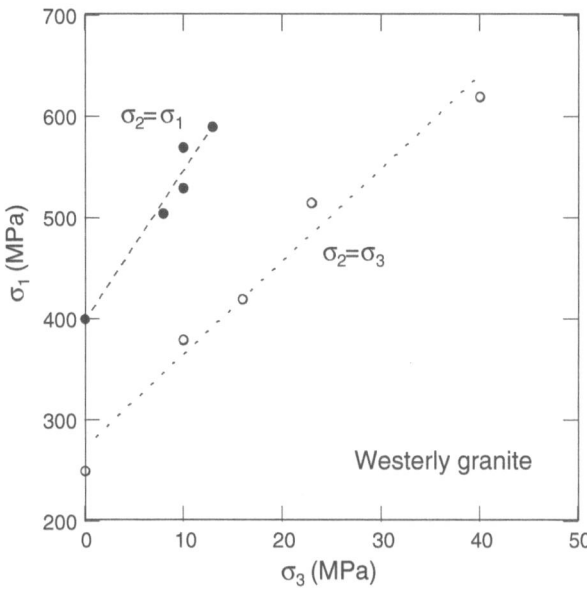

Figure 4

MOGI's (1967) improved conventional triaxial testing results in Westerly granite subjected to relatively low confining pressure σ_3, under both compression and extension.

piston loading by using lubricants such as copper sheet jacketing and Teflon or thin rubber sheets between specimen faces and pistons. Stress concentration at specimen ends was found to be greatly reduced by the high confining pressure (MOGI, 1966). Silicon rubber jacketing of specimen sides subjected to confining pressure and all around the edges of the piston-specimen contacts prevented the confining fluid from permeating the rock. Pressure generation was supplied by high-capacity hydraulic cylinders. A minimum principal stress reaching 800 MPa could be applied.

Using his ingenious apparatus, MOGI (1971) tested several carbonate and silicate rocks, and presented detailed true triaxial fracture strength results for Dunham dolomite and Mizuho trachyte. He discovered that σ_2 has indeed a significant effect on fracture strength at all levels in-between $\sigma_2 = \sigma_3$ and $\sigma_2 = \sigma_1$. Moreover, the largest effect on strength is reached at a level well inside this range. For example, when Dunham dolomite is subjected to $\sigma_3 = 45$ MPa and $\sigma_2 = 250$ MPa, its strength increases to 650 MPa, which is 160 MPa higher than the strength when $\sigma_2 = \sigma_3$, and 90 MPa higher than the strength when $\sigma_2 = \sigma_1$ (compare with Fig. 5 above). Mogi plotted his results in the form of σ_1 at failure versus σ_2 for different families of tests in which σ_3 was kept constant. A typical example is shown in Figure 6 for Dunham dolomite. Fracture strength increases steadily with the magnitude of σ_2.until a plateau is reached following which strength tends to decline as σ_2 approaches σ_1. The best-fitting curve to experimental data for any given σ_3 is downward concave, but

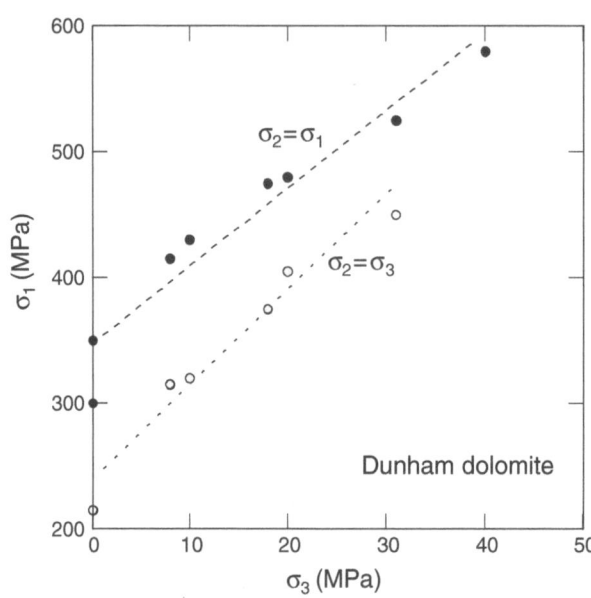

Figure 5
Substantial discrepancy between Dunham dolomite brittle strength under compression ($\sigma_2 = \sigma_3$) and extension ($\sigma_2 = \sigma_1$), using MOGI's (1967) improved conventional triaxial loading apparatus.

strength as σ_2 nears σ_1 remains higher than that when $\sigma_2 = \sigma_3$. Remarkably, this behavior is also predicted by the theoretical model proposed by WIEBOLS and COOK (1968).

In fact, WIEBOLS and COOK (1968) attempted to verify experimentally their own theoretical findings by employing a true triaxial cell in which loading was provided by pressurizing flat jacks inserted between the rectangular prismatic specimen sides and three pairs of rigid anvils. Flat jacks are inflatable bladders consisting of two thin metal sheets that are soldered together around the edges, and which allow hydraulic oil to be pumped in and pressurized against the specimen sides. However, while flat jacks are superior to direct piston loading in terms of the uniformity of the applied stresses, the magnitudes of the stresses must be kept low because of the limited internal pressure accommodated by the soldered sheets. Thus, Wiebols and Cook's experiments in Karroo dolerite lead only to a rather incomplete comparison with their theoretical criterion. It was not until MOGI (1971) that a more convincing agreement between Wiebols and Cook's model and experimental results was achieved in principle for several rocks.

MOGI (1971) also attempted to derive a fitting true triaxial criterion of brittle fracture based entirely on his experimental results. He observed that plotting all his test data for Dunham dolomite and Mizuho trachyte in the NADAI (1950) domain (τ_{oct} versus σ_{oct}, equation 7) did not lead to a uniform criterion. He then rationalized

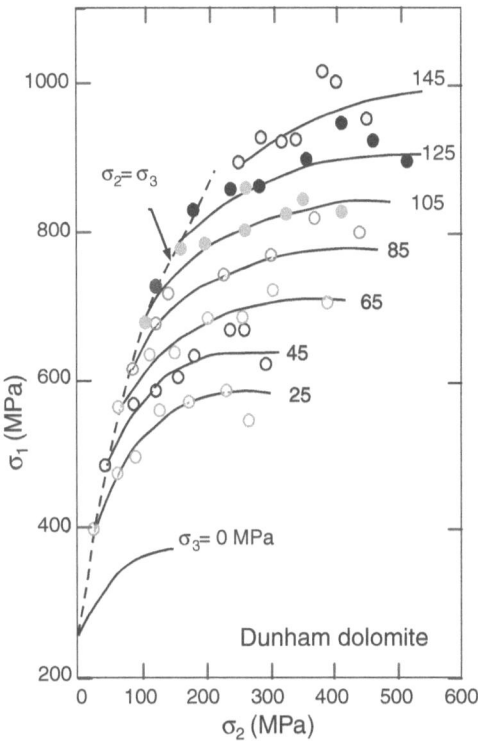

Figure 6
Results of the first comprehensive true triaxial tests conducted by MOGI (1971) using his new apparatus for testing rectangular prismatic specimens. Seven groups of tests are plotted, each for a different σ_3. Within each group a series of tests data are shown, each which is for a different σ_2. Dunham dolomite fracture strength increases in each case when σ_2 is raised above σ_3.

that since brittle fracture was in the form of shearing along one single plane striking in the σ_2 direction, it was more realistic to degenerate the mean normal stress σ_{oct} to just the mean stress on planes striking along σ_2. He found that in both rocks plotting the experimental results as τ_{oct} versus $\sigma_{m,2}$ yielded single monotonically rising curves, best expressed as power functions f_M (Fig. 7):

$$\sigma_{oct} = f_M \sigma_{m,2} \quad \text{where} \quad \sigma_{m,2} = \frac{1}{2}(\sigma_1 + \sigma_3). \tag{11}$$

Contemporaneous with the observations, testing, and theoretical advances related to the role of the intermediate principal stress in rocks, such as those by MURRELL (1963), HANDIN et al. (1967), WIEBOLS and COOK (1968), MOGI (1967, 1971) and others, similar developments were taking place in the field of soil, concrete, and soft-rock mechanics. A series of true triaxial testing machines were developed, typically based on the application of hydraulic pressure through three pairs of flexible

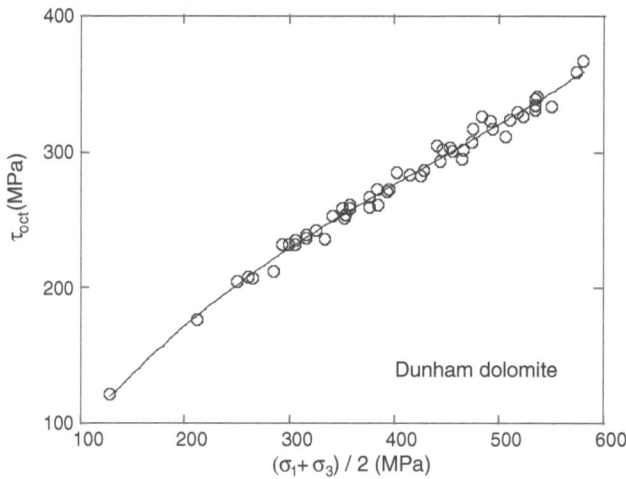

Figure 7
Generalized strength criterion for Dunham dolomite, in which all the experimental data points in Figure 6
align themselves closely with a power function in the octahedral shear stress τ_{oct} versus $(\sigma_1 + \sigma_3)/2$ domain
(after MOGI, (1971).

membranes or fluid cushions to cubical test specimens. KO and SCOTT (1967),
ATKINSON and KO (1973), LADE and DUNCAN (1973), STURE and DESAI (1979), and
DESAI et al. (1982) are among those who developed different designs of true triaxial
testing machines capable of testing soils and weak rocks. The capacity of these
machines was typically less than 200 MPa, and thus not appropriate for loading hard
rocks to failure under high intermediate and least principal stresses.

MICHELIS (1985) modified the flexible membrane design, in order to increase
pressure capacity of the true triaxial cell. His apparatus consisted of a hollow
cylinder that accommodated a prismatic specimen of size $50 \times 50 \times 100$ mm. The
axial (maximum) load was provided by two rigid pistons, while the intermediate and
minimum pressures were supplied by two prismatic PVC bags between specimen and
cylinder walls filled and pressurized by hydraulic oil. The cell allowed pressures
reaching 300 MPa in the lateral directions, and an axial load extending to 1500 MPa.
The initial results in tests conducted on Naxos marble for just one constant σ_3
(13.8 MPa) and different σ_2, showed a clear and substantial effect of the intermediate
principal stress on strength (Fig. 8). As σ_2 was raised from 13.8 MPa to 113.2 MPa,
the true triaxial strength increased from (approximately) 130 MPa to 280 MPa,
which is a 115% rise over the commonly used conventional triaxial strength.

Mogi's pioneering true triaxial testing of hard rocks subjected to high pressures
was followed by TAKAHASHI and KOIDE (1989), who designed and fabricated a near
replica of MOGI's (1971) true triaxial cell. Their equipment accommodated much
larger specimens (reaching $50 \times 50 \times 100$ mm), and was meant specifically for testing

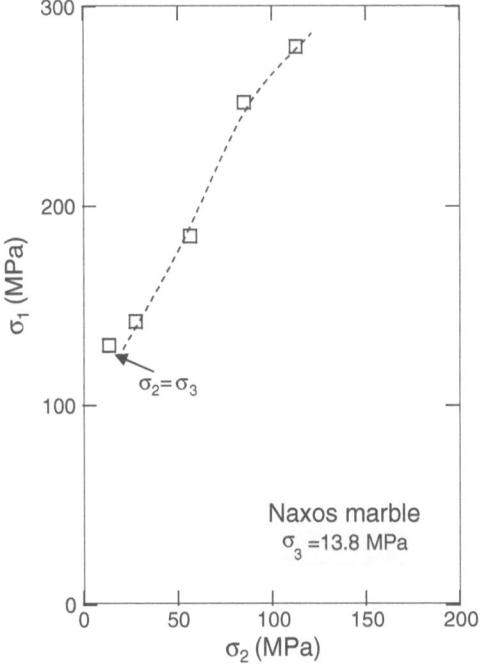

Figure 8
Strength of Naxos marble as a function of σ_2 for a constant and relatively low least principal stress σ_3 in a novel true triaxial apparatus for testing large prismatic rock specimens (after MICHELIS, 1984).

weak rocks such as sandstones and shales. The applied minimum principal stress range was limited to 0–50 MPa, simulating shallow depths, because the interest of these researchers was in rock engineering applications. Again, the strengthening effect of σ_2 as it increased beyond that of σ_3 was evident. Figure 9 reproduces Takahashi and Koide's results in Yuubari shale, which reveals an increase of up to 20% in resistance to brittle fracture as a function of σ_2, for each of the two σ_3 levels tested. TAKAHASHI and KOIDE (1989) also extended WIEBOLS and COOK's (1968) theoretical failure criterion, by rendering it more applicable to true triaxial testing. Their suggested model prediction of the variation of σ_1 as a function of σ_2 for various applied σ_3 matched well their experimental results in Shirahama sandstone.

In the mid-1990s a radically different design of a true triaxial cell was implemented in the form of a conventional triaxial cell in which cylindrical (and not prismatic as in most other designs) rock specimens are subjected to vertical loading applied through pistons at both ends of the cell, and two unequal lateral stresses provided by differential radial pressures (SMART, 1995). Radial pressure variation is achieved through an array of 24 PVC tubes aligned with the specimen long axis and trapped between the specimen and the triaxial cell inner wall. Each tube

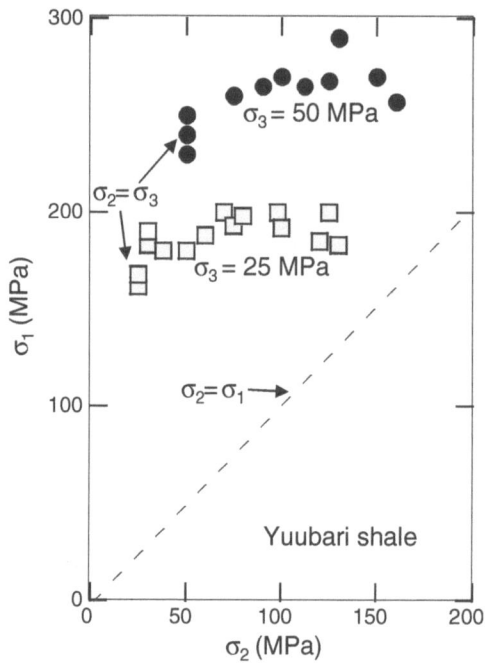

Figure 9
True triaxial testing in a Mogi designed apparatus modified to accept larger samples, shows that the strength of rocks such as shales is also controlled by the level of σ_2, not just σ_3 (after TAKAHASHI and KOIDE, 1989).

is filled with oil and pressurized, and by coupling opposing groups of tubes to independent hydraulic circuits, the application of different lateral pressures to the cylindrical specimen is enabled. The practicality of this cell is obvious, since it does not require tedious six-sided specimen preparation and makes use largely of existing equipment, with some modifications. The limitations of the design are the relatively low absolute and differential pressures that the PVC tubes can tolerate (50 and 7 MPa, respectively) and the intuitive uncertainty regarding the uniformity of the three principal stresses throughout the specimen. The cell cannot be used for a complete series of true triaxial tests since the maximum difference between σ_2 and σ_3 cannot exceed 14 MPa, but important information can still be drawn from tests performed at relatively low intermediate and minimum principal stresses. CRAWFORD et al. (1995) tested two sandstones in this apparatus and found that their fracture strengths were strongly affected by the magnitude of σ_2. Observations of failed sandstone specimens revealed that the brittle fracture mechanism involved the development of stress-induced dilatant microcracks extending parallel to σ_2 and opening against σ_3, progressively coalescing as σ_1 rose, and forming a through-going fault when peak σ_1 was reached.

Another major development in true triaxial testing was the design and fabrication of the Sandia loading facility, which incorporates a conventional triaxial pressure vessel and high-pressure hydraulic rams inside the vessel (WAWERSIK et al., 1997). The cell resembles Mogi's, in that two of the principal stresses are applied through steel pistons, with the third one applied by the confining pressure in the vessel. The innovation here is that the pistons and the hydraulic jacks for the lateral intermediate stress are contained inside the vessel and react against the vessel's inside wall. Bending moments generated by uneven loading are minimized by the use of rubber sheets between the specimen faces and the pistons. Specimen size accommodated by the cell is $57 \times 57 \times 25$ mm or $76 \times 76 \times 178$ mm. The cell is intended specifically for testing high-porosity rocks that undergo large deformations during loading. Unfortunately, beyond some initial true triaxial testing in Castlerock sandstone, no published data on additional experimental results are available.

True Triaxial Testing at the University of Wisconsin

KTB Amphibolite and Westerly Granite

In 1994 the rock mechanics group at the University of Wisconsin embarked on a multi-year true-triaxial research project. This was motivated at first by the need to determine the true triaxial strength criterion of the KTB amphibolite, in order to enable the estimation of the maximum horizontal in situ stress from the logged breakout dimensions in the KTB ultra deep hole. The KTB hole was drilled to a depth of 9100 m in Bavaria by the German Continental Deep Drilling Program to study the properties and processes of the lower continental crust (EMMERMANN and LAUTERJUNG, 1997). One of the main objectives was the measurement of the in situ stress. The vertical stress (σ_v) was assessed from the weight of the overlying strata, and the two horizontal principal stresses (σ_h and σ_H) were estimated using the hydraulic fracturing method. However, below 3000 m only two partially successful hydraulic fracturing tests were carried out (at 6000 m and 9000 m), which yielded estimates of just the least horizontal stress magnitude (BRUDY et al., 1997). Fortunately, borehole breakouts were logged in the amphibolite zone of the KTB hole (3000–7000 m depth). They provided the required information on in situ stress directions. Moreover, the dimensions of the logged breakouts at the borehole wall, together with knowledge of the vertical and least horizontal stresses, were used in an attempt to estimate the maximum tectonic horizontal stress. The model assumed that the stress condition at the points where breakouts intersect the borehole reached the strength criterion of the rock. However, use of criteria based on conventional triaxial tests was considered inadequate for deriving reasonable maximum stress estimates because the principal stresses at the borehole wall are noticeably differential (VERNIK and ZOBACK, 1992). BRUDY et al. (1997) employed the theoretical true triaxial strength criterion suggested by WIEBOLS and COOK (1968) to estimate σ_H. Subsequently, we undertook to re-estimate σ_H by first deriving an empirical true triaxial strength criterion. In the process we conducted a comprehensive experimental study of the true triaxial mechanical behavior of the KTB amphibolite, emphasizing strength and deformability.

The first step was to design and fabricate a true triaxial testing system capable of applying three mutually perpendicular high compressive loads to rectangular prismatic rock specimens (HAIMSON and CHANG, 2000). Our goal was to construct a cell similar to the one used by MOGI (1971), but keep it compact, and make use of existing equipment as much as possible, since our funding was limited. What emerged was a true triaxial loading system, which consists of a polyaxial pressure vessel inside an existing biaxial loading apparatus (Fig. 10). Three independent and mutually orthogonal pressures are generated in this system and applied to a rectangular prismatic specimen. The biaxial apparatus facilitates the application of two pressures, one in the axial (σ_1) and the other in one of the two lateral directions (σ_2) of the specimen. These two pressures are transmitted from the biaxial cell to the rock specimen via two perpendicular pairs of pistons mounted in the pressure vessel. The third pressure (σ_3) is applied directly to the second pair of specimen lateral faces by the confining hydraulic pressure inside the pressure vessel. The loading system was thoroughly calibrated using a strain-gaged aluminum sample of known elastic properties (HAIMSON and CHANG, 2000). The maximum stresses that this loading system can apply to a rock specimen of dimensions $19 \times 19 \times 38$ mm are 1600 MPa in the two piston-loading directions and 400 MPa in the third direction. This high capacity enables the equipment to bring very strong rocks to failure under realistic stress conditions of the order encountered in the earth's crust at depths exceeding 10 km.

In the first series of tests on KTB amphibolite, dry jacketed specimens were loaded to failure under the most general state of stress ($\sigma_1 > \sigma_2 > \sigma_3$). These experiments were intended to provide fundamental data on the mechanical behavior of the amphibolite. The immediate observation in our true triaxial tests was that brittle fracture in the form of a steeply inclined through-going fault plane is generally similar to that observed in conventional triaxial tests. However, the true triaxial loading confirmed the expectation that the dip direction of the fault plane is aligned with σ_3 (CHANG and HAIMSON, 2000). Test results showing the state of stress at failure were plotted graphically as the true triaxial strength σ_1 versus the applied σ_2 for different constant values of σ_3 (Fig. 11). The dependence of σ_1 at failure on the intermediate principal stress σ_2 is obvious. As σ_2 increased beyond σ_3 magnitude, the compressive strength was always higher than that when $\sigma_2 = \sigma_3$. Despite some considerable scatter (the amphibolite tested came from a depth of 6300 m and may have some inhomogeneity and anisotropy), the increase in strength as a function of σ_2 for constant σ_3 is substantial, and for example, as much as 50% or more over the commonly used conventional triaxial strength at $\sigma_3 = 30$ MPa. The effect of σ_2 on

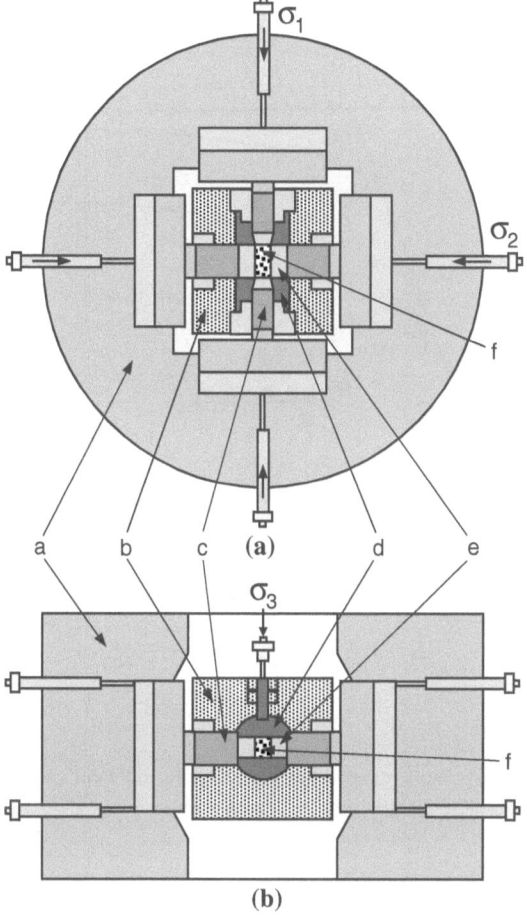

Figure 10
The University of Wisconsin true triaxial testing system: (a) Cross section, (b) profile. In this diagram *a* is
the biaxial loading apparatus; *b* is the polyaxial pressure vessel; *c* is the loading pistons; *d* is the confining
fluid; *e* is the metal anvil; and *f* is the rock specimen.

rock strength appears to weaken as the level of σ_3 rises, but remains significant (10%
higher at $\sigma_3 = 150$ MPa). Thus, conventional triaxial tests provide only a lower
bound of strength for a given least principal stress in KTB amphibolite. It is also
noted that at some level of σ_2 a plateau is reached where strength appears to level off.
This is similar to previous findings in softer rocks by MOGI (1971) and TAKAHASHI
and KOIDE (1989), and in support of WIEBOLS and COOK (1968) theoretical
prediction.

    In order to define a fitting strength criterion to the experimental results in
Figure 11, we replotted the data in the domains used by NADAI (1950), MURRELL

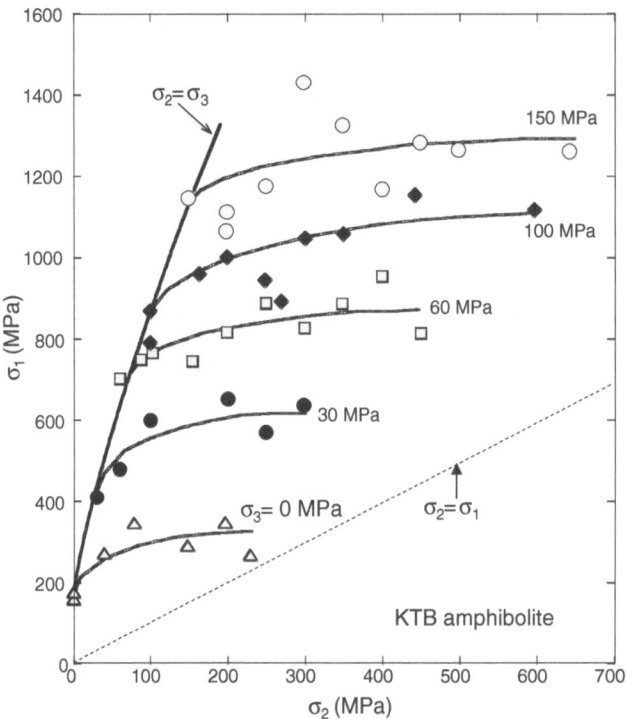

Figure 11

Variation of peak compressive stress σ_1 as a function of σ_2 for different constant values of σ_3 in KTB amphibolite (after CHANG and HAIMSON, 2000).

(1963), and MOGI (1971) for their respective strength criteria (equations 6, 7, 11). The only domain in which our experimental data points fit along one single curve with minimum deviation was that suggested by MOGI (1971) for brittle failure, i.e., the octahedral shear stress τ_{oct} as a function of the mean normal stress acting on the failure plane $\sigma_{m,2}$ (equation 11). The best fitting curve for KTB amphibolite was a monotonically increasing power function of the form (Fig. 12):

$$\tau_{\text{oct}} = A\sigma_{m,2}^{b}, \tag{12}$$

where A and b are two parameters that are rock-specific.

Our testing system enabled the measurement of strain in all three principal directions (HAIMSON and CHANG, 2000). The stress-strain behavior observed, revealing accelerated extension in the σ_3 direction as σ_1 increased, suggests that most of the induced and reopened microcracks eventually leading to brittle failure are aligned with the $\sigma_1 - \sigma_2$ plane and open up in the σ_3 direction. For each set of stress-strain curves we calculated the stress-volumetric strain and marked the onset of dilatancy, the stress level at which the constant rate of volumetric decrease starts

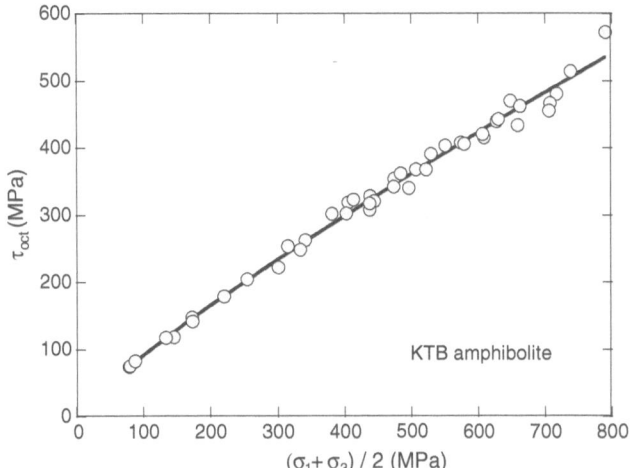

Figure 12
A true triaxial strength criterion for KTB amphibolite, based on all the experimental results shown in
Figure 11, in terms of τ_{oct} versus $(\sigma_1 + \sigma_3)/2$ (after CHANG and HAIMSON, 2000).

reversing itself. Dilatancy has been correlated to internal microcracking responsible
for enlarging specimen volume and for leading eventually to brittle fracture (BRACE
et al., 1966). We determined that dilatancy is more pronounced at low σ_2 magnitudes
but diminishes at higher σ_2 levels, supporting results obtained by TAKAHASHI and
KOIDE (1989) in Shirahama sandstone. The onset of dilatancy for a given σ_3 generally
increases with the magnitude of σ_2 (Fig. 13) similar to observations by MOGI (1971)
in Mizuho trachyte. These findings are significant in that previous conventional
triaxial tests could only identify a unique dilatancy onset for a certain σ_3 in a given
rock. The other significance is that higher intermediate principal stress magnitudes
appear to extend the elastic range of the stress-strain behavior for a given σ_3, and by
doing so it retards the onset of the failure process. The micromechanics leading to
brittle fracture under true triaxial stress conditions begins at dilatancy onset, when
microcracks, subparallel to the major principal stress σ_1, develop. As σ_1 increases
microcracks grow and localize creating a shear-band dipping in the σ_3 direction.
Upon brittle fracture the shear band fails, forming a fault (Fig. 14).

    Using the same true triaxial system, we also conducted an extensive series of tests
in Westerly granite, which had been thoroughly studied in rock mechanics
laboratories in the U.S., and which is well known for its homogeneity and isotropy.
The results were remarkably similar to those obtained in KTB amphibolite, both
with respect to the effect of σ_2 on brittle fracture and the true triaxial strength
criterion (Figs. 15 and 16), as well as dilatancy onset and the micromechanics of
failure. From this analogous behavior between Westerly granite and KTB amphibolite we inferred that it represents typical true triaxial mechanical characteristics of
fine- to medium-grained crystalline rocks.

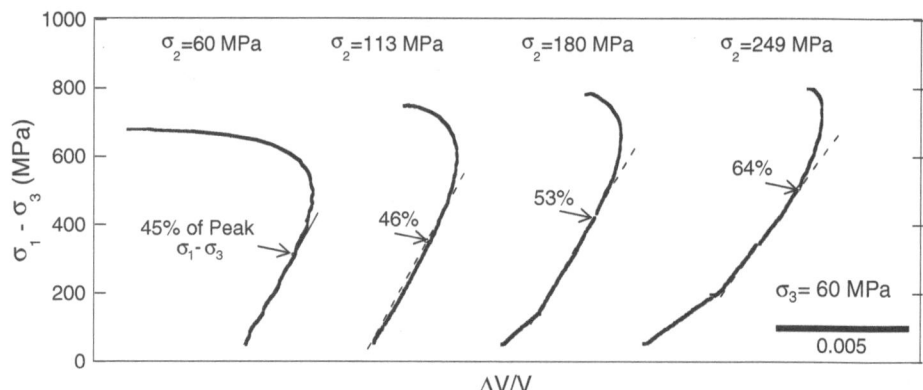

Figure 13
Differential stress ($\sigma_1 - \sigma_3$) versus volumetric strain (ΔV/V). showing that for a constant σ_3 (60 MPa) the onset of dilatancy rises with the increase in σ_2.

Unjacketed Amphibolite and the State of in situ Stress at KTB

A second series of true triaxial tests was conducted on KTB amphibolite, specifically for the purpose of estimating *in situ* stress magnitudes from breakout dimensions logged in the ultra deep hole. These tests were intended to simulate field conditions, in which the least principal stress at the borehole wall equals the radial pressure applied by the borehole fluid to the exposed rock, and where owing to the extremely low porosity and permeability of the KTB amphibolite, the pore pressure in the rock is assumed to be negligible (BRUDY et al., 1997). Thus, in this series of tests samples were initially dry, but the confining-pressure fluid (kerosene), which applies σ_3, was in direct contact with two opposing specimen faces that were left unjacketed. The results of this series of tests were substantially different from those obtained in the jacketed specimens (HAIMSON and CHANG, 2002).

Since KTB amphibolite is a nearly impermeable, practically no confining pressure fluid penetrated the rock prior to the opening of microcracks. However, fluid infiltration upon dilatancy onset completely changed the failure mode. Brittle fracture occurred at or soon after dilatancy onset, roughly at 50–60% of σ_1 at failure under jacketed conditions, and resulted from the development of densely spaced through-going extensile cracks adjacent and subparallel to one of the unjacketed faces (Fig. 17). Confining fluid apparently intruded newly opened microcracks upon dilatancy onset, and facilitated their extension into long tensile fractures subparallel to σ_1–σ_2 plane. For any given least principal stress, the compressive strength typically increased with the rise in the intermediate principal stress, but overall it maintained its much lower magnitude than that of jacketed dry samples of the same rock (Fig. 18). The true triaxial strength criterion of amphibolite under borehole wall conditions is not well represented by the criterion

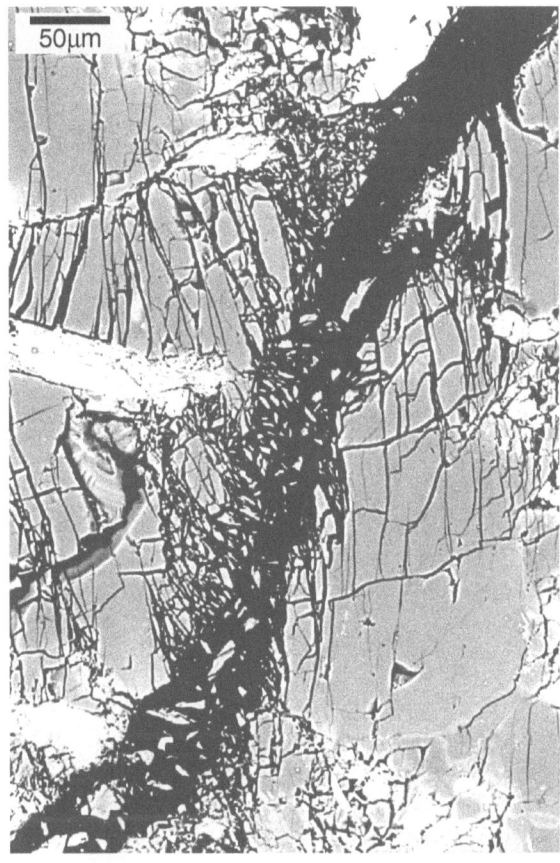

Figure 14

SEM micrograph of a failed KTB amphibolite, showing the microcrack localization that preceded the fault, which is steeply inclined and dips in the σ_3 (acting laterally in this image) direction. Scattered grain debris within the fault is inferred as a result of shear slip.

given in equation (12), but can be expressed as a linear relationship between the octahedral shear stress and the octahedral normal stress at failure (equation 8; Fig. 19). The discrepancy can be explained based on the fact that failure occurs immediately upon dilatancy onset, which is approximately equal to the yield point in this brittle rock and which occurs throughout the specimen and not just along the fracture plane (MOGI, 1971).

Using this criterion together with all the other known data from the KTB hole (BRUDY *et al.*, 1997), we computed the estimated magnitude of the maximum horizontal *in situ* stress there (HAIMSON and CHANG, 2002). Since the dimensions of the laboratory specimens were close to those of the breakouts in the KTB hole, scale effect was considered negligible. Our results (Fig. 20) show that σ_H increases steadily

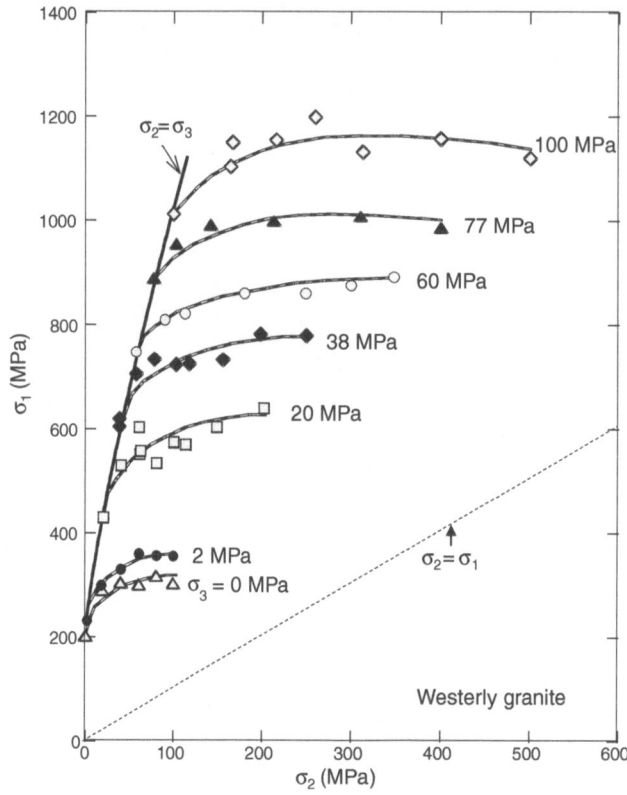

Figure 15
True triaxial test results in Westerly granite, plotted as the peak σ_1 versus σ_2 for several constant σ_3 magnitudes (after HAIMSON and CHANG, 2000).

with depth, within a relatively narrow band of uncertainty, confirming previous assessments of a strike-slip stress regime by BRUDY *et al.* (1997).

Long Valley Hornfels and Metapelite

More recent experiments have shown that the true triaxial mechanical behavior described above cannot be generalized to all brittle rocks. True triaxial strength experiments in two ultra fine-grained brittle rocks, hornfels and metapelite, which together are the major constituent of the Long Valley (California, U.S.A.) caldera basement in the 2000–3000 m depth range, exhibit a behavior unlike that previously observed in other crystalline and clastic rocks under similar testing conditions (CHANG and HAIMSON, 2005). For a given magnitude of σ_3, compressive strength σ_1 does not vary significantly in either hornfels or metapelite, regardless of the applied σ_2, suggesting little or no intermediate principal stress effect on brittle fracture (Fig. 21).

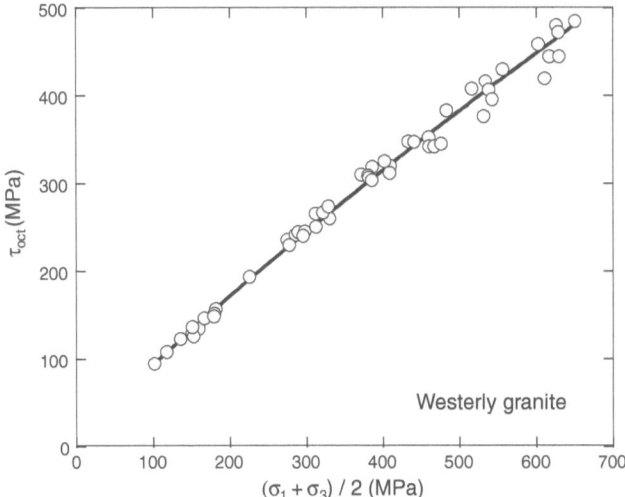

Figure 16

True triaxial strength criterion for Westerly granite, resulting from plotting all the experimental results shown in Figure 15 as τ_{oct} versus $(\sigma_1 + \sigma_3)/2$ (after HAIMSON and CHANG, 2000).

Measured σ_1 versus volumetric strain is linear almost to the point of rock failure, suggesting the absence of dilatancy. This unique mode of compressive failure that is not preceded by dilatancy was not previously detected in true triaxial tests. SEM inspection of failed specimens corroborates the observed nondilatant deformation by failing to reveal microcrack development prior to the emergence of the through-going steeply dipping shear failure plane (Fig. 22). This recent study implies that the commonly accepted mechanism of growth, localization, and coalescence of stress-induced extensile microcracks preceding brittle failure is not universally applicable to all hard rocks. It also indirectly confirms that the onset of microcracking is responsible for both dilatancy and the strengthening effect of the intermediate principal stress.

The mechanism that culminates in compressive failure of the hornfels and metapelite is not clearly understood. One attractive speculation is that minute shear microcracks, which have been found to induce brittle fracture in a serpentinite and a syenite without discernible volumetric increase (ESCARTIN et al., 1997; KATZ and RECHES, 2000) are responsible for the nondilatant deformation and σ_2-independent compressive strength of the Long Valley rocks. However, no such discontinuities were detected in our microscopic study.

Summary, Conclusions and Future Work

In this paper I attempted to demonstrate the ever increasing evidence accumulated over the last century of the effect of the intermediate principal stress on brittle

Figure 17
Cross section of failed unjacketed KTB amphibolite specimen along a $\sigma_1 - \sigma_3$ plane, showing a cluster of through-going extensile cracks parallel to the unjacketed face subjected to σ_3. In this figure σ_1 acts vertically.

fracture and failure criteria of rock. The Coulomb, Mohr, and Griffith criteria have been universally used to depict brittle failure based only on knowledge of the least and largest principal stresses. This approach enables the use of uncomplicated conventional triaxial cells to yield expressions for failure criteria. However, the intermediate principal stress effect on strength and deformability has proven over and over to be substantial. Strength differences of 10 to 50% have been observed in most rocks tested thus far, with the exception of the ultra fine-grained Long Valley hornfels and metapelite. In addition, the angle of fracture orientation with respect to

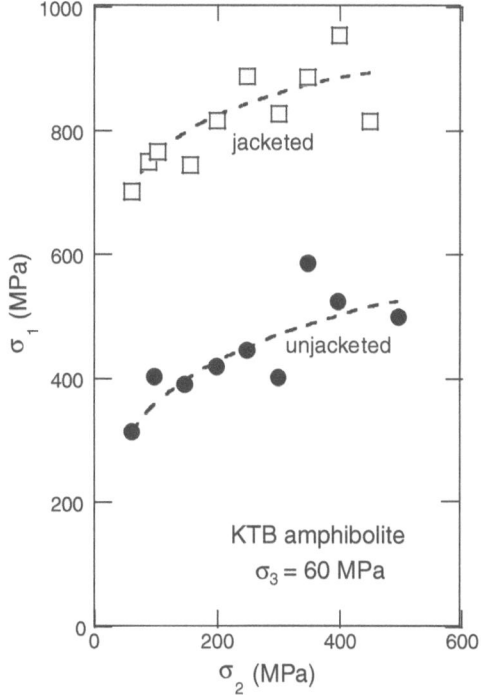

Figure 18

Compressive strength of unjacketed KTB amphibolite (blocked circles) as a function of σ_2 for a constant σ_3 (60 MPa). For comparison the strength of the same rock and for the same σ_3 under jacketed conditions is also plotted (open squares). The increase in strength with σ_2 is similar, but the unjacketed rock is about half as strong as the jacketed.

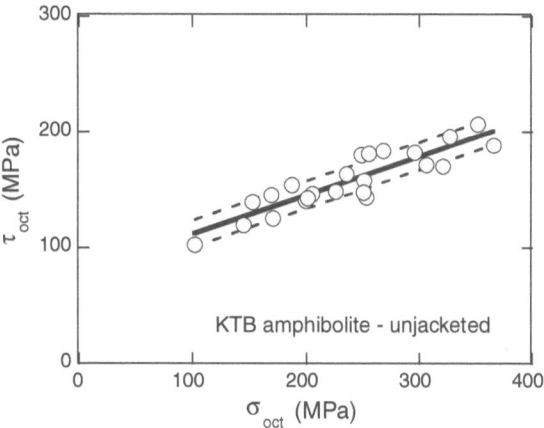

Figure 19

True triaxial strength criterion of unjacketed KTB amphibolite is best described as a linear relationship between the octahedral shear stress τ_{oct} and the octahedral normal stress σ_{oct}.

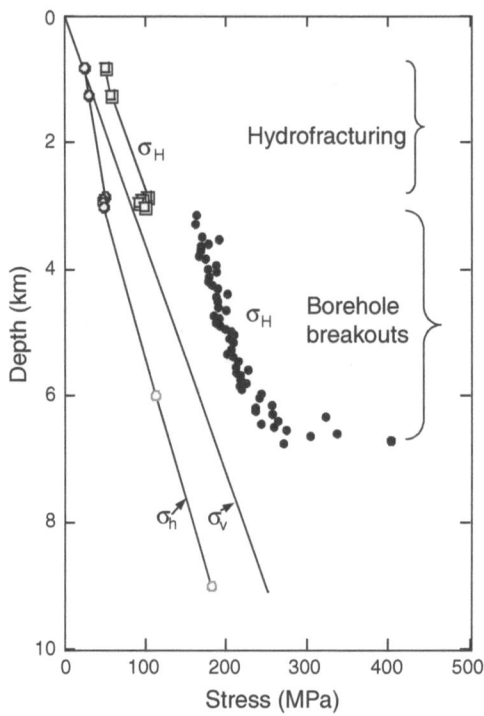

Figure 20
The estimated state of stress around the KTB ultra deep hole. In the 3 to 7 km range hydraulic fracturing yielded only an estimate of the least horizontal stress σ_h (see open circle at 6.0 km depth). The vertical stress σ_v was assumed to equal the weight of the overlying rock (BRUDY et al., 1997). The major horizontal principal stress σ_H was computed from knowledge of the logged breakout dimensions and the true triaxial strength criterion of unjacketed amphibolite (HAIMSON and CHANG, 2002).

the directions of the principal stresses has been found to depend not only on σ_3, but also on σ_2. Also, the onset of dilatancy was shown to increase sharply with the rise in σ_2 in tests run under a constant σ_3.

Employing Coulomb, Mohr or other criteria derived from them (such as the 'Mohr-Coulomb' criterion) for assessing rock strength yields only the lower limit of this parameter, appropriate when σ_2 is equal or only slightly larger than σ_3. For larger intermediate principal stresses a true triaxial strength criterion is recommended, based on true triaxial tests conducted on rectangular prismatic specimens. Several examples are scattered in the literature in which it is reported that using the conventional strength criterion (equation 3) in some specific applications leads to nonsensical results. VERNIK and ZOBACK (1992) found that use of the Mohr-Coulomb criterion in relating borehole breakout dimensions to the prevailing *in situ* stress conditions in crystalline rocks did not provide realistic results. They suggested the use of a more general criterion that accounts for the effect of the intermediate

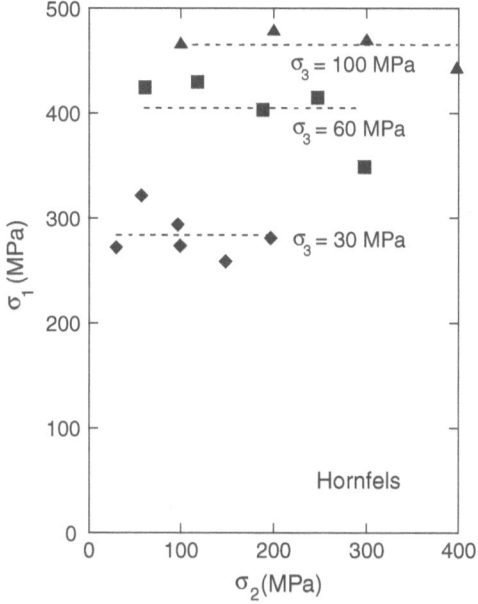

Figure 21
True triaxial strength σ_1 as a function of σ_2 for three separate magnitudes of σ_3 in Long Valley Caldera hornfels extracted from a depth of 2,250 m. The relationship is best represented by a constant (CHANG and HAIMSON, 2005).

principal stress on strength. EWY (1998) reported that for the purpose of calculating the critical mud weight necessary to maintain wellbore stability, the Mohr-Coulomb criterion is too conservative because it neglects the strengthening effect of the intermediate principal stress. Ewy considered using the DRUCKER and PRAGER (1952) criterion or a modified Lade criterion (LADE and DUNCAN, 1973) to take into account the σ_2 effect.

The importance of incorporating all principal stresses in determining experimentally the mechanical behavior of rocks has contributed to our present involvement in conducting true triaxial tests on extracted core from two major site investigations, the San Andreas Fault Observatory as Depth (SAFOD), California, U.S.A., and the Taiwan Chelungpu-fault Drilling Project (TCDP), both programs being under the auspices of the International Continental Scientific Drilling Program (ICDP).

Acknowledgements

The work performed at the University of Wisconsin and the preparation of this paper were funded by National Science Foundation grants nos. EAR-9418738 and

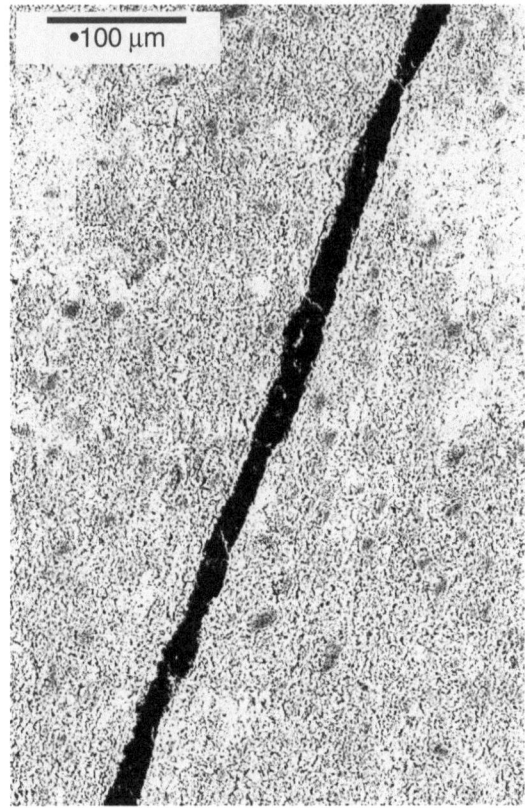

Figure 22

SEM micrograph of a cross section of a failed Long Valley hornfels specimen along a $\sigma_1 - \sigma_3$ plane, showing a steeply inclined shear fault, but lacking any visible microcracks. (Compare with the KTB amphibolite, Fig. 14.) The lack of microcracks is inferred as the reason for the nondilatancy in hornfels. Long Valley metapelite failure mechanism is identical to that of the hornfels (CHANG and HAIMSON, 2005).

EAR-0346141. C. Chang carried out all the reported University of Wisconsin experimental work. I. Song and R. Sandberg assisted in the design of the University of Wisconsin true triaxial cell.

REFERENCES

ATKINSON, R.H., and KO, H.Y. (1973), *A fluid cushion, multiaxial cell for testing cubical rock specimens*, Intl. J. Rock Mech. and Mining Sci. *10*, 351–361.

BÖKER, R. (1915), *Die Mechanik der bleibenden Formänderung in kristallinisch aufgebauten Körpern*, Verhandl. Deut. Ingr. Mitt. Forsch. *175*, 1–51.

BRACE, W.F., PAULING, B.W., and SCHOLZ, C.H. (1966), *Dilatancy in the fracture of crystalline rocks*, J. Geophys. Res. *71*, 3939–3953.

BRESLER, B. and PISTER, K.S. (1957), *failure of plane concrete under combined stresses*, Trans. Am. Soc. Civ. Engrs. *122*, 1049–1068.

BRUDY, M., ZOBACK, M.D., FUCHS, K., RUMMEL, F., and BAUMGÄRTNER, J. (1997), *Estimation of the complete stress tensor to 8 km depth in the KTB scientific drill holes: implications for crustal strength*, J. Geophys. Res. *102*, 18,453–18,475.

CHANG, C. and HAIMSON, B.C. (2000), T*rue triaxial strength and deformability of the KTB deep hole amphibolite*, J. Geophys. Res. *105*, 18,999–19,014.

CHANG, C. and HAIMSON, B.C. (2005), N*ondilatant deformation and failure mechanism in two long valley caldera rocks under true triaxial compression*, Intl. J. Rock Mech. and Mining Sci. *42*, 402–414.

CRAWFORD, B.R., SMART, B.G.D., MAIN, I.G., and LIAKOPOULOU-MORRIS, F. (1995), S*trength characteristics and shear acoustic anisotropy of rock core subjected to true triaxial compression*. Int. J. Rock Mech. and Min. Sci. *32*, 189–200.

DESAI, C.S., JANARDAHONAN, R., and STURE, S. (1982), *High capacity multiaxial testing device*, Geotech. Testing J. *5*, 26–33.

DRUCKER, D.C., and PRAGER, W. (1952), *Soil mechanics and plastic analysis or limit design*, Quart. Appl. Math. *10*, 157–165.

EMMERMANN, R., and LAUTERJUNG, J. (1997), *The german continental deep drilling program KTB: Overview and major results*, J. Geophys. Res. *102*, 18,179–18,201.

ESCARTIN, J., HIRTH, G., and EVANS, B. (1997), *Nondilatant brittle deformation of serpentinites: Implications for Mohr-Coulomb theory and the strength of faults*. J. Geophys. Res. *102*, 2897–2913.

EWY, R.T., *Wellbore stability predictions using a modified lade criterion*. In *Rock Mechanics in Petroleum Engineering*, vol. 1, Proc. Eurock 98 (Society of Petroleum Engineers, 1998), pp. 247–254..

FREUDENTHAL, A., *The inelastic behavior and failure of concrete*. In Proc. I (ASME, New York, 1951), pp. 641–646.

HAIMSON, B., and CHANG, C. (2000), *A new true triaxial cell for testing mechanical properties of rock, and its use to determine rock strength and deformability of westerly granite*, Int. J. Rock Mech. Min. Sci. *37*, 285–296.

HAIMSON, B., and CHANG, C. (2002), *True triaxial strength of the KTB amphibolite under borehole wall conditions and its use to estimate the maximum horizontal in situ stress*, J. Geophys. Res. *107(B10)*, ETG 15-1 to 14.

HANDIN, J., HEARD, H.C., and. MAGOUIRK, J.N. (1967), *Effect of the intermediate principal stress on the failure of limestone, dolomite, and glass at different temperature and strain rate*, J. Geophys. Res. *72*, 611–640.

JAEGER, J.C., and COOK, N.G.W., *Fundamentals of Rock Mechanics*, 3rd ed., (Chapman and Hall, London 1979) 593 pp.

JAEGER, J.C., and HOSKINS, E.R. (1966), *Rock failure under the combined Brazilian test*, J. Geophys. Res. *71*, 2651–2659.

KATZ, O. and RECHES, Z. (2000), *micro- and macro-structural analysis of small faults in a quartz-syenite intrusion: Faulting of a brittle rock without microcracking?* EOS Transactions, AGU *81*, F1121.

KO, H.Y. and SCOTT, R. F. (1967), *A new soil testing apparatus*, Geotechnique *17*, 40–57.

LADE, P.V. and DUNCAN, J.M. (1973), *Cubical triaxial tests on cohesionless soil*, J. Soil Mech. and Foundation Div., ASCE *99*, 793–812.

MICHELIS, P. (1985), *A true triaxial cell for low and high-pressure experiments*. Int. J. Rock Mech. Min. Sci. *22*, 183–188.

MOGI, K. (1966), *Some precise measurements of fracture strength of rocks under uniform compressive strength*, Rock Mech. Engin. Geology *4*, 51–55.

MOGI, K. (1967), *Effect of the intermediate principal stress on rock failure*, J. Geopys. Res. *72*, 5117–5131.

MOGI, K. (1971). *Fracture and flow of rocks under high triaxial compression*, J. Geophys. Res. *76*, 1255–1269.

MURRELL, S.A.F., *A criterion for brittle fracture of rocks and concrete under triaxial stress, and the effect of pore pressure on the criterion*. In Proc. *Fifth Symp. on. Rock Mechanics*, (Pergamon Press, 1963), pp. 563–577.

NADAI, A., *Theory of Flow and Fracture of Solids*, vol. 1 (McGraw-Hill, New York, 1950).

SMART, B.G.D. (1995), *A true triaxial cell for testing cylindrical rock specimens*, Int. J. Rock Mech. and Min. Sci. *32*, 269–275.

STURE, S., and DESAI, C.S. (1979), *Fluid cushion truly triaxial or multiaxial testing device*, Geotech. Testing J. *2*, 20–33.

TAKAHASHI, M., and KOIDE, H., *Effect of the intermediate principal stress on strength and deformation behavior of sedimentary rocks at the depth shallower than 2000 m*, In *Rock at Great Depth* (eds. V. Maury and D. Fourmaintraux) (Balkema, Rotterdam, 1989), pp. 19–26.

VERNIK, L., and ZOBACK, M.D. (1992), *Estimation of maximum horizontal principal stress magnitude from stress-induced well bore breakouts in the cajon pass scientific research borehole*, J. Geophys. Res. 97, 5109–5119.

VON KARMAN, T. (1911), *Festigkeitsversuche unter all seitigem Druck*, Z. Verein Deut. Ingr. *55*, 1749–1759.

WAWERSIK, W.R., CARLSON, L.W., HOLCOMB, D.J., and WILLIAMS, R.J. (1997), *New method for true-triaxial rock testing*, Int. J. Rock Mech. and Min. Sci. *34*, Paper no. 330.

WIEBOLS, G.A. and COOK, N.G.W. (1968), *An energy criterion for the strength of rock in polyaxial compression*, Int. J. Rock Mech. Min. Sci. *5*, 529–549.

(Received April 21, 2005, revised September 9, 2005, accepted September 12, 2005)

 To access this journal online:
http://www.birkhauser.ch

Pure appl. geophys. 163 (2006) 1131–1151
0033–4553/06/061131–21
DOI 10.1007/s00024-006-0067-5

© Birkhäuser Verlag, Basel, 2006

❚ Pure and Applied Geophysics

Discrete Element Modeling of Stress and Strain Evolution Within and Outside a Depleting Reservoir

HAITHAM T. I. ALASSI,[1] LIMING LI,[2] and RUNE M. HOLT[1,2]

Abstract—Stress changes within and around a depleting petroleum reservoir can lead to reservoir compaction and surface subsidence, affect drilling and productivity of oil wells, and influence seismic waves used for monitoring of reservoir performance. Currently modeling efforts are split into more or less coupled geomechanical (normally linearly elastic), fluid flow, and geophysical simulations. There is evidence (from e.g. induced seismicity) that faults may be triggered or generated as a result of reservoir depletion. The numerical technique that most adequately incorporates fracture formation is the DEM (Discrete Element Method). This paper demonstrates the feasibility of the DEM (here PFC; Particle Flow Code) to handle this problem. Using an element size of 20 m, 2-D and 3-D simulations have been performed of stress and strain evolution within and around a depleting reservoir. Within limits of elasticity, the simulations largely reproduce analytical predictions; the accuracy is however limited by the element size. When the elastic limit is exceeded, faulting is predicted, particularly near the edge of the reservoir. Simulations have also been performed to study the activation of a pre-existing fault near a depleting reservoir.

Key words: Reservoir geomechanics, numerical modeling, discrete element method, reservoir compaction, surface subsidence, stress path, fault.

Introduction

Petroleum reservoir depletion leads to stress alteration within and outside the reservoir. During recent years it has become evident that such stress changes can have a profound impact on reservoir management (e.g., TEUFEL *et al.*, 1991; ADDIS, 1997; KENTER *et al.*, 1998; HOLT *et al.*, 2004). Not only do they control purely mechanical deformation (reservoir compaction and surface subsidence), but they also impact petroleum recovery through compaction drive and through possible permeability changes. Furthermore, stress changes may affect the ability to drill stable wells, and the risks for onset of particle production or casing collapse throughout the life of the reservoir. In some cases, depletion-induced stress changes

---

[1] NTNU Norwegian University of Science and Technology, Trondheim, Norway.
E-mail: alassi@ntnu.no; rune.holt@ntnu.no
[2] SINTEF Petroleum Research, Trondheim, Norway. E-mail: liming.li@iku.sintef.no

may be large enough to cause seismicity by activation of existing or generation of new faults. This may be utilized as a tool for reservoir performance monitoring (MAXWELL and URBANCIC, 2001). The main purpose of reservoir monitoring is to identify which parts of the reservoir that are produced, so that the production strategy can be tailored to the behavior of the reservoir. Today reservoirs are frequently monitored by "4-D" (also called time-lapse) seismic surveys. Clearly, stress sensitive wave velocities within a depleting reservoir or its surroundings may cause time-shifts that can be used as indicators of reservoir performance (KENTER *et al.*, 2004).

The economic impact of the issues above calls for modeling tools that can predict the evolution of stresses as a result of pore pressure changes associated with fluid extraction from the reservoir. Further, models need to be available that can also predict associated strains (compaction, subsidence, casing deformations), associated seismic velocity changes, and associated seismicity risk. There is currently no model that can be used to predict all these facets of the problem. Geomechanical simulators (PANDE *et al.*, 1990; ZIENKIEWICZ, 1991; JING and HUDSON, 2002) addressing large-scale problems like those described above are most often based on Finite Element (FEM) formulations, and are inherently static in the sense that they do not predict dynamic features like faulting. They do however predict plastic strain occurrence, but need to be re meshed in order to account for faulting. Although full poromechanical coupling is available (SETTARI and MOURITS, 1994; GUTIERREZ and LEWIS, 1998; LEWIS *et al.*, 2003; KOUTSABELOULIS and HOPE, 1998; OSORIO *et al.*, 1998; LONGUE-MARE *et al.*, 2002) in such models, at least in a staggered manner, a further link to seismic modeling is as yet absent.

The motivation behind the work presented here is to explore the feasibility of applying an inherently dynamic model to this problem, namely a discrete element (DEM) approach. The DEM used here is the Particle Flow Code[1] (PFC), which is available in 2-D and 3-D formulations, and which has been applied with success at grain scale (CUNDALL and STRACK, 1979; POTYONDY and CUNDALL, 2004), and also has been refined to incorporate poromechanical coupling (SHIMIZU, 2004; LI and HOLT, 2004) and elastic wave propagation (LI and HOLT, 2002). Clearly, this model may have severe limitations for a reservoir or even basin-scale application as outlined here — the elements in the model can no longer be particles, but must be several meter large circular or spherical grid blocks. Conversely, the potential of the DEM to study localized failure, as demonstrated by LI and HOLT (2002), makes it attractive for the purpose of studying the impact of inelasticity which has not been properly addressed by other tools.

A key subject in reservoir geomechanics is the reservoir stress path as defined in the next Section, and how the stress path may be linked to the production strategy of

---

[1]Trademark of Itasca c.g., Minneapolis, U.S.A.

the field. We then proceed to describe the basic principles of the DEM used in this work (PFC). It is important to validate such an approach: Since direct experimental calibration is not possible, our validation strategy has been to determine if results of analytical elastic modeling can be reproduced. We will therefore show a comparison between predictions of the DEM and the classical Geertsma theory (GEERTSMA, 1973), both for 2-D and 3-D cases. We then proceed to address the case in which the elastic limit is exceeded somewhere in the model, leading to damage, in the form of fault generation. Finally, we demonstrate how DEM may be used to analyze the circumstances in which a pre-existing fault may be activated as a result of reservoir depletion.

Geomechanics of Depleting Reservoirs

The reservoir stress path is defined through the following parameters (HETTEMA et al., 1998)

$$\gamma_v = \frac{\Delta\sigma_v}{\Delta p_f}; \ \gamma_h = \frac{\Delta\sigma_h}{\Delta p_f}. \tag{1}$$

Here Δ_v and Δ_h denote vertical and horizontal stress path coefficients, representing the change in total vertical and horizontal stresses ($\Delta\sigma_v$ and $\Delta\sigma_h$) with change (Δ_{pf}) in reservoir pore pressure. Notice that the γ – parameters are valid within the reservoir as well as in the surrounding rock volume, but the pore pressure change always refers to the reservoir.

If there is no stress arching so that the full weight of the overburden is carried by the reservoir, then $\gamma_v = 0$. If in addition the reservoir compacts (linearly) elastically with no lateral strain, then

$$\gamma_h = \alpha\frac{1 - 2v_{fr}}{1 - v_{fr}}, \tag{2}$$

where α is the poroelastic (Biot) coefficient and v_{fr} is Poisson's ratio for the drained reservoir rock. Since $\gamma_h > 0$ and the pore pressure decrease is negative, Eqs. (1) and (2) imply that the total horizontal stress is reduced.

It is evident from field experience (TEUFEL et al., 1991; ADDIS, 1997; KENTER et al., 1998) and also from theoretical considerations (RUDNICKI, 1999; SEGALL and FITZGERALD, 1998) that the stress path in a general case will deviate from that above. If the reservoir is drained in such a way that the drained volume cannot be approximated as a flat "pancake"–like object, then stress arching will occur. Also, a stiff (compared to the draining rock volume) overburden will promote stress arching. One consequence of stress arching is that stress changes occur more in the overburden than within the reservoir.

GEERTSMA (1973) used the so-called "nucleus of strain" method to calculate an analytical solution for displacements as well as changes in the stress field for a depleting disk-shaped reservoir. His solution is limited by the assumption of uniform elastic properties of the sedimentary basin, including the reservoir and the surrounding rock.

In order to solve this problem for realistic field cases, where the shape of the reservoir differs from the idealized cylindrical geometry, where there is elastic contrast between the reservoir and its surroundings, and where the reservoir may be tilted, numerical techniques must be used. The Finite Element Method (FEM) has been applied to this problem by e.g. KOSLOFF *et al.* (1980); MORITA *et al.* (1989); BRIGNOLI *et al.* (1997); GAMBOLATI *et al.* (1999; 2001) and MULDERS (2003).

As an example of the outcome of such simulations, Figure 1 shows the stress path coefficients obtained on the basis of FEM simulations (MAHI, 2003) vs. depth for a case of elastic match between reservoir and surrounding rock. Results are shown for two different radii of drainage. γ_v is positive, which means that the reservoir compacts (as a response to effective stress change) less than it would if arching was not present. Outside the reservoir, where the pore pressure is not expected to change much as a result of depletion, the positive γ_v value corresponds to vertical decompression. The other stress path coefficient, γ_h, is positive within the reservoir (reduced total but increased effective horizontal stress), and negative above and below, implying horizontal compression in those areas. Reducing drainage area is

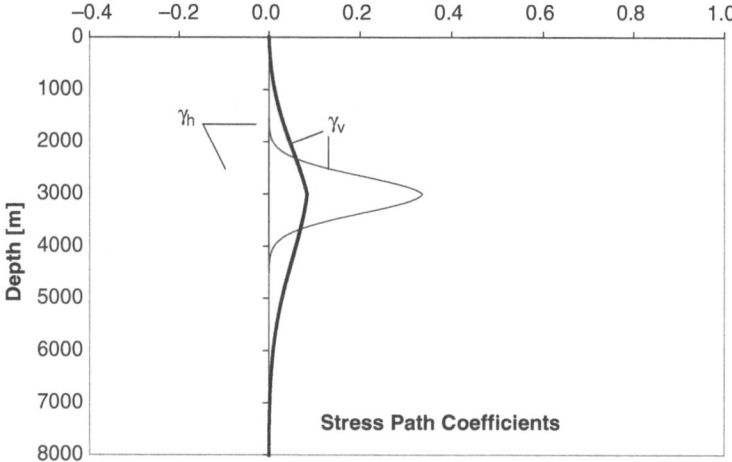

Figure 1
Vertical and horizontal stress path coefficients along a vertical line through the reservoir center, calculated based on FEM simulations (after MAHI, 2003). The computations are performed for a disk-shaped (500 m thick) reservoir centered at 3000 m depth, with a drainage radius of 2000 m (bold curves) and 500 m (narrow curves). Approximate solutions are shown for the case of elastically matched reservoir and surroundings (Young's modulus = 12 GPa; Poisson's ratio = 0.20). Notice that these curves are reproduced as mathematical approximations to FEM simulations.

seen to cause increased arching within and near the reservoir, although the influenced zone is shrinking. This situation may correspond to an early development phase or production of an isolated reservoir compartment. Note that the zone affected by stress alteration as a result of depletion in both cases extends 1000 m or more above and below the reservoir.

An important observation from FEM simulations as well as from analytical computations (SEGALL and FITZGERALD, 1998) is that the vertical stress is strongly increased near the edge on the outside of the reservoir, while the horizontal stress is reduced. This stress alteration may exceed the elastic limit of the rock around the reservoir, and the edge zone is therefore where fault generation or fault activation most likely will take place.

Discrete Element Modeling

We have in this work applied a Discrete Element Method (DEM) named PFC ("Particle Flow Code") (CUNDALL and STRACK, 1979; POTYONDY and CUNDALL, 2004), which is widely used to model the mechanical behavior of rock and other granular materials. The material is represented by discrete particles, basically disks (in 2-D) or spheres (in 3-D) which interact with each other through a user-defined (usually linear) force-displacement contact law, using a soft contact (overlapping particles) approach. Within a calculation cycle, the values of forces and displacements are calculated, and the law of motion is applied to each particle to update position and velocity. Bonds can be inserted at the contacts to represent cementation in rocks. The model is fully dynamic, and hence able to describe complex phenomena like rock failure. One significant point in PFC is that elastic energy can be tracked during simulation, which allows the user to monitor the energy release during crack development and fault sliding. Additionally, wave propagation simulations can be easily performed (LI and HOLT, 2002) since PFC is a dynamic program.

In the subsequent sections of this paper we will use bonded models to simulate reservoir depletion and fault activation.

Elastic Case: Comparison with Geertsma's Analytical Model

Bonded particles can be used to model continuum media, similar to other numerical methods like FEM. The main purpose of the work presented in this section, is to discern to what extent PFC performs as a continuum model. To do this, a set of simulations has been performed both with two-dimensional (PFC^{2-D}) and three-dimensional (PFC^{3-D}) DEM models, and then compared to analytical predictions based on GEERTSMA (1973). Thus; the model and the boundary conditions have been constructed so that no interparticle bonds break, i.e., the model material is linearly elastic.

GEERTSMA (1973) used the center of dilatation ("nucleus of strain") concept to calculate displacements and stress changes associated with depletion of a disk-shaped reservoir in an elastically homogeneous half-space. His analytical solutions are valid for 3-D, making it necessary to derive similar analytical solutions for the center of dilatation (represented as disks) approach in 2-D (the derivation is not shown in this paper). Also, instead of using analytical integrals as done by Geertsma for the disk-shaped reservoir, numerical integral is incorporated to solve the problem of other 3-D reservoir shapes.

Modeling of Depletion for a Rectangular Reservoir Using PFC^{2-D}.

PFC is suitable for grain-scale modeling, where recent studies indicate that a good qualitative and close to quantitative match between modeling and experiment can be obtained (HOLT *et al.*, 2005). Since here we use PFC for modeling of large-scale behavior, the particle size must be chosen large (typically 20 m radius in this work) as well, to keep reasonable computational time. No controlled experiment is possible, consequently, validation is performed by comparison with an analytical model as described above.

In order to make the PFC model most comparable to continuum models, the particle packing should be chosen as compact as possible. In the work presented here, a hexagonal packing of uniformly-sized particles is used. This leads to anisotropy, which creates difficulty in finding suitable linear elastic parameters for the model when comparing it to isotropic analytical theory. An alternative would be to choose a

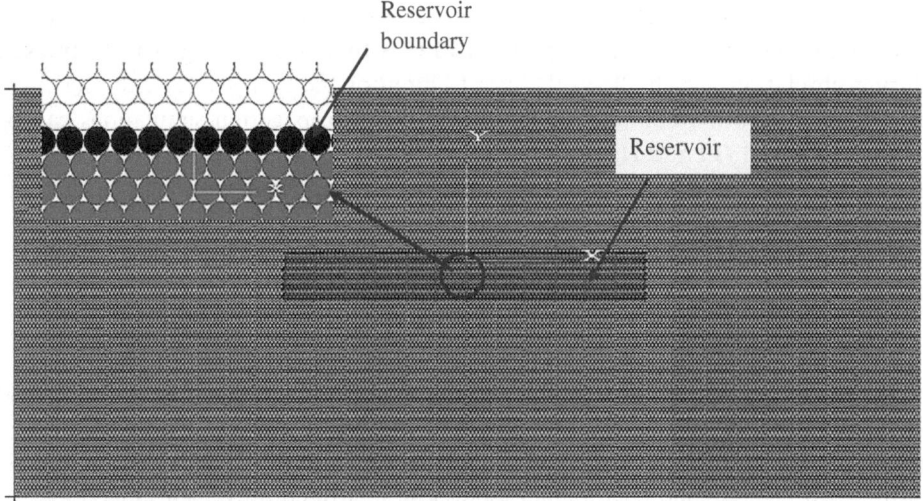

Figure 2
PFC^{2-D} geomechanical model used for modeling reservoir depletion. The black particles along the reservoir boundary denote where forces are applied to simulate depletion.

broad particle-size distribution. Further, force transmission in granular materials is different from that in continua. The force chain pattern depends not only on the elastic parameters of the system, but also on the contact law that governs the relation among the neighboring particles (linear or nonlinear), and the packing of the particles.

The model is 10 km wide and 4.3 km deep. It is composed of a hexagonal packing of 31250 equally sized (radius $=$ 20 m) particles. After packing, gravitational force is added under zero lateral strain (fixed walls) boundary conditions. Finally, parallel bonds are inserted at all interparticle contacts. The tensile as well as the shear strength of the bonds are set equal to 5 MPa. Figure 2 shows the model that is used during the simulation, including a rectangular reservoir inserted at 2000 m depth from the surface. Table 1 shows the model properties. Note that the reservoir parameters do not represent any real reservoir, since the main purpose of this study is to demonstrate feasibility of DEM for reservoir and basin scale studies.

The reservoir is depleted uniformly, with no drainage to the surroundings. Under this assumption, the pore pressure gradient on the boundary will be very large, whereas inside the reservoir it will be zero. In FEM modeling this problem may be solved using a technique presented by GAMBOLATI et al. (2001). They let the pore pressure decrease from p_f to zero on a string of elements around the reservoir. In our model we similarly apply these forces to all particles at the reservoir boundary. The accuracy of our solution will hence depend on the element (i.e., particle) size, which is linked to computational time.

Using this method, the reservoir is depleted by a pore pressure change $\Delta p_f = -10$ MPa. The reservoir has been placed at different depths c within the model basin. Young's modulus and Poisson's ratio of the reservoir material (as listed in Table 1) were determined by performing a biaxial test on a sample with the same PFC parameters as the reservoir. In the reservoir model there is however a stress gradient, therefore elastic parameters are also expected to change with depth. No bonds were broken in the model during this simulation, meaning that the PFC material behaves perfectly elastic. The resulting compaction and subsidence are plotted in Figure 3 together with the analytical solution obtained from Geertsma's

Table 1

Model properties for the PFC^{2-D} simulations

Properties	Values
Model dimensions [km]	10 * 4.3
Reservoir dimension [m]	4000*500
Particle radius [m]	20
Interparticle normal stiffness [GN/m]	24
Interparticle shear stiffness [GN/m]	12
Interparticle normal and shear bond strength [MPa]	5
Young's modulus [Gpa]	30
Poisson's Ratio [-]	0.14

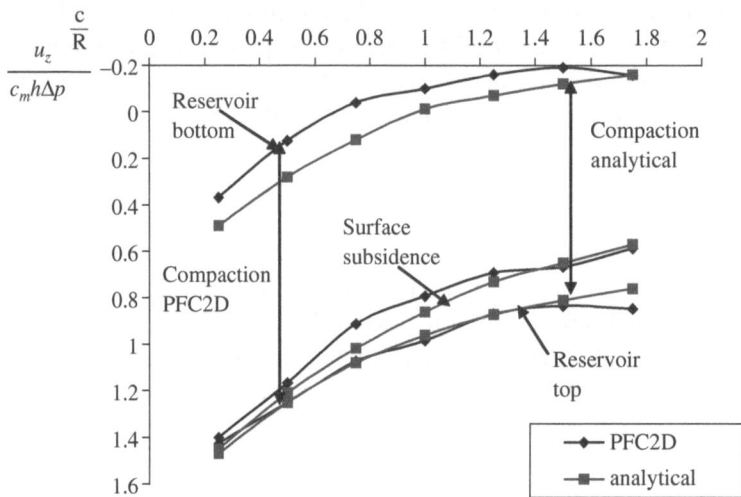

Figure 3
Surface subsidence and displacement at the top and the bottom of a rectangular (4000 × 500 m) reservoir, simulated with a PFC[2-D] model and those obtained by analytical solution. Results are shown for different reservoir depths. The reservoir is depleted with Δp_f 10 MPa. u_z is the vertical displacement, c_m is the uniaxial compaction coefficient, h, R, and c are reservoir thickness, radius (= 2000 m), and depth, respectively. Reservoir compaction equals the difference between reservoir top and bottom displacements.

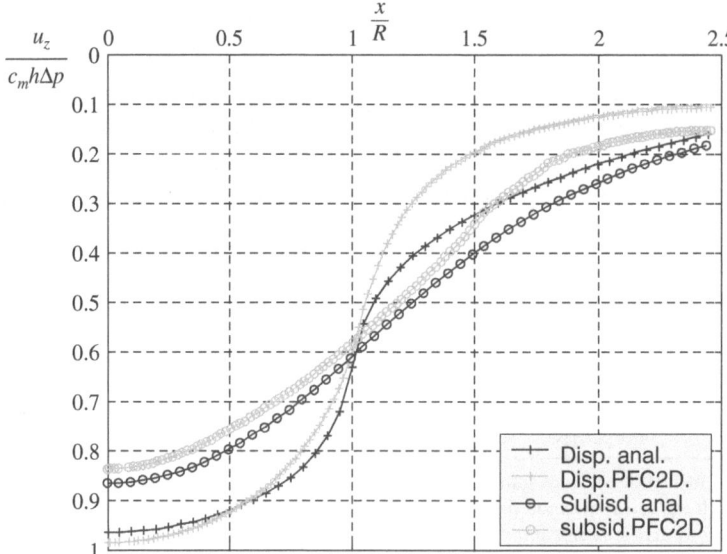

Figure 4
Comparison between PFC[2-D] modeled and analytically calculated reservoir displacement (at the top of the reservoir) and surface subsidence along the x axis (lateral direction). Model parameters are listed in Table 1, reservoir depth is 2000 m, uz is the vertical displacement. cm is the uniaxial compaction coefficient, h is reservoir thickness, R is the reservoir radius.

method, adapted to 2-D. As depth c increases, the values of subsidence and also the displacement of the top of the reservoir decrease, given that the reservoir dimensions are kept unchanged. It can also be seen that for shallow depths ($c/R < 0.5$; R is reservoir radius) the value of subsidence becomes closer to that of vertical displacement at the top of the reservoir. Satisfactory agreement is obtained between the numerical and the analytical solutions. Figure 4 shows a similar comparison between PFC^{2-D} and the Geertsma 2-D solution of the subsidence and compaction bowls in the case of a reservoir placed at 2000 m depth. The agreement is again acceptable.

The PFC^{2-D} simulation permits determination of the stress path coefficients (Eq. (1)) throughout the model. The changes in vertical and horizontal stresses are measured after depleting the reservoir by 10 MPa. The arching coefficients obtained from PFC and analytical solutions are shown vs. depth through the reservoir center in Figure 5, and in the lateral direction just above the top of the reservoir in Figure 6. Note that the discrete element model predicts an increase in the horizontal stress path coefficient with distance from the center of the reservoir towards the edge, as was found also in the finite-element simulations of a disk-shaped reservoir by MULDERS (2003). On the other hand, there is a significant difference between the results of the PFC simulation and the analytical solution: While the trends are the same, the values

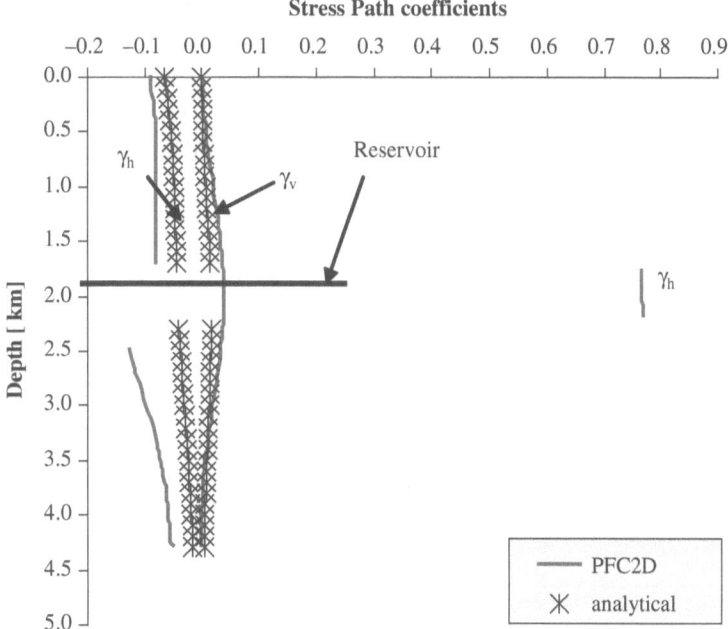

Figure 5

Vertical and horizontal arching coefficients versus depth, from PFC^{2-D} simulation. Reservoir depth = 2000 m.

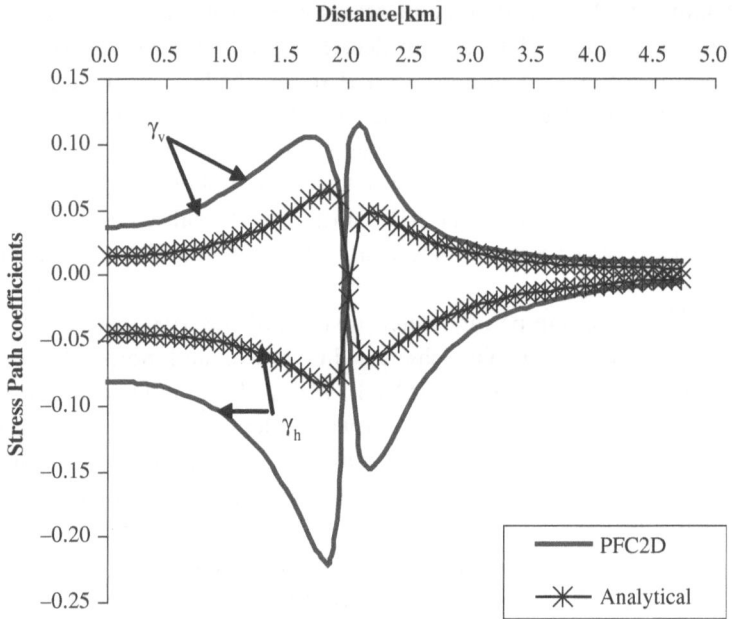

Figure 6

Vertical and horizontal arching coefficients along x axis, from PFC$^{2\text{-}D}$ simulation. Reservoir depth = 2000 m.

of the stress path coefficients differ significantly. This is related to element size and texture as mentioned above, and particularly to choosing the appropriate elastic parameter for the analytical computation. The difference also depends on the method used to measure the stress in PFC: To date the stress is assumed to exist only in the particles (or disks). The boundary conditions also highly contribute to the difference, as can be seen in Figure 4, where the discrepancy between the analytical and the numerical solution increases with distance from the reservoir boundary towards the model boundary.

Table 2

Model properties for the PFC$^{3\text{-}D}$ simulations

Properties	Values
Model dimensions [m]	1600 * 1600*800
Reservoir dimension [m]	800*800*120
Particle radius [m]	20
Young's modulus [GPa]	12
Poisson's Ratio [-]	0.0

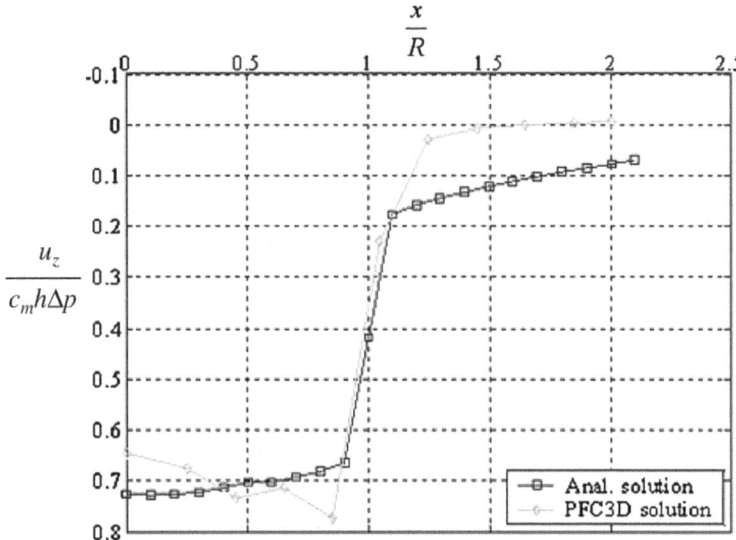

Figure 7
Comparison between PFC$^{3\text{-D}}$ modeled and analytically calculated (with the nucleus of strain model; GEERTSMA, 1973) compaction (at the top of the reservoir). Model parameters are listed in Table 2, reservoir depth is 400 m.

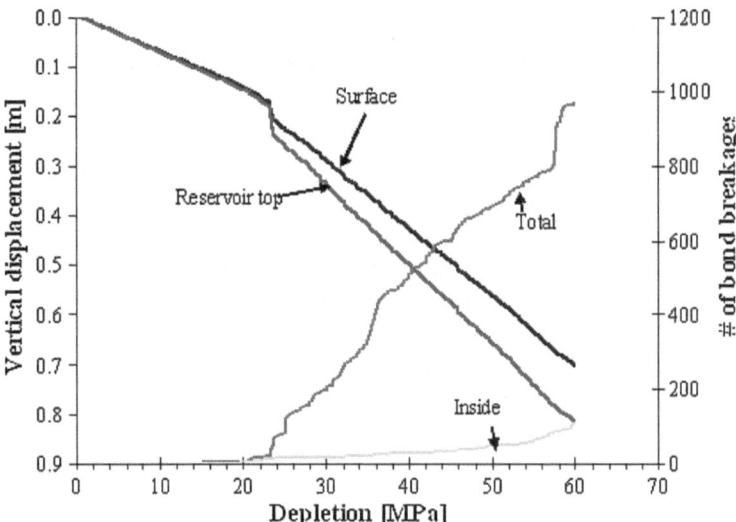

Figure 8
Vertical displacements at the surface and at the top of the reservoir during simulated depletion, using a PFC$^{2\text{-D}}$ model as described in the text. Also shown are recorded numbers of broken bonds between elements within the reservoir and in the full model.

Modeling of Depletion Using PFC^{3-D}

A PFC^{3-D} model consisting of 32,000 spherical particles is constructed, using a cubic packing (see Table 2 for model description). The element (particle) size was kept the same as in the 2-D modeling (20 m). In order to limit computational time, the model size is considerably reduced (1600 × 1600 × 800 m). The reservoir thickness is 120 m, and it is inserted at a depth of 400 m. Depletion of the reservoir is again simulated by applying normal forces to the boundary particles (as was done above).

Figure 7 shows a comparison between PFC^{3-D} modeling and the Geertsma [3-D] solution for compaction at the top of the reservoir. Again, a triaxial test was performed to establish Young's modulus and Poisson's ratio for the reservoir material. As in the 2-D case, the fit is acceptable, but not perfect. The reasons for not accomplishing perfect matching are the same as above: The size of the model relative to the particle size is even smaller in this case, which is a primary source of error. Again; rock properties in the PFC model are expected to vary with depth, and the cubic packing also introduces a slight anisotropy. Nevertheless, a main conclusion is that both 2-D and 3-D PFC simulations with perfectly elastic (no bond breakage) material produce results which are fairly close to analytical predictions.

Beyond Elasticity: Fault Initiation within and outside a Depleting Reservoir

As can be depicted from the previous sections, the stresses evolving during reservoir depletion may exceed the elastic limit; within the reservoir, as well as outside of it. This may lead to the formation of localized deformation bands, or activation of pre-existing faults. In order to study faulting, the PFC^{2-D} model created in the previous section was used, with a significantly larger reservoir depletion (= 60 MPa).

Figure 8 shows the modeled surface subsidence and displacement at the top of the reservoir. The rate of compaction increases with increasing depletion, and the increased reservoir compressibility can be directly linked to damage inside the reservoir as measured by the number of bond breakages. The vertical displacement on the surface of the model shows a similar trend. Obviously, the increased compaction within the reservoir contributes to this. The change in subsidence to compaction ratio is small, in spite of significant bond breakage also in the surrounding material, in particular near the reservoir edges, as illustrated in Figure 9. The localized failure zone near the reservoir edge seems to have little impact on the surface subsidence, at least as long as they do not reach the surface. The observed failure pattern is in agreement with expectations based on analytical computations, finite-element simulations (e.g., BRIGNOLI *et al.*, 1997), as well as laboratory modeling (PAPAMICHOS *et al.*, 2001). While bonds fail largely in shear within the reservoir, tensile bond failures dominate outside. This is partly a result of the somewhat arbitrary choice of tensile vs. shear bond strengths. Notice that

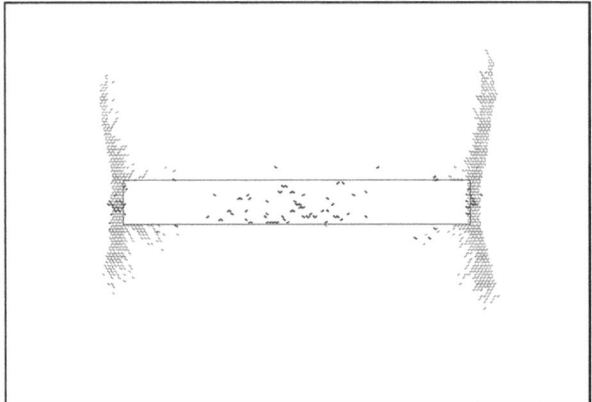

Figure 9
Positions of bond breakages after depleting the PFC$^{2\text{-}D}$ model shown in Figure 2 with 60 MPa. The black line segments indicate local shear failures, while the grey ones represent tensile failures.

the failure pattern in this simulation corresponds largely to that seen in a previous PFC$^{2\text{-}D}$ simulation (HOLT et al., 2004), but differs in details: In that case, significant bond breakage occurred above the reservoir as well as near the edges. The difference is mainly caused by the difference in particle-size distribution and texture. Figure 10 also shows bond breakages after continued depletion to 100 MPa (notice that the values are arbitrarily chosen and do not represent a real case — in reality, the level of depletion should be compared to the strength parameters of the surrounding and reservoir rock). Cracks are seen to propagate to

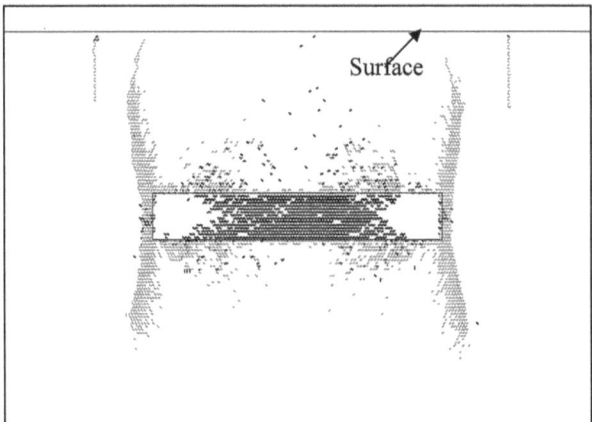

Figure 10
Positions of bond breakages after depleting the PFC$^{2\text{-}D}$ model shown in Figure 2 with 100 MPa. The black line segments indicate local shear failures, while the grey ones represent tensile failures.

the surface and the number of cracks inside and outside the reservoir increase significantly. Although this is not a realistic case, it shows a similar trend to that obtained from laboratory modeling (PAPAMICHOS *et al.*, 2001).

DEM Modeling with a Pre-existing Fault

The simulations shown in the previous section demonstrate that the DEM is able to simulate faulting during depletion of an initially intact reservoir embedded in initially intact surroundings. One may however question if this fault pattern is realistic or not – it is clearly limited by the resolution of the simulation (particle size), which limits the possibility for a fault to localize within the model. In reality, faults may also exist before the reservoir is depleted. The positions of these faults may be seen from seismics, and then it makes no sense to use a numerical model to attempt reproduction of their formation.

These considerations triggered a study of how PFC may be used to embed and simulate the behavior of an existing fault, and to explore the feasibility of studying fault response to reservoir depletion.

To create a fault in PFC^{2-D}, the same model as in previous sections is applied, but with specific properties assigned to a group of particles along a pre-defined fault plane. Table 3 presents the fault properties. Recognize that since the hexagonal packing is used, the fault takes a straight shape because of the chosen dipping angle (60°). Irregular packing may also be used, however then smaller particle sizes need to be created in the fault zone. Slip may be initiated in different ways. A triggering process driven by a high shear stress is mimicked by reducing the friction coefficient between the fault particles and neighboring particles. If the process is triggered by high normal stress, fault activation may be simulated by slightly reducing the size (by 1%) and stiffness (see Table 3) of the fault particles.

After the fault is initiated, the model is run to equilibrium, where the unbalanced force is reduced to a minimum value, and no further fault slipping occurs. In our case, a normal fault is developed according to both scenarios above, since the model

Table 3

Fault properties used in PFC simulations. These properties are assigned to all the particles that compose the fault

Properties	Values
Normal stiffness k_n [GN/m]	1
Shear stiffness k_s [GN/m]	0.5
Friction coefficient μ [-]	0.3
Fault length [m]	1480
Fault dip angle [°]	60

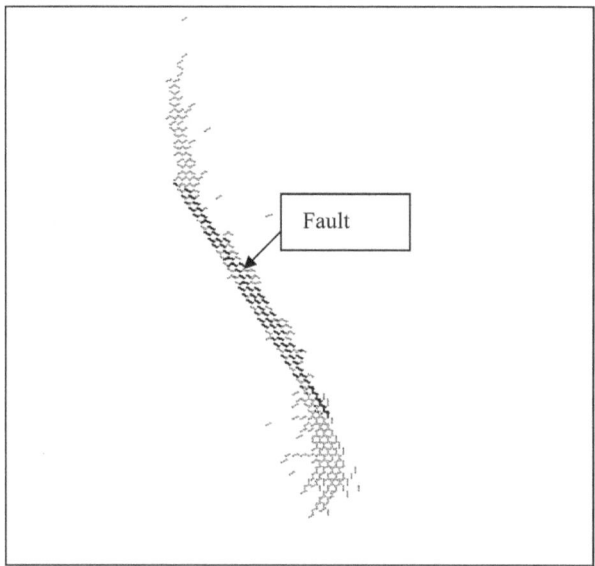

Figure 11
Bond breakages during fault sliding, triggered by reducing the interparticle friction coefficient. Note that all bond failures are tensile (gray color), except at the fault face, which shows failures in shear (black).

is in a normal-faulting environment (vertical > horizontal stress). The hanging wall slips downward and the foot wall slips upward.

The shear-induced fault (Fig. 11) extends in the direction of the maximum principal stress by development of wing cracks. Damage is mainly located of the tip

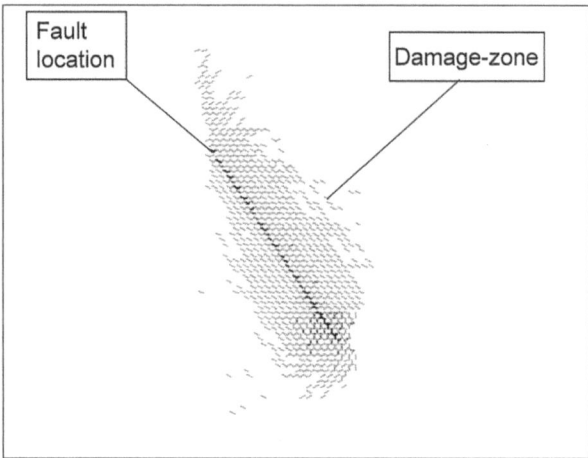

Figure 12
Bond breakages during fault sliding, triggered by reducing particle size and stiffness. Note that all bond failures are tensile (gray color), except at the fault face, which shows failures in shear (black).

regions of the fault. The compaction-induced fault, on the other side, develops a more extended damage zone (Fig. 12).

Fault sliding alters the stress distribution of the model, leading to stress concentrations at the tips of the fault. On one side of the tip, the stress increases (more compression), while on the other side of the same tip, there is an area where the stress decreases (becomes more tensile). Within the stress reduction zones, bonds may break in tension. Cracks grow during sliding of the fault as a result of more stress concentration, and the stress re-distribution caused by bonds breakage. Eventually the cracks that nucleate at different places will coalesce with each other forming a damage-zone around the fault. The cracks do not only start at the tips of the fault, but also along the fault plane, because of the stress disturbance caused by a sudden change of the stiffness and the size of the particles that form the fault. Figure 12 shows the tensile breakages of the parallel bonds at the end of the simulation. ODED *et al.* (2002) presented a fault deformation model which predicts damage (cracks) not only at the fault tips, but also along the fault plane, as is also seen from the PFC model with particle shrinkage as the fault trigger.

Reservoir Depletion, with Fault on the Side of the Reservoir

To study the effect of reservoir depletion on re-activation of a fault, a reservoir is inserted to the left of the fault created previously (see Fig. 13). The size of the reservoir is (arbitrarily chosen) 2500 *500 m. Again an undrained boundary condition is assumed. According to SEGALL and FITZGERALD (1998), normal faults that lie on the side of the reservoir will be re-activated under such circumstances,

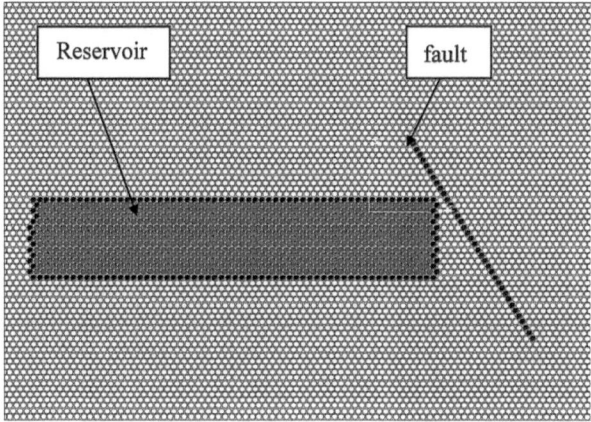

Figure 13
A normal fault is placed to the right of a reservoir. The model is used to simulate the re-activation behavior of the fault due to reservoir depletion.

given a sufficient pore pressure reduction. A simulated reservoir depletion of 40 MPa causes slipping of the fault, the hanging wall moves downward, while the foot wall follows the movement of the reservoir boundary. It can be seen that the deformation of the lower boundary increases the amount of slip, while the deformation of the upper boundary of the reservoir decreases the slip between the two fault faces. This behavior differs from that of a typical normal fault, in which the foot wall is expected to move upward.

The slip or frequently called RSD (relative shear displacement) is plotted in Figure 14 versus depth after 20 and 40 MPa depletion. This value represents the relative displacement between the two sides of the fault in the dipping direction. Reactivation causes new bond breakage in the area of stress concentration; in this case at the tensional side of the fault tips. Figure 15 depicts the new cracks that are developed due to reservoir depletion. The increasing tension on the sides of the reservoir as a result of depletion causes creation of a tensile-normal fault in the direction perpendicular to minimum horizontal stress (SEGALL and FITZGERALD, 1998; FERRILL and MORRIS, 2003). Since in our model the minimum stress is horizontal, the created tensile fault has a dip angle = 0 (vertical fault).

Discussion

The simulations presented here demonstrate the feasibility of using a discrete element model to simulate the geodynamics of a depleting reservoir. The strength of

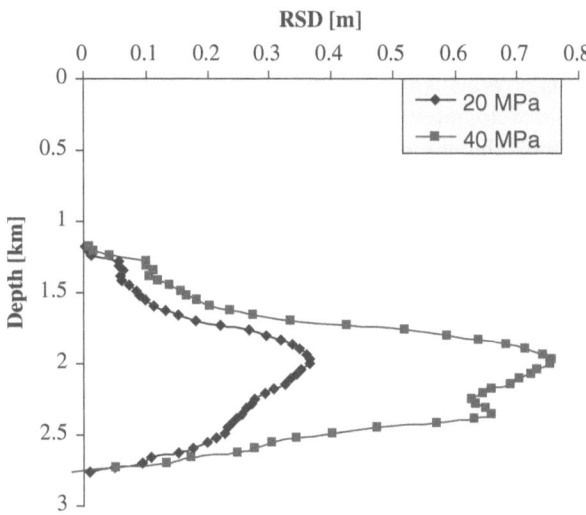

Figure 14
Slip between the fault faces (see Fig. 12) after 20 and 40 MPa depletion.

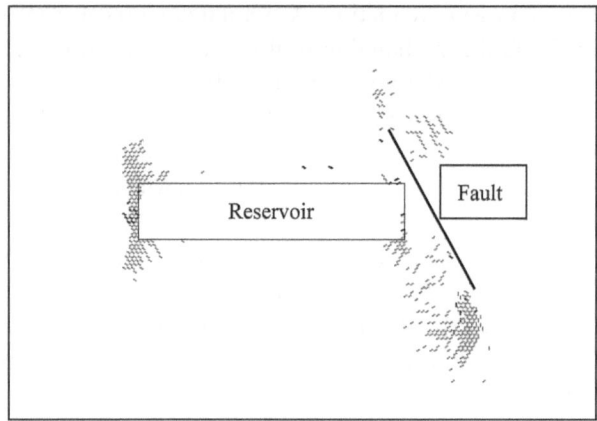

Figure 15
Bond breakages developed after depleting the reservoir in Figure 12 by 40 MPa. Note the concentration of the cracks at the tips of the fault and also at tips of the reservoir, as a result of stress concentration in those areas.

DEM is the ability to simulate faulting and fault activation in a dynamic manner, where natural complexity emerges from simple contact laws. A dynamic approach is therefore beneficial when fracturing is expected to take place. Consequently traditional finite-element solutions suffer, mainly from the need to continuously remesh as a fracture grows, but also because of limitations in path conformability and element size. However; there will be a multitude problems where FEM solutions are sufficient, and these solutions are more efficiently obtained than DEM solutions.

Element size also is a main restriction for DEM. Within limits of current computer technology, element size cannot be reduced to the size of physical particles. Rules need to be developed to guide the choice of particle size distribution and packing, and to guide the choice of parameters for contact laws between elements. Notice that disks or spheres as used here are basic building blocks which may be grouped into clusters or "clumps" to generate elements of various shapes (POTYONDY and CUNDALL, 2004; LI and HOLT, 2002). Micromechanical calibration (as in HOLT *et al.*, 2005) cannot be expected to provide a complete answer here, and the approach must be based largely on field experience, geological considerations, and comparison to theory or other modeling tools. Improved resolution may however be obtained by using small particles in parts of the model where the dynamic feature is most required. This may be achieved with PFC by utilizing a recent option for automatic linkage of the DEM to a continuum (e.g., FEM or finite-difference) model.

In the PFC simulations shown here, poromechanical coupling was (for simplicity) not applied. This is however possible (SHIMIZU, 2004; LI and HOLT, 2004), and would permit more realistic treatments of pressure gradients within a reservoir compartment and across faults. This also permits well drainage to be part of the model. Currently, only single phase fluid flow has been coupled to PFC, nonetheless this is not a fundamental limitation. Also, since wave propagation can be performed relatively easy with PFC (LI and HOLT, 2002), direct simulations of seismic surveys as well as induced seismicity (HAZZARD and YOUNG, 2000) may be incorporated within the same scheme as the geomechanical and fluid flow simulations.

Conclusions

We have demonstrated the feasibility of a Discrete Element Model (PFC) to simulate stress evolution and associated displacements resulting from pore pressure depletion of a producing reservoir. The model was calibrated in 2-D as well as 3-D for a case of perfect elasticity, when comparison could be made to analytical calculations by the nucleus of strain theory (GEERTSMA, 1973). The accuracy of the DEM is limited by element size, which here was 20 m (given by the radius of disk elements in 2-D; spheres in 3-D). While calculated compaction and subsidence were in good agreement with theory, the scatter in stress calculations was more significant. The results are also sensitive to particle size distribution and packing, indicating that more work is required to optimize the choices of these parameters and parameters controlling the contact law between particles. Also as with other numerical methods, the results are largely affected by the boundary conditions. Therefore the model must be refined to achieve better results.

The simulations performed illustrate the ability of the DEM to generate localized faults when the elastic limit is exceeded somewhere in the model. As one would expect from analytical stress calculations and from previous numerical work, faulting is likely to take place in the surrounding near the edge of a depleting reservoir. When faults are known to exist prior to depletion and can be identified from seismic images, they may be embedded in the DEM model by selecting an array of particles with properties different from the surroundings. In our case, two options for numerical simulation of fault activation were considered; (i) reduced friction; (ii) reduced particle size and stiffness. Further work is required to find a geologically representative formulation for a fault in the DEM.

We conclude that DEM, such as PFC, may provide useful insight into the dynamic behavior of a rock mass such as in the case of a depleting reservoir embedded in a sedimentary basin. In principle, fluid flow and elastic wave propagation may also be incorporated in this model. Only when faulting is expected to take place will DEM be beneficial compared to more traditional simulation approaches (like FEM). Improvements include reducing particle (element) size, particularly in zones where failure may occur. Linking of DEM to a continuum model appears to be a promising tool for this.

Aknowledgement

The authors acknowledge the Norwegian Research Council for financial support to this work through the Strategic University Program "ROSE – Improved Overburden Characterization combining Seismics and Rock Physics" at NTNU (Norwegian University of Science and Technology).

REFERENCES

ADDIS, M.A. (1997), *The stress-depletion response of reservoirs*, SPE 38720, 11 pp.

BRIGNOLI, M., PELLEGRINO, A., SANTARELLI, F.J., MUSSO, G. and BARLA, G. (1997), *Continuous and discontinuous deformations above compacting reservoirs; consequences upon the lateral extension of the subsidence bowl*, Int. J. Rock Mech. and Min. Sci. *34* (3–4).

CUNDALL, P.A. and STRACK, O.D.L. (1979), *A discrete numerical model for granular assemblies*, Géotechnique *29*(1), 47–65.

GAMBOLATI, G., TEATINI, P., and TOMASI, L. (1999) *Stress-strain analysis in productive gas/oil reservoirs*, Int. J. Numer. Anal. Meth. Geomech. *23*, 1495–1519.

GAMBOLATI, G., FERRONATO, M., TEATINI, P., DEIDDA, R., and Lecca, G. (2001), *Finite element analysis of land subsidence above depleted reservoirs with pore pressure gradient and total stress formulations*, Int. J. Numer. Anal. Meth. Geomech. *25*, 307–327.

GUTIERREZ, M. and LEWIS, R.W. (1998), *The role of geomechanics in reservoir simulation*, SPE/ISRM 47392. In Proc. EUROCK'98, vol. II, pp. 439–448.

GEERTSMA, J. (1973), *A basic theory of subsidence due to reservoir compaction: The homogeneous case*, Trans. Royal Dutch Soc. Geol. and Mining Eng. *22*, 43–62.

HAZZARD, J.F. and YOUNG, R.P. (2000), *Simulating acoustic emissions in bonded-particle models of rock*, Int. J. Rock Mech. and Min. Sci. *37*, 867–872.

HETTEMA, M.H.H., SCHUTJENS, P.M.T.M., VERBOOM, B.J.M., and GUSSINKLO, H.J. (1998), *Production-induced compaction of sandstone reservoirs: The strong influence of field stress*, SPE50630, 8 pp.

HOLT, R.M., FLORNES, O., LI, L., and FJÆR, E. (2004), *Consequences of depletion-induced stress changes on reservoir compaction and recovery*, In Proc. Gulf-Rock (eds. Arma/Narms) 04–589, ARMA/NARMS, 10 pp.

HOLT, R.M., KJØLAAS, J., LARSEN, I., LI, L., PILLITTERI, A.G., and SØNSTEBØ, E.F. (2005) *Comparison between controlled laboratory experiments and discrete particle simulations of rock mechanical behaviour*, Int. J. Rock Mech. and Min. Sci. *42*, 985–995.

JING, L. and HUDSON, J.A. (2002), *Numerical methods in rock mechanics*, Int. J. Rock Mech. and Min. Sci. *39*, 409–427.

KENTER, C.J., BLANTON, T.L., SCHREPPERS, G.M.A., BAAIJENS, M.N. and RAMOS, G.G. (1998), *Compaction study for Shearwater Field*, SPE/ISRM 47280. In Proc. EUROCK'98; Vol. II, pp. 63–68.

KENTER, C.J., BEUKEL, A. v. d., HATCHELL, P., MARON, K. and MOLENAAR, M. (2004), *Evaluation of reservoir characteristics from timeshifts in the overburden*, In Proc. Gulf-Rock (eds. Arma/Narms) 04–627, ARMA/NARMS.

KOSLOFF, D., SCOTT, R.F., and SCRANTON, J. (1980), *Finite element simulation of Wilmington oil field subsidence: 1. Linear modelling*, Tectonophysics *65*, 339–368.

KOUTSABELOULIS, N.C. and HOPE, S.A. (1998) *"Coupled" stress / fluid / thermal multi-phase reservoir simulation studies incorporating rock mechanics*, SPE/ISRM 47393. In Proc. EUROCK'98 Vol. II, pp. 449–454.

LEWIS, R.W., MAKURAT, A. and PAO, K.S. (2003), *Fully coupled modeling of subsidence and reservoir compaction of North Sea oil fields*, Hydrogeology J. *11*, 142–161.

LI, L. and HOLT, R.M. (2002), *Particle scale reservoir mechanics*, Oil and Gas Science and Technology - Revue de l'Institut Français du Pétrole *57*, 525–538.

LI, L. and HOLT, R.M., *A study on the calculation of particle volumetric deformation in a fluid coupled PFC model*, In *Numerical Modeling in Micromechanics via Particle Methods – 2004* (eds. Shimizu, Hart and Cundall) (A. A. Balkema 2004) pp. 273–279.

LONGUEMARE, P., MAINGUY, M., LEMONNIER, P., ONAISI, A., and GÉRARD, C. (2002), *Geomechanics in reservoir simulation: Overview of coupling methods and field case study*, Oil and Gas Science and Technology - Revue de l'Institut Français du Pétrole *57*, 471–483.

MAHI, A. (2003), *Stress path of depleting reservoirs*, MSc Thesis, NTNU (Norwegian University of Science and Technology).

MAXWELL, S.C. and URBANCIC, T. (2001), *The role of passive microseismic monitoring in the instrumented oil field*, The Leading Edge, 636–639.

MORITA, N., WHITFILL, D.L., NYGAARD, O., and BALE, A. (1989), *A quick method to determine subsidence, reservoir compaction, and in-situ stress induced by reservoir depletion*, JPT Jan.89, pp. 71–79.

MULDERS, F.M.M. (2003), *Modelling of stress development and fault slip in and around a producing gas reservoir*, Ph.D. Thesis, TU Delft, Netherlands.

ODED, K., ZE'EV, R., and GIDON, B. (2003), *Faults and their associated host rock deformation: Part I. Structure of small faults in a quartz–syenite body, southern Israel*, J. Struct. Geol. *25*, 1675–1689.

OSORIO, J.G., CHEN, H-Y., TEUFEL, L.W. and SCHAFFER, S. (1998), *A two-domain, 3-D, fully coupled fluid-flow / geomechanical simulation model for reservoirs with stress-sensitive mechanical and fluid-flow properties*, SPE/ISRM 47397. In Proc. EUROCK'98, Vol. II, pp. 455–464.

PANDE, G.N., BEER, G. and WILLIAMS, J.R., *Numerical Methods in Rock Mechanics* (Wiley 1990).

PAPAMICHOS, E., VARDOULAKIS, I., and HEIL, L.K. (2001), *Overburden modeling above a compacting reservoir using a trap door apparatus*, Phys. Chem. Earth (A) *26*, 69–74.

POTYONDY, D.O. and CUNDALL, P.A. (2004), *A bonded particle model for rock*, Int. J. Rock Mech. and Min. Sci. *41*, 1329–1364.

RUDNICKI, J.W. (1999), *Alteration of regional stress by reservoirs and other inhomogeneities: Stabilizing or destabilizing?* Proc. Int. Congress on Rock Mechanics (eds. G. Vouille and P. Berest), ISRM, Vol. 3; pp. 1629–37.

SEGALL, P. and FITZGERALD, S.D. (1998), *A note on induced stress changes in hydrocarbon and geothermal reservoirs*, Tectonophysics *289*, 117–28.

SETTARI, A. and MOURITS, F.M. (1994) *Coupling of geomechanics and reservoir simulation models*, Computer Methods and Advances in Geomechanics, 2151–2158.

SHIMIZU, Y., *Fluid coupling in PFC^{2-D} and PFC^{3-D}. In Numerical Modeling in Micromechanics via Particle Methods – 2004* (eds. Shimizu, Hart and Cundall) (A. A. Balkema 2004), pp. 281–287.

TEUFEL, L.W., RHETT, D.W., and FARRELL, H.E. (1991), *Effect of reservoir depletion and pore pressure drawdown on in situ stress and deformation in the Ekofisk field*. In *Rock Mechanics as a Multidisciplinary Science* (ed. J.C. Roegiers) (A.A. Balkema 1991), pp. 63–72.

ZIENKIEWICZ, O.C., *The Finite Element Method in Engineering Geoscience* (4th edition). (McGraw-Hill 1991).

(Received April 5, 2005, revised September 19, 2005, accepted October 6, 2005)

To access this journal online:
http://www.birkhauser.ch

Pure appl. geophys. 163 (2006) 1153–1174
0033–4553/06/061153–22
DOI 10.1007/s00024-006-0055-9

© Birkhäuser Verlag, Basel, 2006

❙ Pure and Applied Geophysics

Comparison of Numerical and Physical Models for Understanding Shear Fracture Processes

JOHN NAPIER,[1] and TOBIAS BACKERS[2]

Abstract—An understanding of the formation of shear fractures is important in many rock engineering design problems. Laboratory experiments have been performed to determine the Mode II fracture toughness of Mizunami granite rock samples using a cylindrical 'punch-through' testing device. In this paper we attempt to understand and interpret the experimental results by numerical simulation of the fundamental shear fracture initiation and coalescence processes, using a random array of displacement discontinuity crack elements. It is found that qualitative agreement between the experimental and numerical results can be established, provided that shear-like micro-scale failure processes can be accommodated by the failure initiation rules that are used in the numerical simulations. In particular, it is found that the use of an exclusively tension-driven failure initiation rule does not allow the formation of macro-shear structures. It is apparent, also, that further investigation is required to determine how consistent rules can be established to link micro-failure criteria to equivalent macro-strength and toughness properties for a macro-shear slip surface.

Key words: Fracturing, numerical modelling, Punch-Through Shear test, Mode II.

1. Introduction

Shear fracture and shear band localisation are ubiquitous processes occurring in earthquake mechanics, mining-induced fracture processes and laboratory scale testing of rock samples. In many geotechnical engineering applications, shear strength properties must necessarily be assumed for the completion of required design exercises. However, it is generally recognised that the description of brittle or quasi-brittle failure mechanisms that occur during the failure of bonded granular media, such as rock, are very complex. In particular, the detailed mechanisms of shear rupture sequences are, arguably, still only incompletely understood. A central part of this difficulty is to distinguish the role played by fundamental tensile fracture processes and alternative shear failure mechanisms. The aim of the current

[1] School of Computational and Applied Mathematics, University of the Witwatersrand, Johannesburg, South Africa. E-mail: john.napier@pixie.co.za
[2] GeoFrames GmbH, Telegrafenberg D421, 14473 Potsdam, Germany. E-mail: backers@geoframes.de

paper is to attempt to gain further insight into the basic nature of shear fracturing by comparing experimental observations of so-called Punch-Through Shear (PTS-) tests of rock samples, to numerical simulations of approximately equivalent shear loading tests. This highlights basic issues relating to the feasibility of choosing equivalent material properties for the simplified representation of shear fracture in rock.

2. Laboratory Experiment – The Punch-Through Shear Test

This section introduces the Punch-Through Shear (PTS-) test (BACKERS et al., 2002a, 2004; BACKERS, 2005) and reports typical results from experiments on rock and special results on Mizunami granite.

The granite is from a borehole sunk for the Mizunami Underground (MIU) Research Laboratory project in Japan. It is classified as biotite granite (~50% quartz, ~40% feldspar and ~10% others). The medium grained granite is from ~200 m to ~ 500 m below surface and the average grain size is about 0.7 mm (BACKERS, 2005). From experiments the uniaxial compressive strength was found to be UCS = 166 ± 35 MPa, tensile strength σ_T = 9 ± 2 MPa, Young's modulus E = 50 ± 8 GPa and Poisson' ratio v = 0.37 (JNC DEVELOPMENT INSTITUTE REPORT, 2003; Backers, 2005; and references given therein).

All results and observations described in this section are related to the Mizunami granite although, in principle, also hold for other rock types. This section explains features of the PTS- test procedure to facilitate the comparison of the results of these tests to numerical experiments. For detailed results and interpretations for other rock types, reference should be made to BACKERS et al. (2002a, b, 2004) and BACKERS (2005).

2.1 Experimental Setup and Specimen Preparation

The PTS- test uses cylindrical samples with circular, centred, notches drilled into the end surfaces where the notches serve as a friction free initiation locus for fractures. It is expected that the fractures will connect the notches in a vertical (i.e., shear) direction. The suggested sample geometry, its dimensions and principle loading directions are shown in Figure 1.

Specimens are prepared from core bits (diameter D = 50 mm). The end surfaces are ground with a lathe to provide flat end surfaces perpendicular to the core mantle. Surface roughness is approximately 100 μm. The sample is fixed and centred using a large diameter chuck for manufacturing the notches of the inner diameter, ID, as depicted in Figure 1. The circular notches of width t = 1.5 mm are drilled using standard drill bits. The depths of the notches are indicated as a = 5 mm (top) and b = 30 mm (bottom).

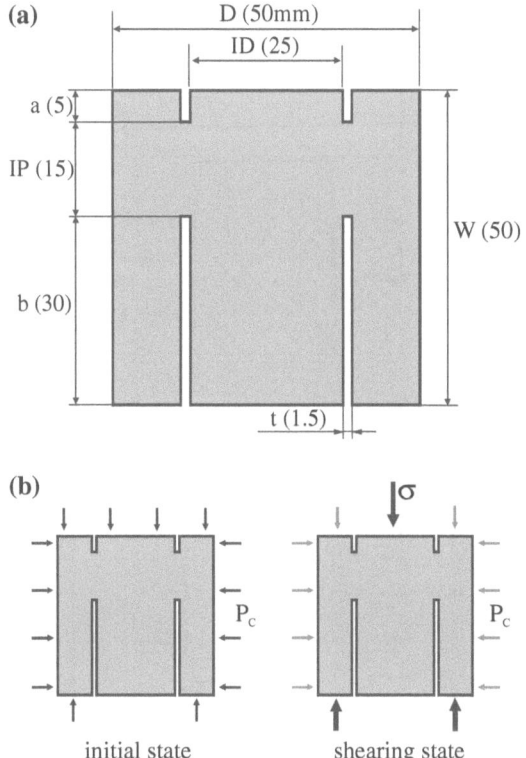

Figure 1
Sample geometry, principle loading, and dimensions for the Punch-Through Shear (PTS-) test. (a) Suggested sample geometry and dimensions (in [mm]). W: sample height, D: sample diameter, ID: inner notch diameter, a: upper notch depth, b: lower notch depth, IP: intact rock portion and t: notch width. (b) Loading. σ: axial stress, and P_C: confining pressure. The left-hand side indicates the initial stress state due to confining pressure (initial state) and the right-hand side indicates the loading regime during application of shear stress (shearing state). The confining pressure applies a normal load to the future fracture plane and the inner part is punched through the sample by an axial load.

The prepared sample is placed between a bottom support and loading stamp assembly (Fig. 2). Special devices were developed for the PTS- test. The bottom support is a solid cylinder with a centred 5-mm deep circular gap at its top surface. This device supports the outer bottom 'ring' of the sample during testing, while the central bottom part is unconstrained.

The hardened stamp is guided by a hollow-cylinder. A combination of Teflon ring and high pressure sealing supports the stamp and prevents migration of oil into the sample during testing. The setup is covered by a rubber sleeve to prevent intrusion of the confining pressure medium into the sample. The ends of the sleeve are clamped and the entire assembly is placed inside the pressure vessel.

The experiments are performed in a stiff ($1.1 \cdot 10^{10}$ N/m) servo-controlled loading frame (MTS, Material Test Systems Corporation, Minneapolis MI, USA; model-no.:

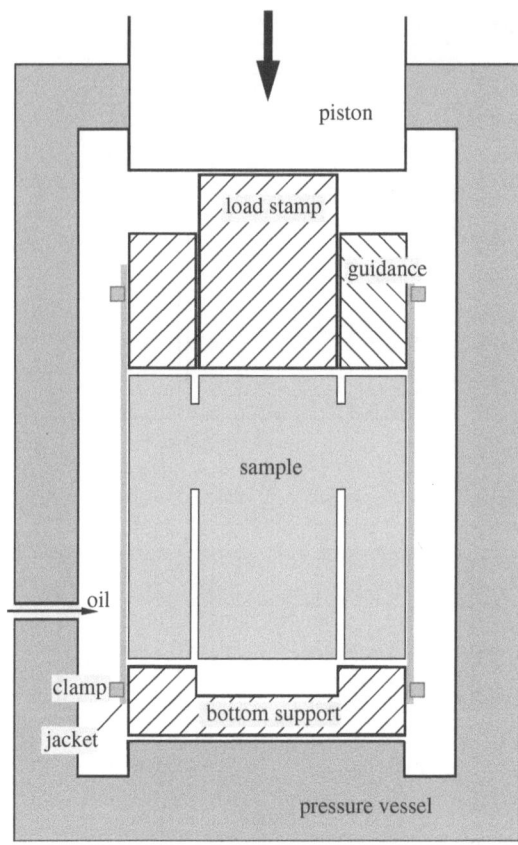

Figure 2
Punch-Through Shear test setup. The sample stack is placed in the pressure vessel (schematic, crosscut view)

815-315-03) with a 200 MPa confining pressure system. A high accuracy load cell with a range of 0–1000 kN (calibration error < 1%; sensitivity = ± 0.05 kN) is used.

Prior to the actual testing procedure, a small preload (typically ~2 kN) is added to the stamp to provide good adjustment and to secure the initial positioning of the assembly. Subsequently, the confining pressure is applied. The inner cylinder of the sample is punched down at a constant displacement rate of $3.3 \cdot 10^{-5}$ m/s. The applied axial stress, σ, generates an increasing shear stress in the intact rock portion between the notches which eventually causes failure. Axial force, confining pressure and axial displacement of the piston are recorded during each experiment.

2.2 Experimental Results on Mizunami Granite

Figure 3 shows typical recordings of axial force vs. axial displacement for the PTS-test performed on Mizunami granite at confining pressures of 30 MPa and 50 MPa.

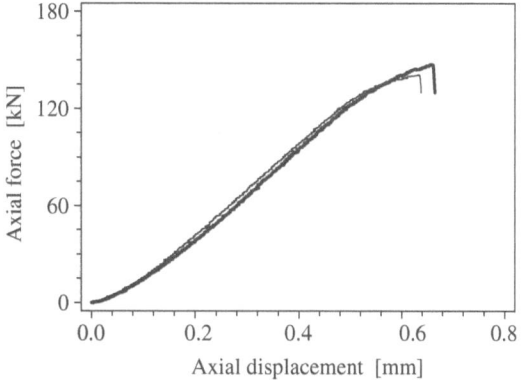

Figure 3

Typical recordings of force vs. displacement for Mode II loading using the PTS- test for confining pressures of 30 MPa (thin line) and 50 MPa (thick line). The recordings show the axial loading history until a significant loss in the load carrying capacity is recorded.

After some stiffening behaviour at low loads, the granite shows an approximately linear response. Prior to peak, some flattening in the force - displacement response slope is observed.

Analysing the peak force data of samples subjected to different levels of confining pressure, it becomes clear that two regimes can be recognised in the strength envelope to the PTS- test data. At low confining pressures the peak load steeply increases with confining pressure, P_C, while at high P_C the load necessary for fracture propagation increases only moderately with the increase in confining pressure. The transition from the steep to the shallow slope is at about 30 MPa. The inferred Mode II fracture toughness, K_{IIC}, shows an analogous behaviour with increasing confining pressure P_C (Fig. 4). The detailed procedure that is used to evaluate K_{IIC} from the

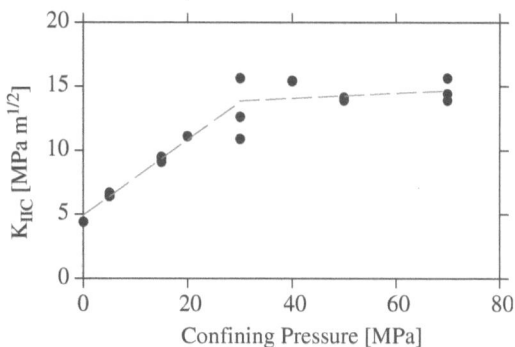

Figure 4

Mode II fracture toughness, K_{IIC}, as function of the confining pressure, P_C. Experiments have been performed up to $P_C = 70$ MPa.

Punch-Through Shear test by a Displacement Extrapolation Technique is described in BACKERS (2005).

It was shown for Mizunami granite and other rock types that a change in loading rate over a range spanning up to six orders of magnitude at zero confining pressure, does not change K_{IIC} (BACKERS, 2005).

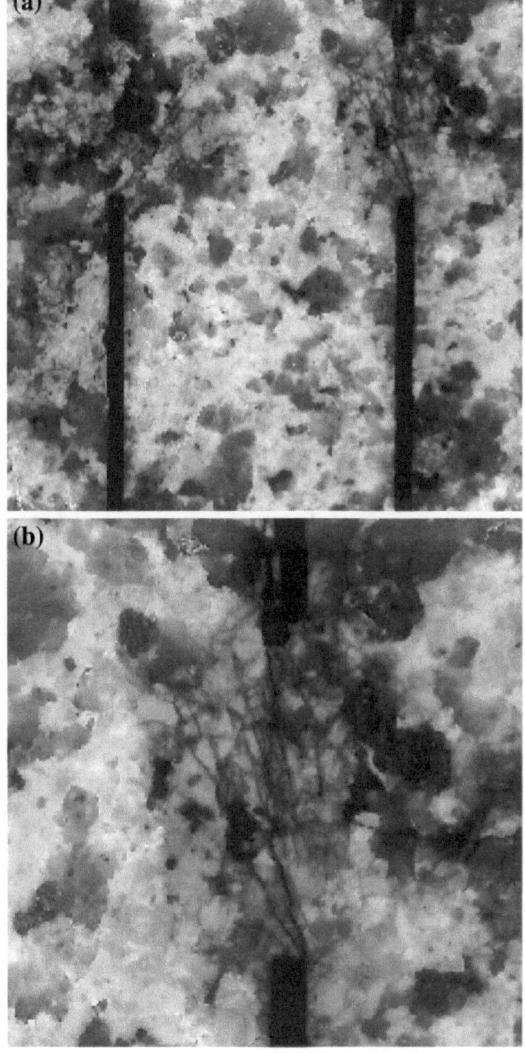

Figure 5
Pictures of a sample subjected to a confining pressure, P_C, of 30 MPa. (a) Grey scale scan of the entire sample cut after testing. Intense microcracking has occurred in a broad band between the upper and lower notches, (b) is a magnification of the area between the notches.

2.3 Microstructural Observations

Subsequent to testing, samples loaded to different percentages of the average peak load were soaked and stabilised with blue epoxy resin and cut in half. The granite develops a wide process zone that is observed to be initiated at the bottom notch at about 30% of the maximum load. In this intensely microcracked zone (fracture process zone, FPZ) the main fracture develops and connects the notches at the peak load (Fig. 5). The rock develops a network of predominately grain boundary but also intragranular cracks.

These observations were confirmed by the analysis of Acoustic Emission (AE) activity (STANCHITS et al., 2004). The located AE data show the development of tensile dominated cracks (derived from polarity analysis) at the bottom notch at an early stage in the loading history (about 30% of the maximum load). On further loading, the clouds of AE events become bigger around the notch tips and close to peak, at about 85% of the peak load, start propagating upwards to connect the notches. Coincidentally, the cracking changes from a domination of tensile cracking to shear and collapse dominated cracking. In the light of this information the force–displacement data from Figure 3 can be interpreted. The flattening behaviour may be dedicated to the extension and propagation of the FPZ and main fracture.

At low confining pressure, P_C (typically below 25 to 30 MPa) wing fractures are initiated at the top and bottom notches. These are increasingly suppressed with higher P_C. Above a certain level of P_C the wing fractures are not initiated at all. This level happens to coincide with the change in slope of the strength envelope. For a detailed description and analysis of the microstructure resulting from the PTS- test, reference should be made to BACKERS et al. (2002b) and BACKERS (2005). However, if the confining pressure is increased, tensile mode wing fractures are suppressed (see, for example, LAWN, 1993). Additionally, it could be shown in a study on the fracturing behaviour of Carrara marble in PTS- testing, that an increase of confining pressure increasingly suppresses the formation of tensile fractures in the FPZ (BACKERS et al., 2002b).

3. Numerical Experiment

3.1 The Numerical Model

General Considerations

A comprehensive review of numerical modelling procedures in rock mechanics has been prepared recently by JING (2003). Most of the available methods can be classified as "continuum" or "distinct element" methods in terms of the basic approach that is taken to represent material deformations once failure is initiated. In the case of continuum plasticity or "damage" models (e.g., SELLERS and SCHEELE, 1996; HAJIABDOLMAJID et al., 2002) it is necessary to define both the failure condition (usually as a function of the principal stress components) and also to prescribe relationships between incremental stress values and incremental plastic strains in the material once failure is initiated (so-called "flow rules"). The formulation of general constitutive relationships to control the evolution of failure patterns in a particulate material such as rock is extremely difficult. An alternative approach is to make some specific assumption about the basic composition of the material at an appropriate fundamental level of detail. For example, it might be assumed that the material can be represented as an assembly of bonded particles that have a mean size that can be linked to the actual granular composition of the material (POTYONDY and CUNDALL, 2004; MORGAN and BOETTCHER, 1999; MORGAN, 1999). The complex evolution of emergent failure patterns is then reduced to an explicit integration of the motion of each particle, subjected to the evolving forces imposed by neighbouring particles. Naturally, the details of the inter-particle contact stiffness and bond strength play a subtle role in determining the overall material deformation and fracture patterning. The choice of the appropriate particle "size" (or size distribution) is also somewhat subjective, depending on the scale of the problem that is to be analysed.

Discontinuity Models

Fracture in rock may be represented by an assembly of interacting discontinuities with specified slip and opening displacement distributions across the discontinuity surfaces. The fundamental "building block" of material failure can then be considered to be a "crack element" rather than a "particle". For computational convenience, it is useful to make the simplifying assumption that the host material is linear elastic. All failure processes are then represented by the interaction of a suitable assembly of mobilised, interacting crack elements. A very convenient method for the resolution of the mutual crack influences is to employ a particular form of the boundary integral method, termed the displacement discontinuity method (see CROUCH and STARFIELD, 1983). Several groups of investigators have followed this approach to simulate failure in rock or rock-like materials. Some examples of this approach are the simulation of rock fracture patterns near deep level mine excavations (NAPIER, 1990; NAPIER and HILDYARD, 1992) and the explicit simulation of crack propagation processes (SCAVIA, 1992, 1995; SHEN and STEPHANSSON, 1993a, b; BOBET and EINSTEIN, 1998a, b). In these studies fractures are simulated as "chains" of connected discontinuity elements that are allowed to propagate from specified points in the medium. The fracture propagation process is controlled by a material failure criterion and by rules that prescribe the direction in which the crack is to be extended. For example, the growth direction of a tensile fracture may be determined by evaluating the stress state at a series of positions that are at a fixed distance from an existing crack tip. The appropriate growth direction is then chosen to correspond to the angular position at which the stress component, normal to the incremental crack surface and joined to the crack tip, has a maximum tension value. Shear growth modes have been simulated as well by SHEN and STEPHANSSON (1993a, b) and by BOBET and EINSTEIN (1998a, b) using composite weightings of Mode I and Mode II stress intensity factors (or corresponding Mode I and Mode II fracture energies) to determine the crack growth direction.

Random Mesh Fracture Model

Some disadvantages of the explicit crack growth procedures are that specific growth sites have to be chosen *a priori* and that complete freedom of fracture initiation within the problem region is curtailed by the selective positioning of these growth sites. In addition, the intersection of growing fractures with existing discontinuities can sometimes lead to numerical difficulties. An alternative method is to consider the problem region to be covered by a random mesh of potential crack segments (NAPIER and PEIRCE, 1995; NAPIER and MALAN, 1997). Each segment of the (two-dimensional) random mesh is assumed to be represented by one or more displacement discontinuity line elements and each element is assigned two internal collocation points at which stress values are evaluated. A fracture simulation exercise is carried out in a series of incremental crack growth cycles. During each cycle the normal and shear stress traction vector components are evaluated at each collocation point of the current population of unmobilised crack elements that exist in the set of defined mesh segment positions. The normal and shear traction components are determined with respect to the local tangent and normal directions of each mesh segment. A choice is then made of the most likely element to fail according to the effective "distance" (in terms of the normal and shear traction components) from a specified failure envelope. Once this choice has been made, the element is added to the population of active crack elements and the entire problem is resolved. The search procedure is then continued for a specified number of boundary loading incremental steps. (Further details of this approach have been reported by NAPIER and MALAN, 1997).

The explicit form of the failure envelope and the definition of the stress "distance" is illustrated in Figure 6 in terms of the normal component, σ_n and shear stress component, τ, acting at a point on a given crack segment. When the normal stress component is tensile (as in the case of point A in Fig. 6), the failure envelope is assumed to follow a power law of the form

$$|\tau| = S_0(1 - \sigma_n/T_0)^\gamma, \tag{1}$$

where S_0 is the intact cohesive strength and T_0 is the intact tensile strength of the material at the collocation point. γ is a dimensionless exponent such that $0 < \gamma \leq 1$. The distance from point A to the failure envelope is measured in the direction of the local origin as depicted in Figure 6. When the normal stress component is compressive ($\sigma_n < 0$), as for point B in Figure 6, the failure envelope is assumed to follow a linear relation ship of the form

$$|\tau| = S_0 - \mu_0\sigma_n, \tag{2}$$

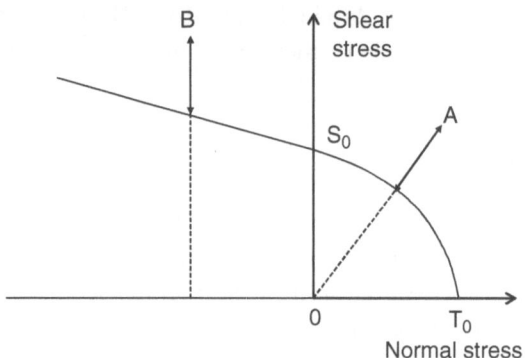

Figure 6

Schematic depiction of the assumed composite failure envelope and definition of the "distance" of a point from the envelope in terms of normal and absolute shear stress "coordinates" with respect to a local discontinuity segment line (the normal stress is assumed to be positive when it is tensile).

where μ_0 is the intact internal coefficient of friction of the material. The distance from point B to the failure envelope is assumed to be the "vertical" distance, parallel to the shear stress axis, as shown in Figure 6.

Once failure is initiated, the cohesive strength and tensile strength are assumed to be reduced to zero as linear functions of the slip and crack opening displacements, D_s and D_n, respectively. The explicit relationships that are assumed are of the following form (see also NAPIER, 2001; ALLODI et al., 2003).

$$S = S_0 - \alpha|D_s|, \tag{3}$$

$$T = T_0 - \beta D_n, \tag{4}$$

where S and T are the current values of the shear strength and the cohesion and α and β are the shear and tension "weakening" parameters, respectively. The values of S and T are set to zero when the maximum slip and opening displacements implied by equations (3) and (4) are exceeded. The composite failure envelope, shown in Figure 6, is updated progressively as failure continues by substituting the current values of S and T for S_0 and T_0 in equations (1) and (2), respectively. It is further assumed that the progressive change in the failure envelope is such that S and T both reach zero simultaneously and that the friction coefficient μ changes congruently from the intact value μ_0 to a specified residual value μ_f as these changes proceed. The failure envelopes in the tensile and compressive regions in Figure 6 are assumed to have the same slope when $\sigma_n = 0$. This, in turn, implies that the exponent γ must satisfy the relationship

$$\gamma = \mu T/S. \tag{5}$$

The slip-weakening concept (see PALMER and RICE, 1973; UENISHI and RICE, 2003) has a number of important implications. In particular, the weakening behaviour imposes an implicit length scale on the material (in some sense analogous to the mean particle size of a particle model) and also implies that numerical experiments of self-similar geometric configurations will exhibit size effects. In addition, it is important to note that the ratio of the host material shear modulus, G, to the shear weakening parameter α is proportional to an implicit slip "nucleation length", h_n for a single discontinuity (UENISHI and RICE, 2003).

Numerical Setup

In the simulations reported in this paper, all element sizes are chosen to be considerably smaller than the nucleation length h_n to ensure that the overall response of the numerical experiments are independent of the random mesh grid size.

It is important to note that the approach that is followed in this paper assumes two-dimensional plane strain conditions and, consequently, is not directly analogous to the loading geometry used in the Punch-Through Shear sample tests. Emphasis is therefore placed on the qualitative nature of the simulated punch experiments rather than attempting to adjust the numerical model parameters to fit the experimental observations directly. Furthermore, from the nature of the slow loading rates applied in the laboratory tests, it is assumed that inertia controlled dynamic effects can be neglected in the simulation of progressive fracture patterns. There is also no evidence from the laboratory experimental results that wave propagation effects play any significant role in determining the observed fracture patterns.

Numerical simulations of the evolutionary development of fracturing between the notches of a plane strain equivalent to the Punch-Through Shear test configuration were carried out. The test material properties are assumed to correspond to the Mizunami granite with a strength envelope of the form shown in Figure 6. The specific material properties that were used are summarised in Table 1.

The weakening parameter values, α and β, are not known for Mizunami granite and were chosen to have a similar order of magnitude to the values inferred in a series of size effect tests on rock carried out by LABUZ and BIOLZI (1991). The relative areas under the cohesion and tension weakening curves can be inferred from the parameters in Table 1 to have a ratio of about 14.8. This ratio can be considered,

Table 1

Summary of material properties used to simulate shear fracture in Mizunami granite

Young's modulus	50000 MPa
Poisson's ratio	0.37
Cohesion, S_0	34.4 MPa
Tensile strength, T_0	8 MPa
Internal friction coefficient, μ_0	tan(45°)
Residual friction coefficient, μ_f	tan(30°)
Cohesion weakening slope, α	100 MPa/mm
Tensile strength weakening slope, β	80 MPa/mm

as well, to reflect the ratio of the intrinsic specific fracture energy required to mobilise shear microcracks compared to the energy required to mobilise tensile microcracks, with an equivalent area. These values can also be related to the intrinsic Mode II and Mode I fracture toughness values of the microcracks in the model although this will not be pursued further in the paper.

It is not considered to be appropriate to include additional parameters such as crack shear and normal stiffnesses or to include an intrinsic dilation angle that is associated with crack sliding movements on crack elements. Such parameters should be introduced only if they can be motivated by suitable experimental observations.

Failure in the test sample was simulated by selecting elements from a random mesh tessellation of potential crack positions, covering the region between the test sample notches (Fig. 7). A number of different tessellation patterns were evaluated,

Random mesh and boundary segments

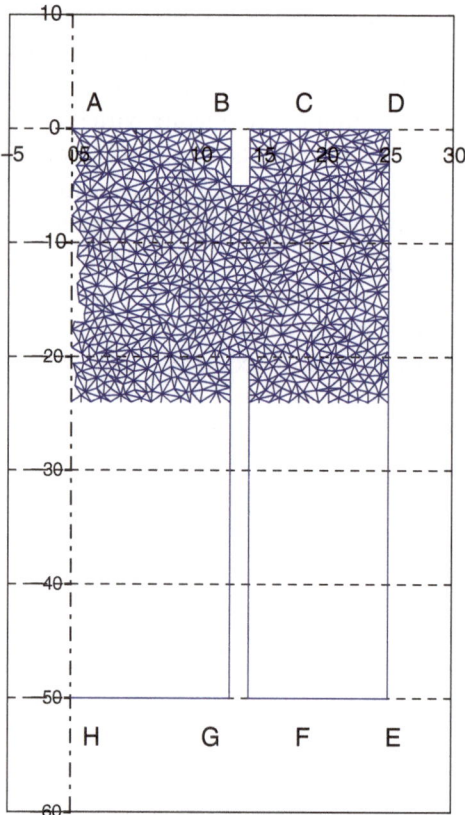

Figure 7
Layout of sample boundaries and internal random mesh used in the numerical simulation of the shear loading test. (The line joining points A to H represents a line of symmetry in the model).

including a random Delaunay triangulation, Delaunay triangulation with internal sub-triangulation and a Voronoi polygon mesh with internal triangulation of the Voronoi cells. It is found that the simulated failure patterns are not particularly sensitive to the specific tessellation structure, provided each junction node in the mesh has on average six impinging arcs. This allows appropriate "wedging" and "splitting" failure mechanisms to develop as well as accommodating general shear failure orientations.

The tessellation pattern depicted in Figure 7 is generated from a Voronoi polygon mesh with internal triangulation of the Voronoi cells. In this case there are 2766 segments in the mesh and the average segment length is 0.89 mm. The detailed distribution of the segment length values is shown in Figure 8. It should also be noted that the nucleation distance, h_n, introduced by UENISHI and RICE (2003), is of the order of 200 mm for the parameter values quoted in Table 1. This is well above the maximum segment length of approximately 2 mm shown in Figure 8. The region covered by the mesh segments is confined to the upper portion of the sample where failure is expected to be initiated. It must be emphasised that the simulated configuration is a plane strain geometry and Figure 7 represents an in-plane sectional view of this configuration. The significant load surfaces are lettered in Figure 7. Segment A-B represents the "platen" surface. A series of vertical displacement increments are imposed on this surface to simulate the physical experiment. The maximum simulated displacement is 0.75 mm which corresponds approximately to

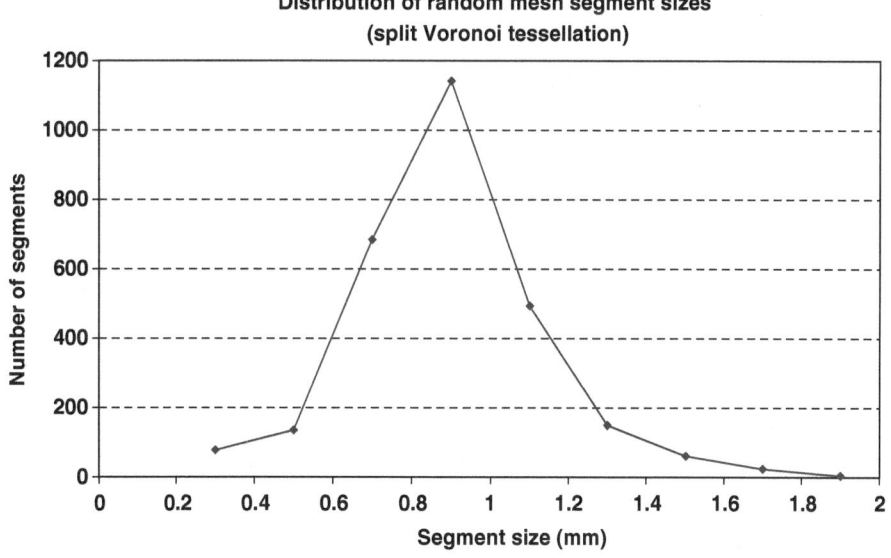

Figure 8
Distribution of the segment lengths in the random mesh depicted in Figure 7. (The total number of segments is 2766 with a mean length of 0.89 mm).

the maximum displacement range applied to the loading platen in the laboratory experiments (see Fig. 3). A confining stress is applied in the normal direction to both the upper surface C-D and to the vertical surface D-E. In addition, the upper surface C-D is constrained to have no lateral (horizontal) movement. It is assumed that no shear stress exists on the vertical surface D-E. The supporting base surface E-F is constrained to have zero horizontal and zero vertical movement. The lower surface G-H and the notch surfaces are all assumed to be stress-free. It is also noted that the numerical simulations were carried out assuming that a vertical line of symmetry is imposed through the central axis of the sample (line A-H in Fig. 7).

3.2 Numerical Results and the Comparison to Physical Experiment

Simulations with 'Mixed-Mode Fracture' Criterion

Figures 9a and 9b show the failure pattern developed at an intermediate stage of loading (platen displacement = 0.28 mm), with two different confining pressures of 30 MPa and 50 MPa applied to the upper loading surface C-D and to the side D-E of the sample as defined in Figure 7. The thick (green) lines in Figures 9a and 9b

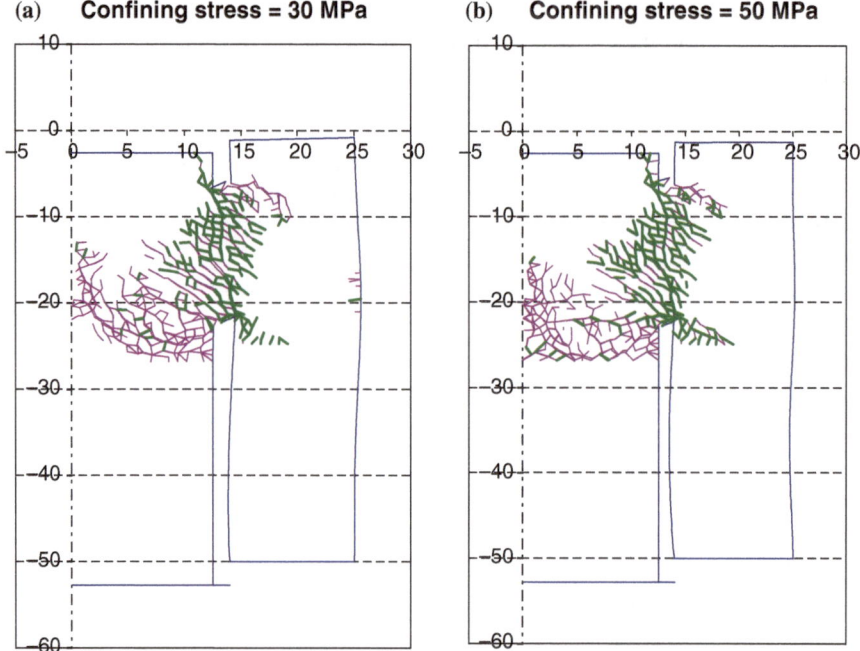

Figure 9

Development of simulated fracture patterns with different confining pressures for a platen displacement of 0.28 mm. Thick lines (green) designate cracks that are initiated in shear mode, thin lines (pink) designate cracks that are initiated in tension. (Voronoi tessellation with internal triangulation: (a) – confining stress = 30 MPa, (b) – confining stress = 50 MPa; displacements are magnified by a factor of 10).

designate crack elements that are initiated in a shear mode. The thin (pink) lines designate crack elements that are initiated when the stress component normal to the crack surface is in tension. It can be observed in Figures 9a and 9b that shear fracture initiation is clustered between the sample notches whereas tensile fractures arise across the sample "diameter". The tensile fracturing is observed also in the physical experiments (see Section 2.3) but is more localised and is suppressed significantly at the higher confining stress of 50 MPa. It must also be recalled that the detailed stress distribution near the physical sample axis is different from the numerically simulated plane strain loading configuration. Figures 9a and 9b also reveal a more intense clustering of shear fracturing between the sample notches as the confining pressure is increased. (It should be noted that the magnitude of the displacement movements across the crack surfaces and on the sample boundary surfaces is magnified by a factor of ten to highlight the simulated deformation magnitudes).

Figures 10a and 10b show the deformation patterns that evolve when the platen displacement is increased to 0.75 mm in incremental steps of 0.05 mm, revealing strong disruption and dilation of the material in the zone between the opposed sample notches. Shear-initiated cracks are again depicted using thick (green) lines.

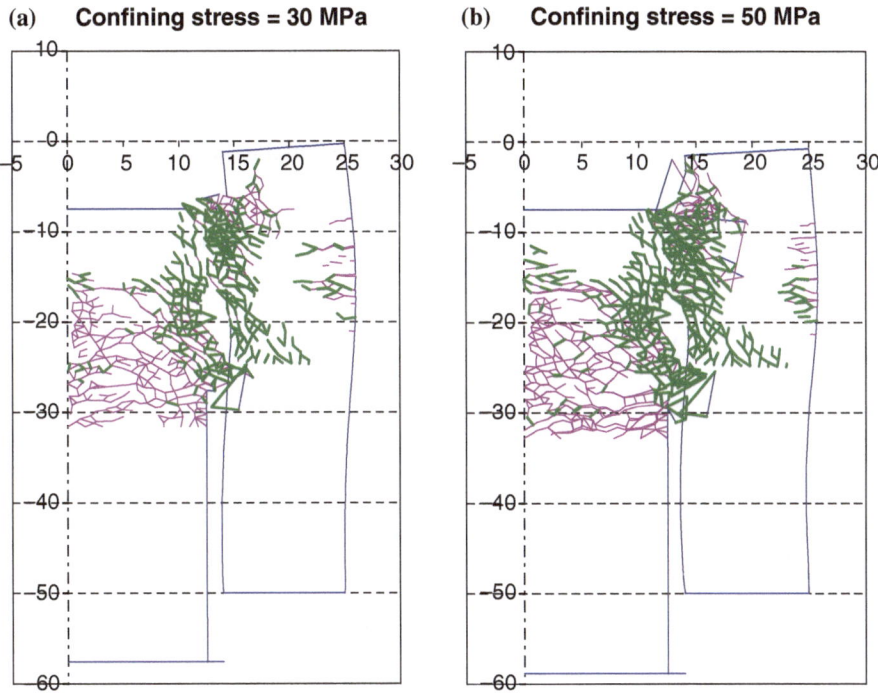

Figure 10

Effect of increased platen displacement on the simulated fracture pattern for confining pressures of (a) 30 MPa and (b) 50 Mpa, respectively. (Platen displacement = 0.75 mm).

Figure 11 is a plot of the fraction of the fracture length that is mobilised in a tension mode during each imposed displacement increment of the loading platen. This shows that when the applied confining stress is 30 MPa, relatively fewer fractures appear to be initiated in tension as the platen displacement is increased. Over 75 percent of the fractures that are mobilised during the first loading increment are initiated in tension. However, when the confining stress is increased to 50 MPa it appears that, apart from the initial load increment, there is no clear trend for the relative fraction of fractures that are initiated in a tensile mode. This behaviour agrees qualitatively with the inferred trends in fracture initiation modes that are observed in the PTS- tests.

It is also of interest to plot the average platen stress simulated in the numerical experiments against the applied platen displacements. The curves marked "A" and "B" in Figure 12 correspond to the simulated platen load-displacement behaviour when using the material properties defined in Table 1. It can be seen that some load shedding occurs as the platen displacement increases beyond 0.5 mm for curve "A" and beyond 0.6 mm for curve "B", respectively. This is not observed in the laboratory experimental force-displacement curves shown in Figure 3. Material properties that may affect this behaviour are the tensile strength, T_0, the residual friction coefficient, μ_f, and the cohesion weakening slope parameter, α. The punch loading tests were repeated with the tensile strength, T_0, increased from 8 MPa to 16 MPa, residual friction coefficient, μ_f, increased from tan(30°) to tan(45°) and cohesion softening slope, α, reduced from 100 MPa/mm to 50 MPa/mm. The

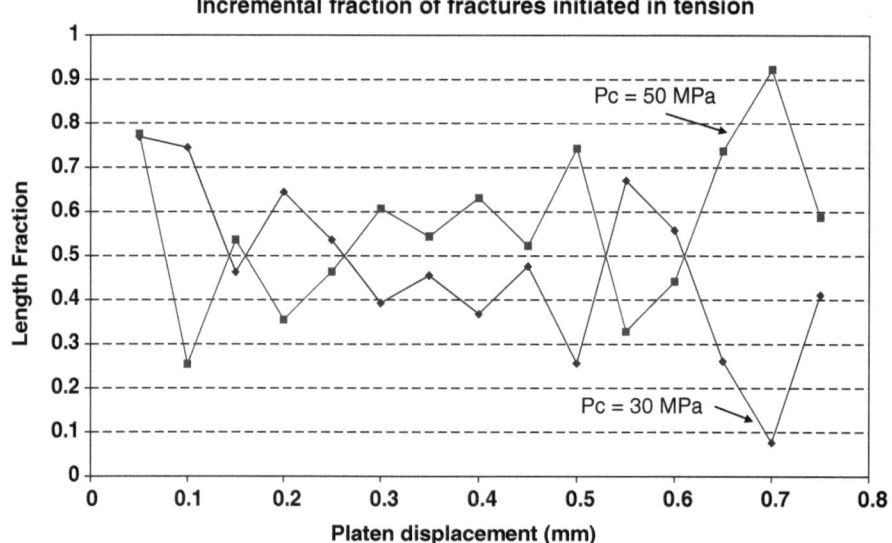

Figure 11
Relative fraction of the fracture length that is mobilised in a tension mode, during each platen displacement increment of 0.05 mm, with applied confining stresses of 30 MPa and 50 MPa.

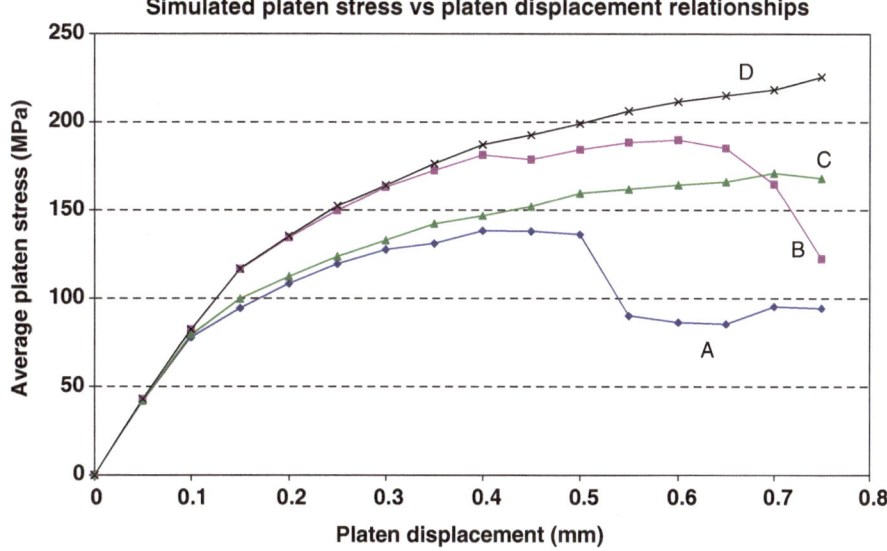

Figure 12
Average platen stress vs. platen displacement for different confining stresses and material properties.
(Curves A and B correspond to the simulated response using the material properties given in Table 1 with
confining stresses of 30 MPa and 50 MPa, respectively. Curves C and D correspond to the case in which
the material properties are adjusted to suppress overall load shedding after failure is initiated. The
confining stresses are equal to 30 MPa and 50 MPa, respectively).

resulting load-displacement curves that arise are marked "C" and "D" in Figure 12
and correspond to confining stresses of 30 MPa and 50 MPa, respectively. It is
observed that with these amended parameter values the load-shedding behaviour is
suppressed in the simulated range of the platen displacement.

The load-response curves shown in Figure 12 can only be claimed to represent a
qualitative agreement to the observed platen load as a function of platen
displacement (see Fig. 3). For example, the peak average platen stress, corresponding
to the central curve "C" in Figure 12, is approximately equal to 170 MPa. This
corresponds to an average force on the loading platen of approximately 83 kN. This
is significantly lower than the observed force of about 150 kN shown in Figure 3 and
may be related to the intrinsically stiffer response of the axi-symmetric laboratory
sample geometry, compared to the simulated plane strain configuration.

Simulations with a 'Maximum Normal Stress' Criterion

It is of considerable interest to examine in more detail the effect of the failure rule,
used in the numerical simulation, on the development of the overall fracture pattern.
In particular, if the criterion for crack activation selection is postulated to be the
maximum normal stress component across each potential crack element (implicitly a
tension-only criterion), then the resulting failure pattern shown in Figure 13 is

Maximum tension growth initiation rule
(Confining stress = 30 MPa)

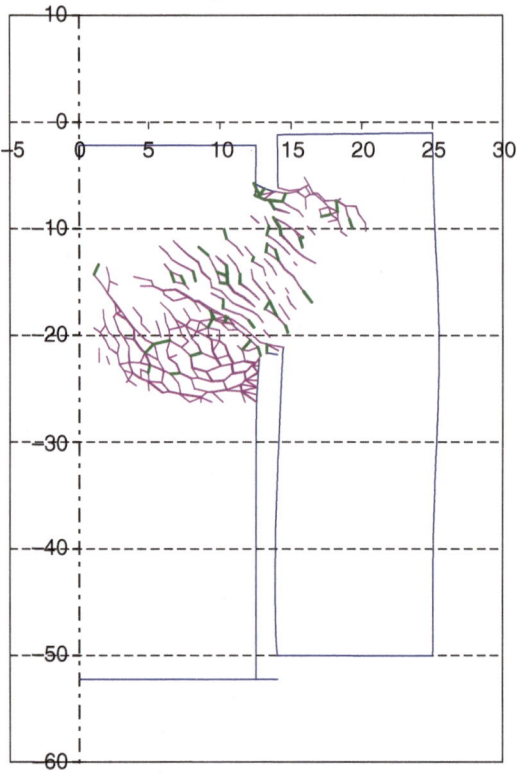

Figure 13
Effect on the simulated fracture pattern of selecting the fracture growth initiation mode according to
the maximum normal stress across the crack plane. (Thick lines designate fracture initiation when the
maximum normal stress across the crack is compressive.)

developed, following a cumulative platen displacement of 0.22 mm. This fracture
pattern reveals no potential evolution of a connecting shear band between the sample
notches. Figure 14 shows the initial load displacement response that is observed
when the tension-only initiation rule is used compared to the corresponding load-
displacement relationship that arises if mixed shear and tension failure is allowed. It
is also found that if the platen displacement is increased further, the load response
corresponding to the tension-only failure rule continues to increase. (At a platen
displacement of 0.60 mm the average platen load is observed to rise monotonically to
294 MPa). The suppression of the macro-shear band structure suggests strongly that
some form of micro-shear mechanism is required in order to enable macro-shear
band development to occur. It does not seem possible to assume a "universal" micro-
failure constitutive rule, based for example on tension only failure, that allows the
formation of macro-shear band structures.

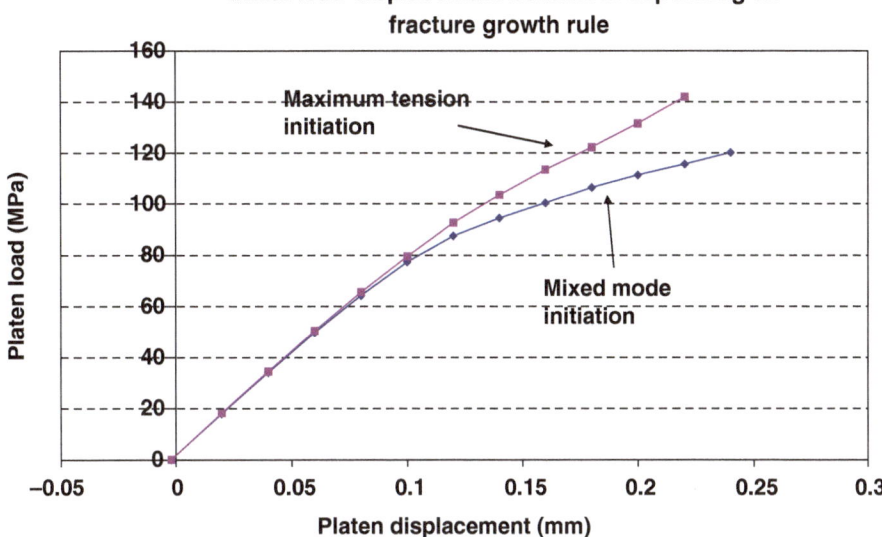

Figure 14
Effect of the failure initiation rule on the simulated load-displacement response.

Single 'Shear' Plane Simulation

Lastly, it is of interest to examine whether the complex tessellation mesh used in the detailed simulation of the Punch-Through Shear test can be replaced by a single "effective" shear plane connecting the sample notches. Figure 15 shows the load-response curves for a single plane, compared to a detailed random mesh simulation in which the intrinsic failure properties of the elements in the random mesh are chosen to be the same as the single shear plane. (In this case the properties are: Confining stress = 30 MPa, Cohesion, $S_0 = 68.8$ MPa; Tensile strength, $T_0 = 16$ MPa; Internal friction angle, $\mu_0 = 45$ degrees; Residual friction angle, $\mu_f = 30$ degrees; Cohesion weakening slope, $\alpha = 100$ MPa/mm; Tension weakening slope, $\beta = 200$ MPa/mm). Clearly, the single "equivalent" shear plane response is substantially different from the detailed response of the tessellation model and reveals a substantially higher initial load build-up followed by rapid load shedding. It is apparent that the use of the same material properties in a "micro" and "macro" level simulation will not, in general, lead to an equivalent overall system response.

4. Conclusions

The broad conclusions of the study of the behaviour of the Punch-Through Shear test are that qualitative agreement, in terms of both fracture patterns and load-response behaviour, can be obtained between the laboratory experiments and

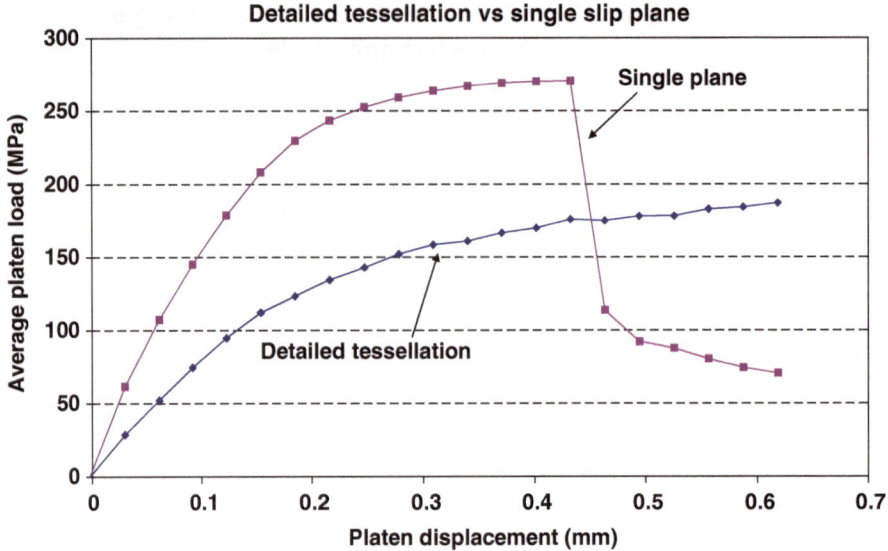

Figure 15
Effect of replacing the tessellation pattern by a single "effective" shear slip plane connecting Punch-Through Shear sample notches. (Strength properties for the random mesh cracks and the effective slip plane are the same. Confining stress = 30 MPa).

the numerical simulation of these tests using a random mesh assembly of displacement discontinuity crack elements. However, significant differences are observed in the shape and magnitude of the overall load-deformation response curves. The numerical experiments also suggest that the relative extent of fracture initiation in tension decreases progressively as the platen displacement is increased when the confining stress is 30 MPa. However, this trend is not displayed when the confining stress is increased to 50 MPa. This corresponds approximately to the apparent suppression of tensile wing cracks, in the physical experiments, when the confining stress is increased to 50 MPa. The computational aspects of simulating rock failure still require a more thorough evaluation and understanding of appropriate intrinsic material properties such as the rate of cohesion and tension weakening as a function of slip and crack opening movements. The material cohesive and tensile strength and the cohesion and tension weakening parameters can be considered to determine intrinsic material properties that are essentially equivalent to Mode II and Mode I fracture toughness parameters, respectively. The study reveals as well that it is not straightforward to infer equivalent macro-failure constitutive properties from micro-failure constitutive specifications. This, in turn, suggests that further investigation is required to define an appropriate framework that enables engineering "strength" properties of rock mass structures to be inferred consistently at different length scales.

The main highlight of the present study is to emphasise the role of micro-level shear fracturing as a building block for the formation of a macro-shear band structure such as the shear zone that is observed in the Punch-Through Shear test. The numerical experiments reveal that the assumption of fundamental failure rules such as tension-only failure at a "micro" scale will not, in general, support the formation and growth of complex shear structures.

Acknowledgements

Stefan Gehrmann (GFZ Potsdam) assisted with sample preparation. His contribution is acknowledged. We appreciate the detailed and helpful comments of two reviewers.

REFERENCES

ALLODI, A., CASTELLI, M., MARELLO, S., and SCAVIA, C. (2003), *Shear band propagation in soft rocks: Numerical simulation of experimental results.* ISRM 2003 Proceedings - Technology Roadmap for Rock Mechanics, South African Institute of Mining and Metallurgy, pp. 23–28.

BACKERS, T. (2005), *Fracture toughness determination and micromechanics of rock under Mode I and Mode II loading,* Doctorate Thesis, Potsdam University.

BACKERS, T., DRESEN, G., RYBACKI, E., and STEPHANSSON, O. (2004), *New Data on Mode II Fracture Toughness of Rock from Punch-Through Shear (PTS) Test,* SINOROCK2004 Paper *1A01,* Int. J. Rock Mech. Min. Sci.

BACKERS, T., STEPHANSSON, O., and RYBACKI, E. (2002a), *Rock Fracture Toughness Testing in Mode II – Punch-Through Shear Test,* Int. J. Rock Mech. Min. Sci. *39,* 755–769.

BACKERS T., RYBACKI E., ALBER M. and STEPHANSSON O. (2002b), *Fractography of rock from the new Punch-Through Shear Test.* In (Dyskin, A.V., Hu, X. and Sahouryeh, E., eds.), *Structural Integrity and Fracture – The International Conference on Structural Integrity and Fracture,* Perth, Australia, pp. 303–308.

BOBET, A. and EINSTEIN, H.H. (1998a), *Fracture coalescence in rock-type materials under uniaxial and biaxial compression,* Int. J. Rock Mech. Min. Sci. *35,* 863–888.

BOBET, A. and EINSTEIN, H.H. (1998b), *Numerical modelling of fracture coalescence in a model rock material,* Internati. J. Fracture. *92,* 221–252.

CROUCH, S.L. and STARFIELD, A.M. (1983), *Boundary Element Methods in Solid Mechanics* (George Allen and Unwin, London).

HAJIABDOLMAJID, V., KAISER, P.K., and MARTIN, C.D. (2002), *Modelling brittle fracture of rock,* Int. J. Rock Mech. Min. Sci. *39,* 731–741.

JING, L. (2003), *A review of techniques, advances and outstanding issues in numerical modelling for rock mechanics and rock engineering,* Int. J. Rock Mech. Min. Sci. *40,* 283–353.

JNC (Japan Nuclear Cycle) DEVELOPMENT INSTITUTE (2003), *Mizunami Underground Research Laboratory Project*; Results from 1996–1999 Period.

LABUZ, J.F. and BIOLZI, L. (1991), *Class I vs. Class II stability: A demonstration of size effect,* Int. J. Rock Mech. Min. Sci. *28,* 199–205.

LAWN, B. (1993), Fracture of brittle solids. Cambridge University Press, Cambridge.

LI, J., NGAN, A.H.W., and GUMBSCH, P. (2003), *Atomistic modeling of mechanical behaviour,* Acta Materialia. *51,* 5711–5742.

MORGAN, J.K. and BOETTCHER, M.S. (1999), *Numerical simulations of granular shear zones using the distinct element method - 1. Shear zone kinematics and micromechanics of localization*, J. Geophys. Res. *104* (B2) 2703–2719.

MORGAN, J.K. (1999), *Numerical simulations of granular shear zones using the distinct element method - 2. Effects of particle size distribution and interparticle friction on mechanical behaviour*, J. Geophys. Res. *104*, B2, 2721–2732.

NAPIER, J. A. L. (1990), *Modelling of fracturing near deep level mine excavations using a displacement discontinuity approach*. Proc. First Internat. Conf. *Mechanics of Jointed and Faulted Rock* (Rossmanith, ed.) Balkema, Rotterdam) pp. 709–715.

NAPIER, J.A.L. and HILDYARD, M.W. (1992), *Simulation of fracture growth around openings in highly stressed, brittle rock*, J. S. Afr. Inst. Min. Metall. *92*, 159–168.

NAPIER, J A L., and PEIRCE, A.P. (1995), *Simulation of extensive fracture formation and interaction in brittle materials*, Proc. Second Internat. Conf. *Mechanics of Jointed and Faulted Rock*, (Rossmanith, ed.), (Balkema, Rotterdam) pp. 63–74.

NAPIER, J.A.L. and MALAN, D.F. (1997), *A viscoplastic discontinuum model of time-dependent fracture and seismicity effects in brittle rock*, Int. J. Rock Mech. Min. Sci. *34*, 1075–1089.

NAPIER, J.A.L. (2001), *Scale effects in the numerical simulation of time-dependent mine seismic activity*, Proc. *38th U.S. Rock Mechanics Symp.* (Elsworth, Tinucci and Heasley, eds.), (Balkema, Rotterdam) pp. 1297–1304.

PALMER, A.C. and RICE, J.R. (1973), *The growth of slip surfaces in the progressive failure of over-consolidated clay*, Proc. Roy. Soc. Lond. A *332*, 527–548.

POTYONDY, D.O. and CUNDALL, P.A. (2004), *A bonded-particle model for rock*, Int. J. Rock Mech. Min. Sci. *41*, 1329–1364.

SCAVIA, C. (1992), *A numerical technique for the analysis of cracks subjected to normal compressive stresses*, Int. J. Numer. Methods Eng. *33*, 929–942.

SCAVIA, C. (1995), *A method for the study of crack propagation in rock structures*, Geotechnique *45*, 447–463.

SELLERS, E. and SCHEELE, F. (1996), *Prediction of anisotropic damage in experiments simulating mining excavations in Witwatersrand quartzite blocks*, Int. J. Rock Mech. Min. Sci. *33*, 659–670.

SHEN, B. and STEPHANSSON, O. (1993a), *Numerical analysis of mixed mode I and mode II fracture propagation*, Int. J. Rock Mech. Min. Sci. and Geomech. Abstr. *30*, 861–867.

SHEN, B. and STEPHANSSON, O. (1993b), *Modification of the G-criterion for crack propagation subjected to compression*, Engin. Fracture Mech. *47*, 177–189.

STANCHITS S., BACKERS T., STEPHANSSON, O. and DRESEN, G. (2004), *Comparison of acoustic emission events and micromechanics of granite under Mode I and Mode II loading*. In (Makurat A. and Curri P., eds.), EURO-Conf. *Rock Physics and Geomechanics – Micromechanics, Flow and Chemical Reactions*. Extended Abstract Volume, 7–11 September 2003, Delft, Netherlands.

UENISHI, K. and RICE, J.R. (2003), *Universal nucleation length for slip-weakening rupture instability under nonuniform fault loading*, J. Geophys. Res. *108*, B1, ESE 17-1–17-13.

(Received April 18, 2005, revised October 7, 2005, accepted November 18, 2005)

 To access this journal online:
http://www.birkhauser.ch

Pure appl. geophys. 163 (2006) 1175–1192
0033–4553/06/061175–18
DOI 10.1007/s00024-006-0053-y

© Birkhäuser Verlag, Basel, 2006

❘ Pure and Applied Geophysics

Upscaling: Effective Medium Theory, Numerical Methods and the Fractal Dream

Y. Guéguen,[1] M. Le Ravalec,[2] and L. Ricard[2]

Abstract—Upscaling is a major issue regarding mechanical and transport properties of rocks. This paper examines three issues relative to upscaling. The first one is a brief overview of Effective Medium Theory (EMT), which is a key tool to predict average rock properties at a macroscopic scale in the case of a statistically homogeneous medium. EMT is of particular interest in the calculation of elastic properties. As discussed in this paper, EMT can thus provide a possible way to perform upscaling, although it is by no means the only one, and in particular it is irrelevant if the medium does not adhere to statistical homogeneity. This last circumstance is examined in part two of the paper. We focus on the example of constructing a hydrocarbon reservoir model. Such a construction is a required step in the process of making reasonable predictions for oil production. Taking into account rock permeability, lithological units and various structural discontinuities at different scales is part of this construction. The result is that stochastic reservoir models are built that rely on various numerical upscaling methods. These methods are reviewed. They provide techniques which make it possible to deal with upscaling on a general basis. Finally, a last case in which upscaling is trivial is considered in the third part of the paper. This is the fractal case. Fractal models have become popular precisely because they are free of the assumption of statistical homogeneity and yet do not involve numerical methods. It is suggested that using a physical criterion as a means to discriminate whether fractality is a dream or reality would be more satisfactory than relying on a limited data set alone.

Key words: Upscaling, Effective Medium, Numerical Methods, Fractals.

1. Upscaling and Effective Medium Theory

At all scales, from local to regional, rocks are heterogeneous. The influence of heterogeneity in rock, as long as it remains moderate, can often be dealt with using the ergodic assumption, considering that the medium is statistically homogeneous. That means that a representative elementary volume (REV) exists (Fig. 1) and that any part of the system with a volume considerably larger than the REV has identical physical properties. In this case, one can state that "upscaling" is equivalent to "homogenizing." The medium can be considered as invariant by translation (for any property

---
[1]Laboratoire de Geologie, Ecole Normale Superieure, Paris, France
[2]Institut Français de Petrole, IFP, France

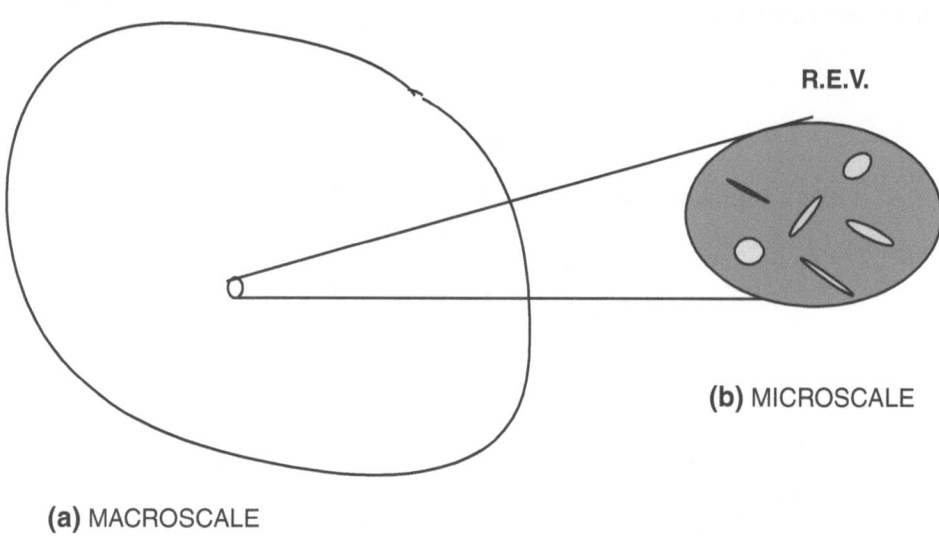

Figure 1
Representative Elementary Volume (REV): Statistical homogeneity (or equivalently translational invariance) assumes that any property averaged over a REV at any place gives the same result.

averaged over a REV). This is obviously applicable to geological objects substantially greater than a REV. In most cases, rock heterogeneity is mainly the result of variable composition and of the presence of pores and cracks. Effective Medium Theories (EMT) have been constructed that allow the successful prediction of rock properties as long as the degree of heterogeneity is not large. There exists in general some critical threshold above which such an approach does not apply. The limits of application of EMT to rocks has been examined by GUÉGUEN *et al.* (1997). We restrict ourselves in the following to the applicability field of EMT. Because there are many close but different EMT schemes, we present in the following a short introduction which is the common basis for all schemes. We then take as an example the elastic properties.

1.1 General Statistical Approach

In order to determine the macroscopic effective properties of a heterogeneous medium, some manner of ensemble averaging is usually performed. To the degree of the property of interest can be considered as a generalized susceptibility χ defined by a linear relation between a generalized force σ (for instance stress, electric field, etc.) and a generalized flux ϵ (for instance strain, electric flux, etc.), the general method is as follows. Let us point out that the properties considered here all belong to that category since elastic moduli, conductivities and permeability enter into that group. In each case, there is also a conservative flux equation (GUÉGUEN and PALCIAUSKAS, 1994) which makes the analogy complete. The flux (i.e., hydraulic, electric, thermal flux or stress) is assumed to be a linear function of the force (i.e., respectively fluid pressure gradient, electric potential gradient, temperature gradient, strain):

$$\sigma(\vec{r}, t) = \int d\vec{r}' \int dt' \chi(\vec{r}, \vec{r}', t - t') \epsilon(\vec{r}', t'). \tag{1}$$

Using the example of elasticity where σ stands for stress and ϵ stands for strain, equation (1) can be simply understood as follows. Any incremental strain $\delta\epsilon$ at a point of coordinates (x', y', z') and time t' produces a stress $\delta\sigma$ at a point of coordinates (x, y, z) and at a later time t. It is assumed that $\delta\sigma$ is a linear function of $\delta\epsilon$, the coefficient of proportionality being the generalized susceptibility χ (in practice the elastic constants, electrical conductivity, etc.). The susceptibility χ is considered to be a function of $(x', y', z'); (x, y, z); (t' - t)$. Summing all contributions $\delta\sigma$, equation (1) results. Let us note that the formal theory can be developed considering σ in terms of ϵ or, equivalently, the reverse.

The Fourier transform equation is

$$\sigma(\vec{q}, \omega) = \int d\vec{q}' \chi(\vec{q}, \vec{q}', \omega) \epsilon(\vec{q}', \omega). \tag{2}$$

If one can define a representative elementary volume which is large compared to any correlation lengths, but small compared to the macroscopic sample of interest, then it can be shown (GUBERNATIS and KRUMHANSL, 1975) that the equation (2) reduces to

$$\sigma(\vec{q}, \omega) = \bar{\chi}(\vec{q}, \omega) \epsilon(\vec{q}, \omega) + \text{small fluctuations}. \tag{3}$$

Depending on the physical property which is considered, σ and ϵ can be respectively stress and strain, fluid flow and pressure gradient, etc. The calculation of the generalized susceptibility χ (which can be elastic modulus, permeability, etc.) can be performed when the specific field equations are introduced as shown later.

Let us point out that frequency effects can be important and in general, the susceptibility χ is frequency-dependent. When laboratory measurements are performed, for instance for elastic properties, ultrasonic waves are used and $\vec{q}$ and ω are non-zero. The elastic moduli may vary with ω. In that case, the physical process behind that is fluid relaxation in pores and cracks (LE RAVALEC and GUÉGUEN, 1996; SCHUBNEL and GUÉGUEN, 2003): high frequency (unrelaxed) moduli differ from low frequency ones (relaxed).

In the simple static case, both $\vec{q}$ and ω are zero. One can show that the volumic averages of static generalized force $\langle\sigma\rangle$ and flux $\langle\epsilon\rangle$ are

$$\langle\sigma\rangle = \sigma(0, 0) \text{ and } \langle\epsilon\rangle = \epsilon(0, 0) \tag{4}$$

so that the effective static susceptibility is

$$\bar{\chi}(0, 0) = \frac{\langle\sigma\rangle}{\langle\epsilon\rangle}. \tag{5}$$

The above relations establish the formal theory.

1.2 *Example of Elastic Properties*

We will focus in this section on elastic properties. Then σ is the stress and ϵ the strain. Both are tensors. We restrict ourselves to the simple static case. However, the practical calculation of effective moduli $\bar{\chi}(0,0)$ remains far from straightforward. This is, because what can be determined usually is not the stress or strain distribution through the rock sample (which are required as input data in the above equation) but the composition and microstructure average distributions. At that point, as noted by GUBERNATIS and KRUMHANSL (1975), and focusing on static moduli, there are two options to determine the unknown fields of σ and ϵ. Interestingly enough, in the case of crustal rocks, the main source of heterogeneity is the presence of pores and cracks. The focus then becomes accountability for pore and crack effects on elastic moduli. In order to simplify the calculations, it can be assumed that either pores or cracks are dominant. The case of cracks is well taken into account by the first option and that of pores by the second one.

1.2.1 Inclusion models

The first option is more or less intuitive and can be split in several subgroups. Variational methods can provide rigorous bounds. HASHIN and SHTRIKMAN (1962) have contributed substantially to provide such useful bounds. We will not attempt to review these results here, and neither attempt to cover all other approaches which belong to that first group. We focus rather on methods following Eshelby's approach for an inclusion for practical reasons which will be explained below. Eshelby's solution provides the stress and strain fields due to a single inclusion. Such an approach has been extensively developed by KACHANOV (1993) within the so-called non-interacting scheme. In that case, the effects of many inclusions are obtained as the superposition of the effect of each individual inclusion. Basically, the elastic modulus M_0 of the uncracked rock is modified due to heterogeneities by ΔM

$$\langle \sigma \rangle = (M_0 + \Delta M)\langle \epsilon \rangle. \tag{6}$$

The unknown quantity ΔM is derived from Eshelby' solution for a single inclusion in a given field. It has been shown by KACHANOV (1992, 1993) that the non-interacting approximation is a very good one as regards cracks, to the degree their spatial distribution is random. This is due to compensating effects between positive and negative interactions. It can be also understood from the Mori-Tanaka scheme. In that scheme, any crack is considered to be located in an effective field which is that of the others cracks. If cracks are introduced in a medium with constant surface conditions, the average stress field remains constant. If we choose, as the simplest approximation, the effective field to be the spatially averaged field, the implication is that the effective field is unchanged whether cracks are present or not. That means that crack interactions cancel.

1.2.2 T matrix model

If, in the case of cracks, the non interacting scheme is optimal, it can be shown (GUÉGUEN et al., 1997) that in the case of round pores, an excellent approximation is obtained with the differential self-consistent method, which is an improved form of the self-consistent scheme. This last scheme can be understood as derived from the so-called T matrix model which represents the second option to determine the unknown fields. This model is very general indeed. It uses integral equations, assuming that stress and strain obey

$$\frac{\partial \sigma_{ij}(\vec{r})}{\partial x_j} = \frac{\partial [C_{ijkl}(\vec{r})\epsilon_{kl}(\vec{r})]}{\partial x_j} = F_i. \tag{7}$$

An important result is that the general solution for the static case can be obtained from equation (3) using integral equations with Green's function kernels. The solution of the integral equation can itself be systematically approximated. Writing the elastic stiffnesses tensor as

$$C_{ijkl} = C^0_{ijkl} + \delta C_{ijkl}, \tag{8}$$

and introducing the tensor Green's function $G_{ijkl}(\vec{r}, \vec{r}')$ for strain, the following equation results (GUBERNATIS and KRUMHANSL, 1975)

$$\epsilon_{ij}(\vec{r}) = \epsilon^0_{ij}(\vec{r}) + \int d\vec{r}' G_{ijkl}(\vec{r}, \vec{r}')\delta C_{klmn}(\vec{r}')\epsilon_{mn}(\vec{r}'). \tag{9}$$

This provides, using the various possible approximations, the solutions for the static field. The iterated solution of equation (9) is

$$\epsilon = \epsilon^0 + G\delta C\epsilon^0 + G\delta CG\delta C\epsilon^0 + \ldots \tag{10}$$

where δC stands for the elastic stiffnesses tensor. The strain field is obtained as a series which can be formally summed so that

$$\epsilon = \epsilon^0 + G\left(\delta C \frac{I}{I - G\delta C}\right)\epsilon^0 = \epsilon^0 + GT\epsilon^0 \tag{11}$$

where $T = \delta C(I - G\delta C)^{-1}$ is called the T matrix in the scattering theory. For a constant strain field ϵ^0, and using the definition of T, the effective susceptibility, i.e., the tensor C^{eff} is obtained

$$C^{\text{eff}} = C^0 + \langle T \rangle \frac{I}{I + \langle GT \rangle}. \tag{12}$$

This gives the exact solution provided that $\langle T \rangle$ and $\langle GT \rangle$ can be computed. At this point various approximations are used. For instance, a particularly important approximation is the self consistent approximation. It assumes $\langle T \rangle = 0$. Because T is a function of C^0, it is possible to make such an approximation which provides a set of implicit equations for C^0. In that particular case, $C^0 = C^{\text{eff}}$. Let us emphasize that C^0 here is no longer the elastic stiffness tensor of the intact (non-porous) rock. It is the effective medium itself. The method is called "self-consistent" precisely because it assumes the effective properties (to be calculated) known from the beginning. The method consists in calculating the effective properties in terms of a set of implicit equations.

2. Upscaling and Numerical Methods

In many cases, the situation is that of a strong heterogeneity. Then one cannot define any REV. The objects of interest are of size smaller than a hypothetical REV. An example of great importance is the oil reservoirs. Any model of fluid flow (or any other process) in such a case cannot rely on the EMT approach. Numerical methods are required to simulate flow at the reservoir scale. This implies that a discrete model of the geological object considered must be constructed. Because such methods can be highly CPU-time consuming, there exists various attempts which aim at the optimum compromise between the quality of the simulation and the required amount of CPU time. In that case, "upscaling" refers to the techniques used to transform a fine-grid model into a more practical, coarser one.

Aquifers or oil reservoirs models are discretized over fine grids populated by permeability values. These models generally contain too many gridblocks to be used directly for flow simulation. Some manner of averaging or upscaling is therefore required to reduce the number of gridblocks and make flow simulation tractable in terms of CPU time.

2.1 From Effective to Equivalent Permeability

As discussed in section 1, the term effective refers to media which are statistically homogeneous on the large scale. In the case of transport properties, the effective permeability is defined from an averaged form of Darcy's law as:

$$K^{\text{eff}} = -\mu \frac{\langle \vec{U} \rangle}{\vec{\nabla} P}. \tag{13}$$

K^{eff} is the effective permeability, μ fluid viscosity, $\langle \vec{U} \rangle$ the mean over the domain of Darcy velocity and $\vec{\nabla} P$ the pressure gradient. This simple expression refers to isotropic media. Once the effective permeability is computed, the finely gridded reservoir model is replaced by a single gridblock with a constant permeability (Fig. 2).

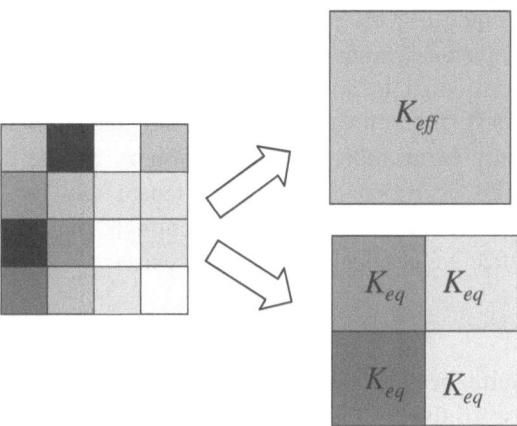

Figure 2
From fine gridblocks to coarse ones.

However, in most engineering situations involving water flow or oil flow, conditions for the emergence of an effective permeability are not met. The studied domains are not large enough to ensure statistical homogeneity. Instead, one focuses on equivalent or block permeabilities. An equivalent permeability is assigned to a group of neighboring fine gridblocks (Fig. 2). In this case, the fine reservoir model is replaced by a set of permeability values instead of a single one. The equivalent permeability must be selected so that flow is possibly least perturbed when replacing the permeabilities of the fine gridblocks by this equivalent permeability. This intuitive concept needs to be mathematically expressed, which entails the definition of equivalence criteria.

As permeability is not an additive property, the equivalent permeability cannot be derived from a simple algebraic expression as the arithmetic mean. Therefore, one instead considers flow rate (CARDWELL and PARSONS, 1945; WARREN and PRICE, 1961; RUBIN and GOMEZ-HERNÁNDEZ, 1990) or dissipated energy (MATHERON, 1967; INDELMAN and DAGAN, 1993). In the first case, the equivalence between the grouped fine gridblocks and the resulting coarse gridblock is expressed in terms of flow rates: Flow rates at the boundaries of the aggregate of fine gridblocks must be the same as those of the coarse gridblock. In the second case, the dissipated energy must be identical for the aggregate of fine gridblocks and the corresponding coarse gridblock. Flow or dissipated energy provide equivalence criteria. Stating equivalence following one of these criteria leads to an estimation of the equivalent permeability of the coarse gridblock. Although both criteria involve distinct formulations, they come down to identical results with periodic conditions applied at the boundaries of the coarse gridblock (BOE, 1994).

The equivalent permeability is generally not a scalar, but a second-order tensor. Its shape motivated many research works. For instance, GELHAR and AXNESS (1983) stated that it is symmetric. This property was actually evidenced by MATHERON (1967) for stationary porous media; that is media with unchanged two-order statistics regardless of the considered location. On the other hand, according to ABABOU (1988) or KING (1993), the equivalent permeability tensor is not symmetric. For engineering problems, permeability is not stationary at the scale of the coarse gridblock. Hence, the equivalent permeability tensor can be considered as non-symmetric. Last, if the coarse gridblock is large enough, the equivalent permeability tends towards the effective permeability, when it exists.

2.2 Numerical Upscaling Techniques

In order to estimate the equivalence criteria, we need to perform flow simulations over the target coarse gridblock. The problem to solve is based upon the combination of Darcy's law and the continuity equation under steady-state conditions

$$\vec{u} = -\frac{k}{\mu}\vec{\nabla}p \quad \text{and} \quad \vec{\nabla} \cdot \vec{u} = 0, \tag{14}$$

where $\vec{u}$ is the filtration or Darcy velocity, k permeability, μ fluid viscosity and p pressure. In this particular case, the continuity equation is written for an incompressible fluid. We also assume that there are no source terms. Combining both equations gives

$$\vec{\nabla} \cdot (k\vec{\nabla}p) = 0. \tag{15}$$

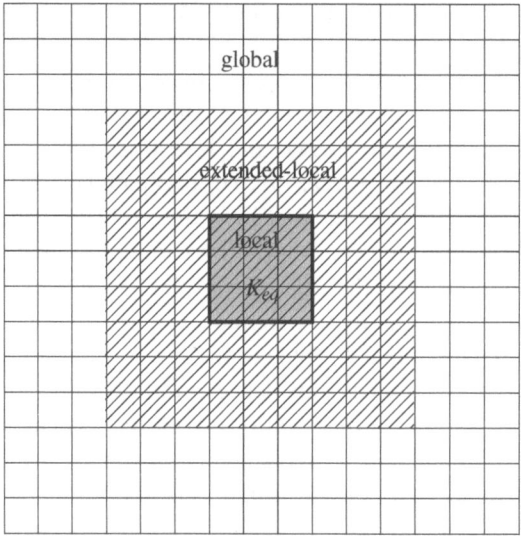

Figure 3
Classification of the upscaling techniques depending on the size of the flow simulation domain.

A variety of upscaling techniques have been developed. They can be classified depending on the size of the domain, over which the above equation is solved (Fig. 3). When this domain is limited to the target coarse gridblock, upscaling techniques are said to be local. When it includes the immediate neighboring gridblocks, upscaling techniques are said to be extended-local. Last, when it extends to the entire fine grid, upscaling techniques are said to be global. In addition, solving the above equations calls for boundary conditions. Most of the time, they are unknown. Subsequently, arbitrary boundary conditions are selected: They are often of the no-flow type or uniform or periodic. Clearly, the computed equivalent permeabilities depend on the boundary conditions considered.

2.2.1 Local Techniques

Local upscaling techniques derive the equivalent permeability of a coarse gridblock from the permeabilities of the fine gridblocks included in this coarse gridblock. First, we assume that we extract the fine gridblocks inside the coarse one from the entire fine grid. Second, we solve equation (15) over the finely gridded target coarse gridblock and we use the solution to compute its equivalent permeability. A flow direction has to be selected: it may be axis X, Y or Z. Boundary conditions must also be prescribed. For instance, flow can be simulated with no-flow (or permeameter) conditions applied at the boundaries of the coarse gridblock (WARREN and PRICE, 1961). In such a case, constant pressures are imposed at inlet and outlet faces and no flow is allowed across the other faces. Let L be the distance between the two faces where pressures are imposed, A the surface of the section crossed by fluid, $\triangle P$ the pressure difference and Q the flow rate across the coarse gridblock. In this case, we consider the equivalence criterion based upon flow equality. Flow rate Q results from the addition of the elementary flow rates at the outlet face. These elementary flow rates are obtained when solving equation (15). Reverting to Darcy's law yields the equivalent permeability of the coarse gridblock

$$K^{\mathrm{eq}} = -\frac{Q}{A}\mu\frac{L}{\triangle P}.\tag{16}$$

When flow is driven in the X direction, the computed equivalent permeability is K^{eq}_{XX}. Repeating the computations for flow along axes Y and Z provides the diagonal terms of the equivalent permeability tensor. Extra-diagonal terms cannot be determined, because no flow is allowed in the direction transverse to the pressure differences.

The major differences in local upscaling techniques are due to their treatment of the boundary conditions. Detailed reviews were published by RENARD and DE MARSILY (1997) and WEN and GOMEZ-HERNÁNDEZ (1996). For instance, GALLOUËT and GUÉRILLOT (1994) used uniform boundary conditions while DURLOFSKY (1991) considered periodic boundary conditions. We note that a complete permeability tensor cannot be computed when assuming no-flow boundary conditions as mentioned above. Also, the upscaled permeability tensor is non-symmetric with uniform boundary conditions and symmetric with periodic ones. PICKUP *et al.* (1994) showed that periodic boundary conditions provide the most reliable results in the example problems considered. The key point is that actual boundary conditions change with time and are unknown.

Local techniques focus on the fine gridblocks inside the coarse gridblock only. They do not account for the effects of the neighboring regions in the estimation of the equivalent permeability. In addition, the computed equivalent permeability tensor strongly depends on the type of boundary conditions. There is no unique equivalent permeability.

Local upscaling techniques provide reliable estimation for reservoir models with smooth permeability variations, but not for reservoir models with complex geological features and strong permeability contrasts.

2.2.2 Extended-local Techniques

The development of extended-local upscaling techniques has been motivated, at least partially, by the work of HOU and WU (1997). These authors obtained improved results by incorporating the effects of neighboring gridblocks in their computations. Extended-local upscaling techniques are very similar to local approaches. However, instead of using the coarse gridblock solely to derive the equivalent permeability, we consider a slightly larger domain consisting of the coarse gridblock plus local border regions (GOMEZ-HERNÁNDEZ and JOURNEL, 1994; HOLDEN and LIA, 1992; WU *et al.*, 2002; WEN *et al.*, 2003). The objective is to obtain more accurate estimations by capturing the effects of larger-scale permeability connectivity and by reducing the influence of boundary conditions into the calculation of the equivalent permeability. It is reasonable to expect that the effect of the boundary specification on the upscaled equivalent permeability will be less when increasing the border regions and that results obtained with different types of boundary conditions will tend to converge.

2.2.3 Global Techniques

Extended-local upscaling techniques yield refined estimations of equivalent permeabilities of coarse gridblocks compared to local techniques. However, in some cases, particularly when applied to reservoirs with complex and tortuous geological features such as channels, extended-local upscaling techniques still need improvement. The leading idea is to better capture the influence of the neighboring geological bodies and to reduce further the dependency on boundary conditions, which are arbitrarily selected. This motivated the development of global upscaling techniques.

With global approaches, the intent is to solve equation (15) on the entire fine grid. The solution is then used to estimate the boundary conditions to be applied to coarse gridblocks when computing equivalent permeabilities. We hope that this choice is

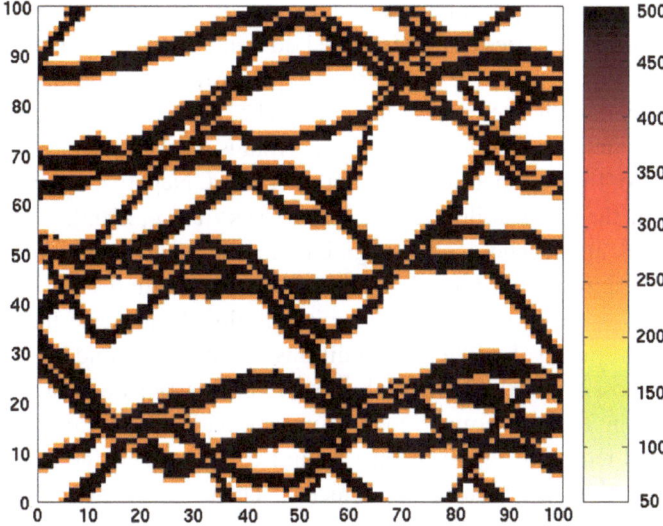

Figure 4
Fine channelized reservoir model.

more realistic than the boundary conditions used with local and extended-local upscaling techniques. Also, considering the flow simulation over the entire fine grid should help to better catch the connectivity of neighboring geological bodies. Because purely global methods are computationally intensive (HOLDEN and NIELSEN, 2000), recent works mainly focus on quasi-global techniques, which decrease the computational overburden by substituting some type of approximate global information in place of global fine scale results. There are two kinds of quasi-global approaches: local-global and global-local.

Local-global methods attempt to estimate the effects of the global flow without actually solving a global fine scale problem. CHEN *et al.* (2003) proposed to first perform global coarse scale flow simulations and then local fine scale flow simulations. The boundary conditions considered for the global flow simulations are arbitrary. These flow simulations provide pressures at the centers of the coarse gridblocks. Then, the domain defined by the target coarse gridblock surrounded by half a coarse

Table 1

Facies properties

Facies	Volume fractions (%)	Permeability (mD)
Foodplain shale	50	50
Channel margin	15	250
Channel sand	35	500

gridblock is extracted. An interpolation process yields the pressures in the fine gridblocks at the boundaries of this extended domain. Thereafter, extended-local flow simulations can be run to derive the equivalent permeability tensor of the target coarse gridblock. The overall method is iterative. The procedure is repeated until the upscaled permeability tensor no longer varies, that is until the result is self-consistent. The suitability of the obtained results is not warranted: Negative terms may appear on the diagonal of the permeability tensor (DURLOFSKY *et al.*, 2003).

Global-local approaches call for global fine scale flow simulations and then local fine scale flow simulations. Unlike local-global methods, the flow problem is first solved over the entire fine grid. This initial global fine scale flow simulation is run with appropriate boundary conditions, that is boundary conditions as consistent as possible with the actual ones. It allows the boundary conditions to be considered in the subsequent local flow simulations. In this special case, no interpolation is required: Pressures are computed at once in the fine gridblocks at the boundaries of the coarse gridblocks. Then, local flow simulations are

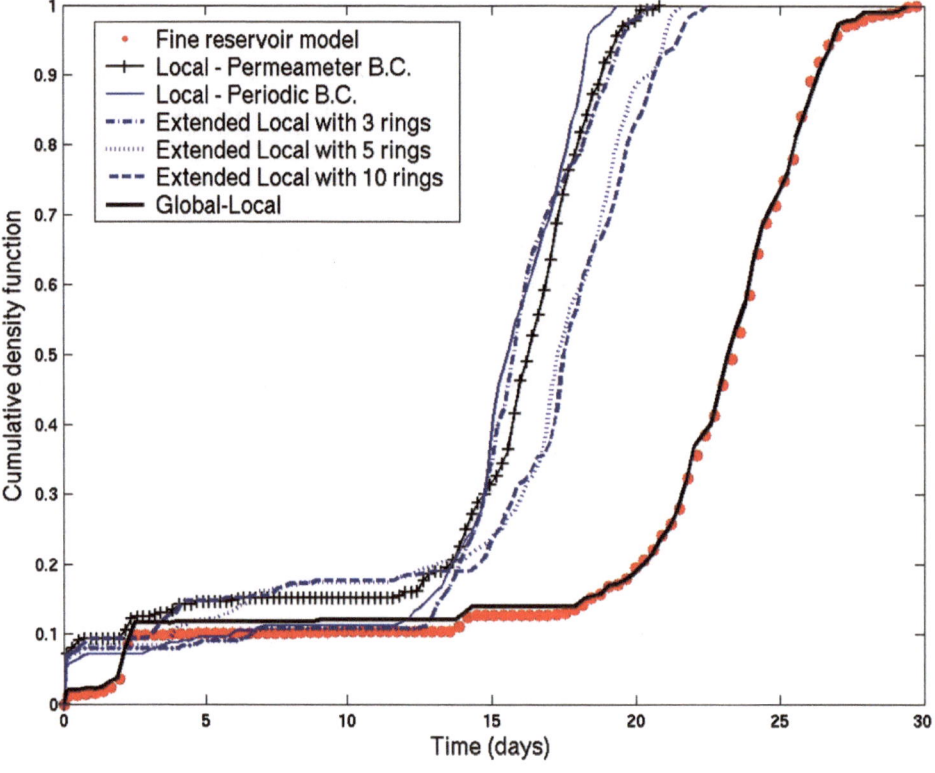

Figure 5
Cumulative density function of the times needed by particles to cross the reservoir.

performed for each coarse gridblock, first with the computed boundary conditions and second with perturbed boundary conditions (PICKUP et al., 1992; RICARD et al., 2005). This results in the estimation of the equivalent permeability tensors of the coarse gridblocks. Contrary to the local-global approach of CHEN et al. (2003), these global-local upscaling methods are not iterative. They provide at once the coarse scale permeability tensors and give no negative diagonal terms. Despite their interesting features, global-local upscaling approaches are not very popular due to the CPU-time required to run a flow simulation over the entire fine grid.

2.3 A Numerical Example

For underground reservoirs with smooth permeability variations, local and extended-local upscaling techniques provide satisfactory approximations. However, they can fail in the case of more complex geological systems such as channelized reservoirs. In the example depicted in Figure 4, the two-dimensional reservoir model is discretized over 100×100 gridblocks. This horizontal layer is made of three facies: A floodplain shale, a channel margin and a channel sand. Their volume fractions and permeabilities are reported in Table 1. For simplicity, porosity is assumed to be 0.25 everywhere. The reservoir is bounded by two horizontal wells: An injector at $x = 1$ and a producer at $x = 100$. A pressure difference of 1 kPa is set between the two wells. Next, inert particles are injected into the injector. Due to pressure difference, they move from the injector to the producer. For the example studied, fluid viscosity is assumed to be 1 cP and the size of each gridblock 1×1 m^2. This numerical tracer test experiment allows for computing the times needed by the particles to cross the reservoir. The cumulative density function of these times is reported in Figure 5. This fine reservoir model is upscaled to 20×20 gridblocks on the basis of an aggregation rate of 5×5. We illustrate the influence of large-scale connectivity on the computation of equivalent permeabilities by applying local, extended-local and global methods. The objective is to perform the same numerical tracer experiment as the one described above for the upscaled reservoir models and to compare the resulting travel times to those determined for the fine reservoir model. Therefore, we hope to assess the quality of the upscaled models. Various upscaling techniques are considered:

1— a local upscaling with permeameter (or no-flow) boundary conditions;
2— a local upscaling with periodic boundary conditions;
3— an extended-local upscaling with target gridblocks surrounded by 3 rings of fine gridblocks and with periodic boundary conditions;
4— an extended-local upscaling with target gridblocks surrounded by 5 rings of fine gridblocks and with periodic boundary conditions;

5— an extended-local upscaling with target gridblocks surrounded by 10 rings of fine gridblocks and with periodic boundary conditions;

6— a global-local upscaling.

The computed cumulative density function of travel times is reported in Figure 5. Clearly, the results derived from local and extended-local upscaling methods are very different from the ones pointed out for the reference finely gridded reservoir. Particles move much faster in the upscaled reservoir than in the fine one. Two main features can be stressed for the example presented here. First, in the case of purely local approaches, the determined cumulative density functions are very close, regardless of the boundary conditions considered. Second, in the case of extended-local approaches, the larger the border region, the better the agreement between the upscaled and fine results. Now, if we examine the cumulative density function of travel times derived from a global-local upscaling technique (RICARD *et al.*, 2005), the improvement is obvious. The cumulative density function of travel times estimated for the upscaled reservoir model fits almost perfectly that one identified for the fine reservoir. This example illustrates the benefit in using global computations, although they depend on flow simulations on the whole fine grid.

3. Upscaling and the Fractal Dream

A particular situation which corresponds to media where statistical homogeneity does not apply, and yet can be handled without the time consuming numerical simulations discussed in the previous section, is that of a fractal geometry. Fractal objects have the property of scale invariance. This obviously makes "upscaling" an irrelevant question. The properties of a fractal medium such as transport properties are scale-dependent through power laws. This makes the situation considerably simpler than that examined in section 2. For all these reasons, the scale invariance property is a very appealing characteristic of fractal media. One could list an impressive number of geological features which have been considered to be fractal and characterized by power laws: Fault length, fault roughness, gouge particle size, slip displacement, pore surface, etc. This raises indeed two questions. First, could one put most geological objects into one of the two following groups: (1) statistical homogeneity, or (2), fractal geometry? The first case corresponds to translational invariance and the second one to scale invariance. Such an alternative would offer a way out of the time-consuming techniques described in section 2. Second, if the answer to the previous question is positive, and pervasive fractality is real and not a dream, what is the reason for a such universal behavior? More specifically, since fractality is a geometrical feature, the basic question is: What are the physical processes which could lead to a fractal geometry? and what are those relevant to geology?

We suggest in this last section that a precise investigation of each situation in which a geological object is considered to be fractal should be conducted. Because most data on possibly fractal objects cover a narrow scale range, and because such geological data are difficult to obtain the desired accuracy, we suggest that a physically based and more restrictive criterion should be used to attest fractality. Only a clear positive answer to the previous question, i.e., identification of the physical mechanism explaining fractality, should be used to identify fractal objects. A multitude of observations and models has been devoted to fractal systems in the past 30 years. We do not attempt to review all of these investigations, however concentrate instead on the few cases where it appears that there are well identified processes which could explain the existence of fractal geological objects. Two examples appear to be good candidates for inclusion into that category. One at macroscopical scale, and the other at microscopical scale. We do not know of possible processes which could explain similarly the hypothetical fractality of other objects, nonetheless we suggest that further investigations should be conducted to seek out physical processes in the other cases.

3.1 Percolation Theory and Fractal Geometry

The first situation pertains to fractal networks of fractures and percolation theory. This is very relevant to upscaling since percolation theory deals with strongly heterogeneous media. As stated in section 1, EMT does apply to statistically homogeneous media only. However, beyond some degree of heterogeneity, i.e. a large number of fractures in the medium in our case, EMT is no longer valid. This is easy to understand as clustering effects, clusters are connected fractures in our case, become increasingly more important. The aim of percolation theory is precisely to investigate such clustering effects and account for the behavior of the medium in that state. The concept of percolation threshold, here it means overall interconnection of fractures providing a connected path, is a key one in this theory (MADDEN, 1983; STAUFFER and AHARONY, 1992). It is well known that the circumstances of such a percolative system at threshold are those of a phase transition. As shown by WILSON (1971), renormalization group theory can explain the particular behavior of physical systems in those conditions. Fluctuations at all scales are present or, stated in different terms, the system can be described as a fractal one. This implies that percolation models provide a simple explanation for the existence of a fractal set. For that reason, it has been suggested that fractal set of cracks, or fault networks, could exist if they were linked to a percolation effect. A possible model is that a fractal set of fractures would be the result of fluid pressure driven mechanisms (GUÉGUEN et al., 1991). If the driving force for fracture propagation is assumed to be fluid pressure, it will be turned off as soon as the percolation threshold is reached because a strong permeability increase would develop then, relaxing the fluid pressure. It follows that the network, in proximity to percolation threshold, is fractal. In terms of phase transition, the system stays very close to its critical point.

Although this mechanism provides a clear explanation for the possible existence of fractal sets of fractures, it remains to be established whether such situations exist or not. At least in such cases, one would expect a fractal distribution. There does not seem to exist other clear processes implying a fractal geometry for fractures.

3.2. Aggregation Processes and Fractal Geometry

Another geological object which can be fractal is pore surface (KATZ and THOMPSON, 1985). Interestingly, it ensues that various growth models can explain the development of fractal surfaces. In particular, Diffusion Limited Aggregation models (DLA theory) provide a means to finish with fractal objects. A common point to those models is the Laplace equation. In rocks, diagenesis, healing and sealing processes could be the mechanisms responsible at the microscopical scale for growth. Although it has been well established that some pore surfaces (KATZ and THOMPSON, 1985), or rough surfaces (KARDAR *et al.*, 1986) are fractal, this is not the general state in geology. Again, it remains for investigation to determine if this is more the exception or the rule.

4. Conclusion

Geological media can be viewed as disordered media. In the simple case where the real system can be approximately described as a homogeneous embedding medium containing various inclusions, Effective Medium Theory (EMT) is an appropriate tool to derive any property of interest (a generalized susceptibility) at large scales. This provides a first possible method for upscaling. However, EMT becomes itself ineffective when the degree of heterogeneity is too large. This is so because basically, EMT is a perturbation method which fails beyond some point. Despite this drawback, the advantage of EMT is that it relies on the assumption of translational invariance: The medium is considered to be statistically homogeneous. That means that, beyond the scale of a Representative Elementary Volume (REV), the averaged susceptibility has the same value everywhere. When this breaks down, one may be tempted to shift from translational invariance to scale invariance: Any kind of invariance is of great interest because it does simplify the description of the system. By itself, scale invariance suppresses the upscaling question. It is a manner of unexpected answer, that we could dream of, to the question examined here. Scale invariance becomes the specific property of fractal systems. This is why fractal models have become so popular. As regards geological systems, fractal geometry appears however to be possible in a small number of specific situations only. For most cases, when Effective Medium Theory becomes ineffective, the only way to address upscaling is to use numerical methods. Considerable progress has been reached in that direction and global techniques allow the calculation of upscaled properties, at the expense of a certain amount of CPU-time.

REFERENCES

ABABOU R. (1988) Three-dimensional flow in random porous media. Ph.D. dissertation, M.I.T., Cambridge, MA, USA.

BOE, O. (1994), *Analysis of an upscaling method based on conservation of dissipation*, Transport in Porous Media *17*, 77–86.

CARDWELL, W., and PARSONS, R.L. (1945), *Average permeabilities of heterogeneous oil sands*, Trans. Am. Inst. Mining. Met. Pet. Eng. 34–42.

CHEN, Y., DURLOFSKY, L.J., GERRITSEN, M., and WEN, X.H. (2003), *A coupled local-global approach for simulating flow in highly heterogeneous formations*, Advances Water Resources *26*, 1041–1060.

DURLOFSKY, L.J. (1991), *Numerical calculation of equivalent grid block permeability tensors for heterogeneous media*, Water Resources Res. *27*, 699–708.

DURLOFSKY, L.J. (2003), *Upscaling of geocellular models for reservoir flow simulation: A review of recent progress*, Proc. 7th International Forum on Reservoir Simulation, Bühl/Baden-Baden, Germany.

GALLOUËT, T., and GUÉRILLOT, D. (1994), *Averaged heterogeneous porous media by minimisation of error on the flow rate*, 4th Europ. Conf. Math. Oil Rec.

GELHAR, L.W., and AXNESS, C.L. (1983), Three-dimensional stochastic analysis of macrodispersion in aquifers, Water Resources Research, 19, 161–180.

GOMEZ-HERNÁNDEZ, J.J., and JOURNEL, A.G. (1994), *Stochastic characterization of grid block permeability*, SPE Form. Eval. *9*(2), 93–99.

GUBERNATIS, J.E., and KRUMHANSL, J.A. (1975), *Macroscopic engineering properties of polycristalline materials: Elastic properties*, J. Appl. Phys. *46*(5), 1875–1883.

GUÉGUEN, Y., CHELIDZE, T., and LE RAVALEC, M. (1997), *Microstructures, percolation thresholds and rock physical properties*, Tectonophysics *2798*, 23–35.

GUÉGUEN, Y., DAVID, C., and GAVRILENKO, P. (1991), *Percolation and fluid transport in the crust*, Geophys. Res. Lett. *18*, 5, 931–934.

GUÉGUEN, Y., and PALCIAUSKAS, V., *Introduction to the Physics of rocks* (Princeton Univ. Press, 1994) 294 pp.

HASHIN, Z., and SHTRIKMAN, S. (1962), *A variational approach to the theory of the elastic behaviour of polycristals*, J. Mech. Phys. Sol. *10*, 343–352.

HOLDEN, L., and NIELSEN, B.F. (2000), *Global upscaling of permeability in heterogeneous reservoirs: The Output Lest Squares (OLS) methods*, Transp. Porous Media *40*, 115–143.

HOLDEN, L. and LIA, O. (1992), *A tensor estimator for the homogeneization of absolute permeability*, Transp. Porous Media *8*(1), 37–46.

HOU, T.Y., and WU, X.H. (1997), *A multiscale finite element method for elliptic problems in composite materials and porous media*, J. Comp. Phys. *134*(1), 169–189.

INDELMAN, P., and DAGAN, G. (1993), *Upscaling of permeability of anisotropic heterogeneous formations, 1- The general framework*, Water Resour. Res. *29*(4), 917–923.

KACHANOV, M. (1992), *Effective elastic properties of cracked solids: critical review of some basic concepts*, Appl. Mech. Rev. *45*, 304–335.

KACHANOV, M. (1993), *Elastic solids with many cracks and related problems*, Advan. Appl. Mechan. *30*, 259–445.

KARDAR, M., PARISI, G., and YI-CHEN ZHANG (1986), *Dynamic scaling of growing interfaces*, Phys. Rev. Lett. *56*, 889.

KATZ, A.J., and THOMPSON, A.H. (1985), *Fractal sandstone pores: Implications for conductivity and pore formation*, Phys. Rev. Lett. *54*, 1325.

KING, M.J., *Application and analysis of tensor permeability to crossbedded reservoirs*, SPE Paper 26118, Western Regional Meeting (Anchorage, Alaska, USA, 1993).

LE RAVALEC, M., and GUÉGUEN, Y. (1996), *High and low frequency elastic moduli for saturated porous cracked rock, differential self consistent and poroelastic theories*, Geophys. *61*, 1080–1094.

MADDEN, T.R. (1983), *Microcrack connectivity in rocks: a renormalization group approach to the critical phenomena of conduction and failure incrystalline rocks*, J. Geophys. Res. *88*, 585–592.

MATHERON, G., *Eléments pour une théorie des milieux poreux* (Masson, eds, Paris, 1967).

PICKUP, G.E., JENSEN, J.L., RINGROSE, P.S., and SORBIE, K.S. (1992), *A method for calculating permeability tensors using perturbed boundary conditions*, 3rd Europ. Conf. Math. Oil Rec. 225–238.

PICKUP, G.E., RINGROSE, P.S., and SORBIE, K.S. (1994), *Permeability tensors for sedimentary structures*, Math. Geol. *26*(2), 227–250.

RENARD, P., and DE MARSILY, G. (1997), *Calculating equivalent permeability : A review*, Advan. Water Resources *20*(5–6), 253–278.

RICARD, L., LE RAVALEC-DUPIN, M., and GUÉGUEN, Y. (2005), submitted.

RUBIN, Y. and GOMEZ-HERNANDEZ, J.J. (1990), *A stochastic approach to the problem of upscaling of conductivity in disordered media: Theory and unconditional numerical simulations*, Water Resources Res. *26*(4), 691–701.

SCHUBNEL, A., and GUÉGUEN, Y. (2003), *Dispersion and anisotropy of elastic waves in cracked rocks*, J. Geophys. Res. *108*, 2101–2116.

STAUFFER, D., and AHARONY, A., *Introduction to Percolation Theory* (Taylor and Francis, London, 1992) 181 pp.

WARREN, J., and PRICE H. (1961), *Flow in heteregeneous porous media*, Soc. Petrol. Engin. J. *1*, 153–169.

WEN, X.H., DURLOFSKY, L.J., and EDWARDS, M.G. (2003), *Use of border regions for improved permeability upscaling*, Math. Geology *35*, 521–547.

WEN, X.H., and GOMEZ-HERNANDEZ, J.J. (1996), *Upscaling hydraulic conductivities in heterogeneous media: An overview*, J. Hydrol. *183*(1–2), ix-xxxii.

WHITE, C.D., and HORNE, R.N., *Computing absolute transmissibility in the presence of fine scale heterogeneity*, SPE 16011, SPE Symp. Reservoir Simulation, San Antonio, TX, 1987).

WILSON, K.G. (1971), *Renormalization group and critical phenomena*, Phys. Rev. *B4*, 3174–3205.

WU, X.H., EFENDIEV, Y.R., and HOU, T.Y. (2002), *Analysis of upscaling absolute permeability*, Discrete Contin. Dyn. Syst., Ser. B *2*(2), 185–204.

(Received April 21, 2005, revised September 29, 2005, accepted September 30, 2005)
Published Online First: May 12, 2006

To access this journal online:
http://www.birkhauser.ch